CRYOGENIC MIXED REFRIGERANT PROCESSES

For other titles published in this series, go to
www.springer.com/series/6086

INTERNATIONAL CRYOGENICS MONOGRAPH SERIES

G. Venkatarathnam

CRYOGENIC MIXED REFRIGERANT PROCESSES

Springer

Gadhiraju Venkatarathnam
Department of Mechanical Engineering
Indian Institute of Technology Madras
Chennai – 600036
India
gvenkat@iitm.ac.in

Series Editors
Klaus D. Timmerhaus
Department of Chemical Engineering
University of Colorado
Boulder, CO 80309-0424
USA
klaus.timmerhaus@colorado.edu

Carlo Rizzuto
Department of Physics
University of Genova
Genova
Italy

ISBN: 978-1-4419-2690-6 e-ISBN: 978-0-387-78514-1
DOI: 10.1007/978-0-387-78514-1

Printed on acid-free paper

9 8 7 6 5 4 3 2 1

springer.com

Dedicated to the loving memory of my father

G. V. Narasimham

Preface

This monograph deals with mixed refrigerant processes that operate at temperatures less than 123 K. Most conventional cryogenic refrigerators and liquefiers operate with pure fluids, the major exception being natural gas liquefiers that use mixed refrigerant processes. The fundamental aspects of mixed refrigerant processes, though very innovative, have not received the due attention in the open literature in view of commercial interests. Hundreds of patents exist on different aspects of mixed refrigerant processes for the liquefaction of natural gas and the composition of mixtures for Joule–Thomson (Linde–Hampson) and other refrigerators. It is difficult to piece together the existing information to choose an appropriate process and an optimum composition for a given application. The main purpose of this monograph is to explain all the aspects of mixed refrigerant processes and the methods for choosing the composition of refrigerants using robust analytical methods based on sound thermodynamic principles.

All concepts required to design and evaluate mixed refrigerant processes including exergy are introduced from first principles in the first chapter. The performance of traditional cryogenic processes that operate with pure fluids such as Linde–Hampson and Kapitza and the reasons for the low exergy efficiency of these processes are also presented in the first chapter.

Cryogenic processes differ from general chemical processes in several ways. The use of multistream heat exchangers with internal pinch points makes it necessary to use somewhat different approaches to simulate mixed refrigerant processes. The methods for simulating and optimizing cryogenic processes using a process simulator are presented in the second chapter.

The need for using refrigerant mixtures over pure fluids is presented in the third chapter with reference to simple refrigeration and gas cooling processes. The more complex refrigeration processes are presented in the fourth chapter. A unified design approach has been evolved for optimizing mixed refrigerant process refrigerators and liquefiers and is presented in the fifth chapter. The different natural gas and nitrogen liquefaction processes are presented in the sixth and seventh chapters, respectively.

Optimum operating pressures and mixture compositions have been determined for a variety of mixed refrigerant process refrigerators and natural gas/nitrogen liquefiers. In some cases, the performance of processes with different alternate mixture compositions and operating pressures has been presented to help understand these processes thoroughly. The performance of different processes is evaluated in terms of exergy losses in different components. Most of the examples and case studies presented are largely unpublished.

The examples presented in this monograph were solved using Aspen Plus, a commercial process simulator, and CRYOSIM, a cryogenic process simulator developed in-house. It should be possible, however, to use any process simulator with optimization capabilities to solve the examples independently. The reader can also use the data presented in the examples as the starting values (estimates) in his or her own optimization studies. Those who are familiar with process optimization will appreciate the value of good estimates that allow the optimizer to start from a feasible point and converge rapidly. Many of the examples provided are nearly optimal solutions. However, students and practicing engineers are encouraged to find even better solutions as a part of their learning exercise. Practicing refrigeration and cryogenic engineers will benefit from this monograph and would be able to apply the methods presented to design optimum mixed refrigerant processes. This monograph can also be used as a textbook for a graduate course on advanced refrigeration or cryogenic processes. No prior knowledge of refrigeration/cryogenic processes is required to read this monograph. Some knowledge of thermodynamics and optimization is helpful, but not necessary. Access to a process simulator is, however, necessary to design mixed refrigerant processes.

The material covered in this monograph has been drawn largely from my research on refrigerant mixtures for over 15 years. I used the material presented in a graduate course on advanced cryogenic systems at IIT Madras.

I am grateful to my senior colleague Prof. S. Srinivasa Murthy of IIT Madras, my friend and collaborator Prof. L. R. Oellrich of Universität Karlsruhe, and my Ph.D. supervisor Prof. Sunil Sarangi of IIT Kharagpur for their support and encouragement at different stages of my career, including the writing of this monograph.

I would like to thank my colleague Prof. K. Ramamurthi, who read the manuscript and offered useful comments. Thanks are also due to Prof. K. D. Timmerhaus, Editor of the International Cryogenics Monograph Series, for reviewing the manuscript critically.

I gratefully acknowledge the contributions of my graduate students and the technical staff of the Refrigeration and Airconditioning Laboratory, IIT Madras, especially Mr. R. Elangovan, in fabricating and testing different mixed refrigerant prototypes over the years. They are too large a number to be acknowledged individually.

My mother, Samrajya Lakshmi, encouraged me to write this monograph and supported me through times of despair. My wife, Suchitra, and son, Akhil, enthusiastically sacrificed all their time and allowed me to work long hours on this monograph for several years. This finalized text would not have been possible without their love and unstinting support.

Contents

Nomenclature

A	Heat transfer area (m^2)
a_i	Constant in Eq. (3.15)
b_i	Constant in Eq. (3.16)
C	Clearance ratio of compressor
COP	Coefficient of performance (–)
c	Constant in Eq. (5.1)
c_p	Specific heat at constant pressure (J/mol·K)
c_v	Specific heat at constant volume (J/mol·K)
ex	Specific exergy of a stream = $[(h - h_o) - T_o(s - s_o)]$ (J/mol)
f	Fugacity (bar)
HX	Heat exchanger
h	Specific enthalpy at pressure p, temperature T (J/mol)
h_o	Specific enthalpy at ambient pressure and temperature (J/mol)
i	Iteration number
LNG	Liquefied natural gas
LPG	Liquefied petroleum gas
MRC	Mixed refrigerant cascade
n	Moles (mol)
$\dot{n}$	Mole flow rate (mol/s)
$\dot{n}_i$	Mole flow rate of the ith component (mol/s)
p	Pressure (bar)
p_c	Critical pressure (bar)
Q	Heat transferred (J)
$\dot{Q}$	Heat transfer rate (W)
$\dot{Q}_o$	Heat transfer rate to ambient (W) ($\dot{Q}_o < 0$)
Q_v	Volumetric cooling capacity (J/mol)
R	Universal gas constant (bar·m^3/mol·K)
s	Specific entropy at pressure p, temperature T (J/mol·K)
s_o	Specific entropy at ambient pressure and temperature (J/mol-K)

T	Temperature (K)
$T_{c,\text{in}}$	Temperature of cold stream at heat exchanger inlet (K)
$T_{c,\text{out}}$	Temperature of cold stream at heat exchanger outlet (K)
$T_{h,\text{in}}$	Temperature of hot stream at heat exchanger inlet (K)
$T_{h,\text{out}}$	Temperature of hot stream at heat exchanger outlet (K)
T_o	Ambient temperature (K)
t_o	Ambient temperature (°C)
U	Overall heat transfer coefficient (W/m^2·K)
$\dot{V}$	Volume flow rate (m^3/s)
$\dot{V}_c$	Displacement rate of compressor (m^3/s)
VLE	Vapor liquid equilibria
VLLE	Vapor-liquid-liquid-equilibria
v	Specific volume (m^3/mol)
W	Work (J)
$\dot{W}$	Power (W)
$\dot{W}_c$	Compressor power (W) ($\dot{W}_c < 0$)
$\dot{W}_e$	Power extracted from an expander (W) ($\dot{W}_e > 0$)
x	Vapor fraction
Y	Liquid yield (flow rate of liquid product/flow rate through compressor, also the fraction of the gas that gets liquefied on expansion)

Greek letters

α	Constant in Eq. (4.10)
β	Constant in Eq. (4.10)
$\Delta c_{p,\text{fusion}}$	$c_{p,\text{liq}} - c_{p,\text{solid}}$ at triple point (J/mol· K)
ΔEx_{loss}	Ratio of exergy loss to power input (-)
Δh_{fusion}	Enthalpy of fusion (J/mol)
Δh_{min}	Specific refrigeration effect (J/mol)
Δs	Specific entropy change (J/mol· K)
ΔT	Temperature difference, temperature approach between hot and cold streams of a heat exchanger (K)
ΔT_{min}	Minimum temperature approach between the hot and cold streams in the heat exchanger (K)
Δv_{fusion}	$v_{\text{liq}} - v_{\text{solid}}$ at triple point (m^3/mol)
ε	Effectiveness of the heat exchanger
γ	Adiabatic index = c_p/c_v
η_{ad}	Adiabatic efficiency (-)
η_{ex}	Exergy efficiency (-)
η_{fc}	Frictional efficiency of compressor (-)
η_m	Efficiency of motor (-)
η_v	Volumetric efficiency (-)
Λ	Ratio of the length of a part of heat exchanger to the total length of the heat exchanger

μ_{J-T}	Joule–Thomson coefficient (K/bar)
Θ	Ratio of the heat transfer over a part of the heat exchanger to the heat transfer over the entire heat exchanger
τ	Dimensionless temperature of streams in a heat exchanger $= (T - T_{c,\text{in}})/(T_{h,\text{in}} - T_{c,\text{in}})$
ρ	Density (mol/l)
ξ	Mole fraction (-)

Subscripts

ad	Adiabatic
c	Compressor, cold stream
comp	Compressor
cb	Cold box
ce	Cold end of heat exchanger
cs	Compressor section (compressor and aftercooler/condenser)
cv	Control volume
e	Expander
ex	Exergy
gas	Process gas being cooled
h	Constant enthalpy, hot stream, high temperature
hp	High-pressure stream
hx	Heat exchanger
i	*i*th component
ideal	Ideal system
l	Liquefier, gas cooler, low temperature
liq	Liquid phase
lm	LMTD
lp	Low-pressure stream
max	Maximum
min	Minimum
N_2	Nitrogen
o	Ambient, surroundings
p	Constant pressure
pre	Precooler
ps	Phase separator
r	Refrigerator
ref	Refrigerant
rev	Reversible
solid	Solid phase
specified	Specified
T	Constant temperature
t	Turbine
tp	Triple point
vp	Vapor pressure
we	Warm end of heat exchanger

1

Fundamental principles and processes

Single-stage mixed refrigerant processes that can provide refrigeration at very low temperatures were first proposed nearly 70 years ago by Podbielniak [69] and were adopted for large-scale liquefaction of natural gas after the pioneering work of Kleemenko [50] of the former Soviet Union in the 1960s. Today most base-load natural gas liquefaction plants operate on mixed refrigerant processes. Mixed refrigerant processes have also been adopted for peak-shaving natural gas liquefaction plants.

Mixed refrigerant processes were also studied in the early 1970s in the former Soviet Union by Brodyanskii and his colleagues for small cryocooler applications [17]. The interest in mixed refrigerant cryocoolers was revived about 10 years ago when DARPA funded projects for the development of low-cost cryocoolers [62].

Currently, there's worldwide interest in using mixed refrigerant processes for the liquefaction of nitrogen and separation of air [25]. Several U.S. patents have been granted during the last five years on the liquefaction of nitrogen using mixed refrigerant processes and two large plants have been built and tested [25].

Refrigeration processes can be divided into two broad groups based on the variation of pressure with time at any location of the process as follows [72]:

- periodic refrigerators in which the pressure at any point of the cycle varies with time, and
- steady-state refrigerators in which the pressure at any point of the cycle is constant and does not vary with time.

Cycles such as the Stirling, Gifford–McMahon, and Pulse-Tube belong to the first group, whereas cycles such as the Linde–Hampson (Joule–Thomson), Kapitza, etc. belong to the second group. The classification is analogous to alternating- and direct-current electrical machines. The theory and fundamental principles of these two groups are different enough to be treated separately. This monograph deals with mixed refrigerant processes in which the pressure at any part of the process is steady and does not vary with time. The flow is also unidirectional and steady in these processes.

G. Venkatarathnam, *Cryogenic Mixed Refrigerant Processes*,
DOI: 10.1007/978-0-387-78514-1_1, © Springer Science+Business Media, LLC 2008

Steady-state processes have been used in cryogenic liquefiers and refrigerators for over a century. Pure (single-component) fluids have traditionally been used in cryogenic refrigerators, whereas the fluid being liquefied is itself used as the refrigerant in the traditional liquefaction processes, the exception being the liquefaction of natural gas using mixed refrigerant processes. There are several advantages in using zeotropic refrigerant mixtures in cryogenic refrigerators and liquefaction systems:

- The exergy efficiency (figure of merit) of refrigeration and liquefaction systems operating with refrigerant mixtures is many times that of systems operating with pure fluids.
- The operating pressure is much lower when refrigerant mixtures are used, compared to pure fluids.
- Refrigeration and liquefaction systems operating with pure fluids operate largely in the superheated vapor region, whereas those operating with refrigerant mixtures operate largely in the two-phase region. Consequently, the heat transfer coefficients in the heat exchangers are much larger in systems operating with refrigerant mixtures compared to those operating with pure fluids, resulting in smaller heat exchangers.
- The degradation of heat exchanger performance due to longitudinal (axial) heat conduction is much smaller due to higher apparent specific heat $[(\partial h/\partial T)_p]$ of refrigerant mixtures in the two-phase region compared to the specific heat at constant pressure (c_p) of pure fluids in the superheated (single-phase) region.

It will be advantageous to replace many traditional processes operating with pure fluids with those operating with refrigerant mixtures.

1.1 Applications

Mixed refrigerant processes are used in numerous applications. Some of them are listed below

Liquefiers

- Liquefaction of natural gas in base-load and peak-shaving plants
- Liquefaction of natural gas derived from capped wells, biogas from landfills, municipal wastes, etc. on a small scale
- Liquefaction of nitrogen
- Liquefaction and separation of air
- Recovery of volatile organic compounds

Refrigerators/Cryocoolers

- Cooling of water traps and cryo vacuum panels in the manufacture of semiconductors, hard disks, LCD displays

- Cooling of telescopes, electronic devices, computers, CCD cameras
- Cooling of infrared, gamma-ray, and X-ray detectors used in a variety of applications
- Cryosurgical devices used in gynecology, cardiac, prostrate and dental surgeries
- Cryofreezers, biofreezers, etc. used in the preservation of cells, tissues and cultures

The main aim of this monograph is to teach the different mixed refrigerant processes and the methods to optimize the composition of refrigerants used in these processes.

It is necessary to understand the fundamental cryogenic processes that operate with pure fluids thoroughly to be able to design mixed refrigerant processes that overcome the deficiencies of conventional systems. The concept of exergy efficiency/exergy loss and the working of fundamental cryogenic refrigeration and liquefaction processes operating with pure fluids are discussed in this chapter.

1.2 Sign convention

The following sign convention is followed in this monograph. The heat transfer to a system (from the surroundings) is considered positive, and heat transfer from the system to the surroundings is considered negative. Similarly, the work delivered by a system is considered positive, and the work done on the system is considered negative. The sign conventions are illustrated in Fig. 1.1 with reference to a heat engine and a refrigerator. Note that $\dot{Q}_o$ and $\dot{W}_c$ are negative, whereas $\dot{Q}_h$, $\dot{Q}_l$, and $\dot{W}$ are positive.

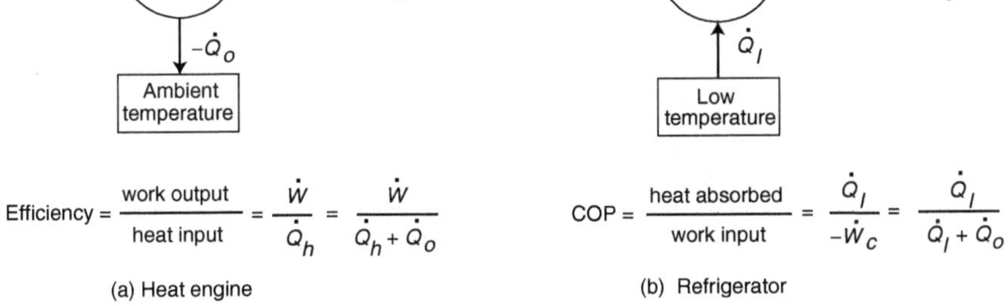

Fig. 1.1. Sign convention used in this monograph.

1.3 Ideal refrigeration and liquefaction processes

Refrigeration systems can be largely divided into two groups based on the refrigeration temperature as follows:

- refrigerators that provide refrigeration over a constant temperature such as Carnot, Stirling, Erricson (Fig. 1.2), and
- refrigerators that provide refrigeration over a range of temperatures such as those required for cooling a gas from room temperature to a low temperature, viz., the reverse Brayton refrigerator (Fig. 1.3).

1.3.1 Ideal constant-temperature refrigeration process

Consider a refrigerator that provides refrigeration over a constant temperature and operates on reversible thermodynamic processes. Such a refrigerator will henceforth

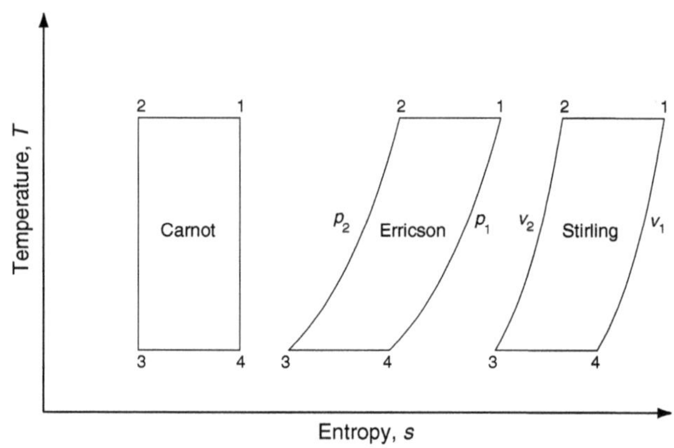

Fig. 1.2. Ideal refrigerator cycles that provide constant-temperature refrigeration (process 3–4).

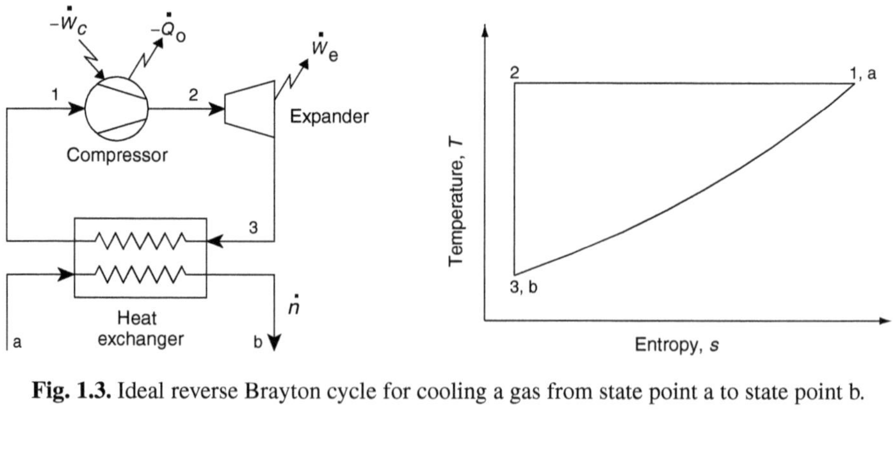

Fig. 1.3. Ideal reverse Brayton cycle for cooling a gas from state point a to state point b.

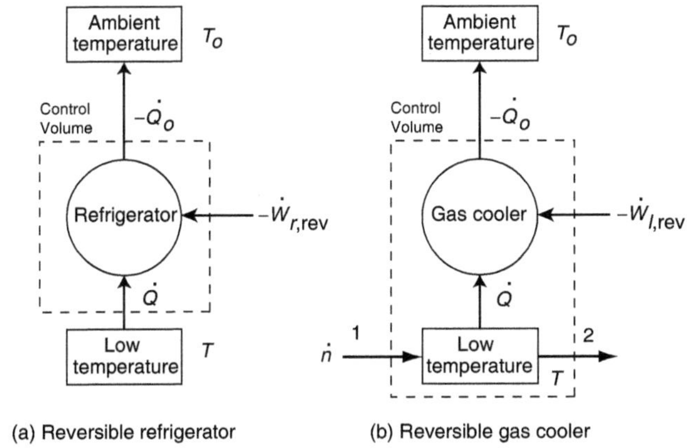

Fig. 1.4. Schematic of (a) a reversible refrigerator and (b) reversible gas cooler for cooling/liquefying a gas from state 1 to state 2.

be called a reversible refrigerator [Fig. 1.4(a)]. Heat is rejected to the surroundings at a temperature T_o and absorbed at a temperature T $(T < T_o)$. The heat transfer between the refrigerator and source/sink is assumed to occur at a zero temperature difference in all reversible refrigerators. The temperature of the refrigerant is therefore the same as that of the ambient (T_o) during the heat rejection process and that of the load (T) during the heat absorption process. The first and second laws of thermodynamics can be written for a reversible refrigerator as follows:

$$\text{first law:} \quad \dot{W}_{r,\,\text{rev}} = \dot{Q} + \dot{Q}_o, \tag{1.1}$$

$$\text{second law:} \quad \frac{\dot{Q}}{T} + \frac{\dot{Q}_o}{T_o} = 0. \tag{1.2}$$

Substituting Eq. (1.2) into Eq. (1.1) gives the expression for the power required by a reversible refrigerator as follows:

$$\boxed{-\dot{W}_{r,\,\text{rev}} = \dot{Q}\left(\frac{T_o}{T} - 1\right)}, \tag{1.3}$$

where T and T_o refer to the refrigeration and ambient temperatures, respectively. $\dot{Q}$ and $-\dot{Q}_o$ are the heat absorbed and heat rejected, respectively.[1]

The coefficient of performance (COP) of any refrigerator is defined as follows:

$$COP = \frac{\text{heat absorbed at low temperature}}{\text{compressor work input}} = \frac{Q}{-W_c} = \frac{\dot{Q}}{-\dot{W}_c}, \tag{1.4}$$

where Q and $-W_c$ refer to the heat absorbed and compressor work input in joules, and $\dot{Q}$ and $-\dot{W}_c$ refer to the heat transfer rate from the low-temperature source and the power supplied to the compressor in watts.

[1] $\dot{Q} > 0, \dot{Q}_o < 0, \dot{W}_{r,\,\text{rev}} < 0.$

The coefficient of performance (COP) of an ideal reversible refrigerator providing refrigeration at constant temperature can be expressed in terms of the temperatures for the heat source and heat sink using Eq. (1.3) as follows:

$$COP_{r,\,\text{rev}} = \frac{\dot{Q}}{-\dot{W}_{r,\,\text{rev}}} = \frac{T}{T_o - T}. \tag{1.5}$$

1.3.2 Ideal gas-cooling/liquefaction process

Figure 1.4(b) shows the schematic of a gas cooler in which the process fluid is cooled from a temperature T_1 to a temperature T_2. The first and second laws of thermodynamics can be written for the control volume of an ideal gas cooler [Fig. 1.4(b)] operating on reversible processes and providing refrigeration over a range of temperatures as follows:

$$\text{first law:} \quad \dot{W}_{l,rev} = \dot{Q} + \dot{Q}_o = \dot{n}(h_1 - h_2) + \dot{Q}_o, \tag{1.6}$$

$$\text{second law:} \quad \dot{n}(s_1 - s_2) + \frac{\dot{Q}_o}{T_o} = 0, \tag{1.7}$$

where $-\dot{W}_{l,\,\text{rev}}$ refers to the power input to the reversible gas cooler and $\dot{n}$ is the mole flow rate of the process fluid.[2]

Substituting Eq. (1.7) into Eq. (1.6) gives the expression for the minimum power required for cooling a gas from state 1 to state 2 as follows:

$$\boxed{-\dot{W}_{l,\,\text{rev}} = \dot{n}[(h_2 - h_1) - T_o(s_2 - s_1)] = \dot{n}(\text{ex}_2 - \text{ex}_1)}\,. \tag{1.8}$$

In the above expression, ex refers to the exergy of the fluid being cooled [$\text{ex} = (\text{h} - \text{h}_\text{o}) - \text{T}_\text{o}(\text{s} - \text{s}_\text{o})$], and T_o is the ambient temperature [see Eq. (1.16)]. It is evident from Eq. (1.8) that the minimum work required to cool a unit mole of a gas using an ideal gas cooler operating on reversible processes is the same as the exergy change of the fluid being cooled and is independent of the process used for cooling.

Consider the gas-cooling process shown in Fig. 1.5. The gas to be liquefied is compressed in an isothermal compressor (process 1–2) and expanded in an isentropic expander (process 2–3) to the required pressure and temperature (state point 3). The net power input to the process is given by the expression

$$-\dot{W}_{\text{net},\,l} = -\dot{W}_c - \dot{W}_e = \dot{n}[(h_2 - T_o\, s_2) - (h_1 - T_o\, s_1)] - \dot{n}(h_2 - h_3). \tag{1.9}$$

Since the entropy at state points 2 and 3 is the same ($s_2 = s_3$), the above expression for the net power required to cool the gas from temperature T_1 to T_3 can be expressed as

$$-\dot{W}_{\text{net},\,l} = \dot{n}[(h_3 - T_o\, s_3) - (h_1 - T_o\, s_1)] = \dot{n}(\text{ex}_3 - \text{ex}_1). \tag{1.10}$$

---

[2] $\dot{Q} > 0$, $\dot{Q}_o < 0$, and $\dot{W}_{l,\,\text{rev}} < 0$.

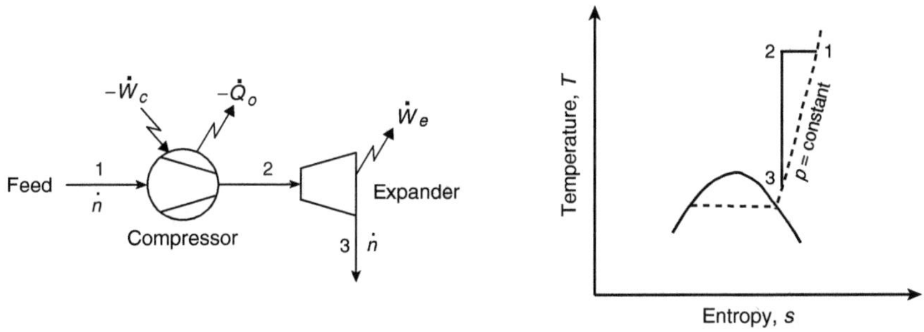

Fig. 1.5. Ideal gas cooling process.

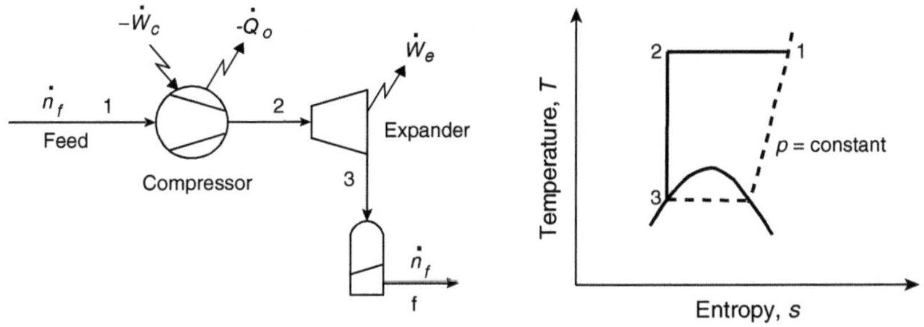

Fig. 1.6. Ideal process for gas liquefaction.

It can be seen that the net power required in the process shown in Fig. 1.5 is the same as that required by a reversible gas cooler for cooling a gas from state 1 to 3 [Eq. (1.8)].

The process shown in Fig. 1.5 is thus a reversible gas cooler. In order to differentiate the process shown in Fig. 1.5 from the generic ideal gas cooler shown in Fig. 1.4(b), the process shown in Fig. 1.5 is termed the ideal gas-cooling process.

Equation (1.8) is also valid when a gas is cooled from ambient temperature to the point where it is a saturated liquid. An ideal process for the liquefaction of a gas is shown in Fig. 1.6. The ideal gas liquefaction process shown in Fig. 1.6, however, has certain limitations:

- The operating pressure required for the reversible process is very high.
- Most expansion machines can tolerate only a very small fraction of liquid, making it difficult to use practical expanders for liquefaction.

Figure 1.7 shows an ideal liquefaction process operating at pressures of 10,000 bar and 1 bar with nitrogen. It can be observed that complete liquefaction of nitrogen is not possible with the ideal liquefaction process, even at an operating pressure of 10,000 bar. The concept of minimum power required for liquefaction using an ideal process, however, is useful to compare the performance of different processes.

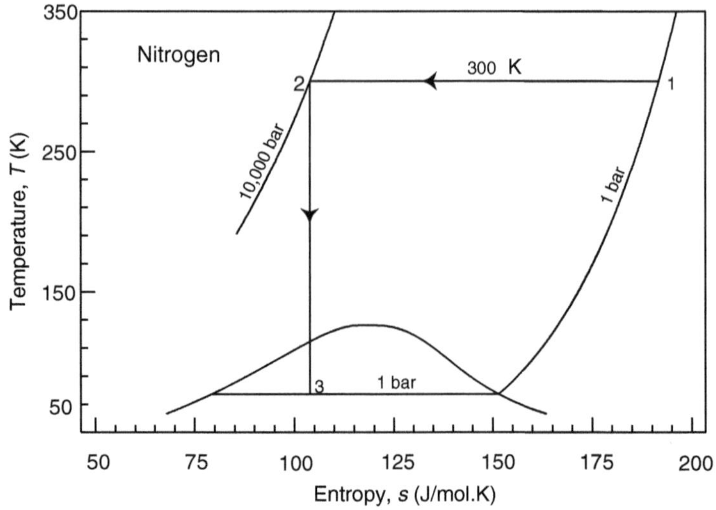

Fig. 1.7. Ideal nitrogen liquefaction process operating between pressures of 10,000 bar and 1 bar.

1.4 Exergy

The performance of a number of refrigeration and liquefaction processes is studied in this monograph. The performance of different systems can be compared on a common-denominator basis using the concept of exergy efficiency. Exergy efficiency is also used as the objective function in the optimization of low-temperature processes discussed in this monograph. A good understanding of the basic concepts of exergy, exergy efficiency, and exergy loss is essential to design efficient mixed refrigerant processes.

Exergy has several definitions [28, 82]. In this section, the concepts of exergy are introduced in relation to refrigerators and liquefiers. The reader may refer to advanced textbooks on exergy for more generalized approaches. The performance of systems that transfer heat to and from the ambient is strongly related to the ambient temperature and pressure. Consider a refrigerator that absorbs a certain quantity of heat Q at a temperature T. The heat absorbed along with the work supplied is rejected to the ambient. The work required by the refrigerator for providing the required refrigeration is therefore related to the ambient temperature. The work required will be minimized when a refrigerator operates on a reversible thermodynamic process, for example, Carnot, Stirling, etc. in which heat transfer between the refrigerant and the surroundings occurs at a zero temperature approach. This minimum work ($-W_{\text{rev}}$) required to provide a refrigeration Q at temperature T while rejecting the heat to ambient at temperature T_o is called exergy.

The coefficient of performance (COP) of a refrigerator operating on reversible processes is related to the temperature at which heat is absorbed (T) and the temperature of the ambient (T_o) as follows (Fig. 1.1):

$$COP_{r,\,\text{rev}} = \frac{Q}{-W_{\text{rev}}} = \frac{T}{T_o - T}. \tag{1.11}$$

The exergy or minimum work ($-W_{rev}$) required to transfer heat Q at temperature T using a reversible refrigerator is given by the expression

$$-W_{\text{rev}} = Q\left(\frac{T_o - T}{T}\right). \tag{1.12}$$

In the case of heat engines, exergy is the maximum work that can be obtained from a reversible heat engine while absorbing heat Q at a temperature T and rejecting part of this heat to ambient at a temperature T_o. It can easily be shown that the above expression relating exergy and the operating temperatures is also applicable for a heat engine (see Fig. 1.8).

The concept of exergy can be extended to other systems as well. Consider a reversible heat engine in which the heat is supplied by a fluid stream as shown in Fig. 1.9. The fluid stream enters at state 1 and leaves at state 2. The first and second laws of thermodynamics can be expressed for a reversible heat engine as follows:

$$\text{first law:}\quad W_{\text{rev}} = Q + Q_o = n(h_1 - h_2) + Q_o, \tag{1.13}$$

$$\text{second law:}\quad n(s_1 - s_2) + \tfrac{Q_o}{T_o} - 0, \tag{1.14}$$

where W_{rev} refers to the work output from the heat engine and n is the number of moles of the fluid stream.

Substituting Eq. (1.14) into Eq. (1.13) gives the expression for the work that can be obtained from a fluid stream as

$$W_{\text{rev}} = n[(h_1 - h_2) - T_o(s_1 - s_2)]. \tag{1.15}$$

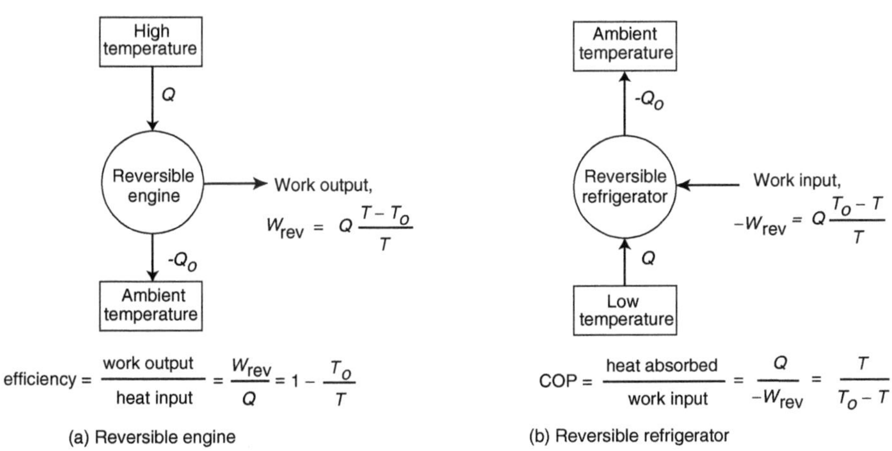

Fig. 1.8. Relationship between (a) work obtained from a reversible engine and (b) work supplied to a reversible refrigerator.

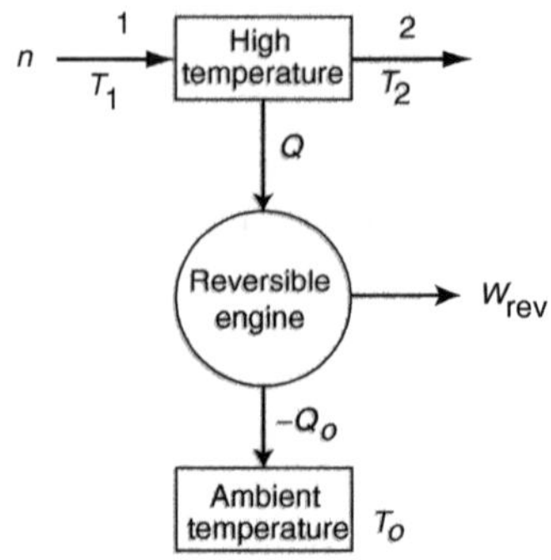

Fig. 1.9. Heat engine receiving heat from a fluid stream.

The above expression gives the relationship between the reversible work that can be extracted from a fluid between state points 1 and 2, the ambient temperature, and the thermodynamic properties of the fluid at the two states. The maximum work is obtained from the fluid stream when it is cooled from state 1 to the ambient temperature and pressure (state 0). The specific exergy of a fluid stream can be defined as the maximum work that can be obtained from the fluid when it follows reversible processes to reach equilibrium with the surroundings and can be expressed as follows:

$$\mathrm{ex} = \frac{W_{\mathrm{rev}}}{n} = (h - h_o) - T_o(s - s_o), \tag{1.16}$$

where ex refers to the specific exergy of the fluid stream at an enthalpy h and entropy s. h_o and s_o are the enthalpy and entropy of the stream at ambient temperature and pressure, respectively.

1.5 Exergy loss and exergy efficiency

Consider the system shown in Fig. 1.10. The first law of thermodynamics (energy balance) can be written as follows:

$$\dot{n}_1 h_1 - \dot{n}_2 h_2 + \dot{Q}_1 + \dot{Q}_2 - \dot{W}_1 - \dot{W}_2 = 0. \tag{1.17}$$

The second law of thermodynamics (Clausius inequality) can be written as follows:

$$\dot{n}_1 s_1 - \dot{n}_2 s_2 + \frac{\dot{Q}_1}{T_1} + \frac{\dot{Q}_2}{T_2} \leq 0. \tag{1.18}$$

Equations (1.17) and (1.18) can be combined after multiplying Eq. (1.18) with the surroundings (ambient) temperature T_o as follows:

$$\dot{n}_1 \{(h_1 - T_o s_1)\} - \dot{n}_2 \{(h_2 - T_o s_2)\} + \dot{Q}_1 \left(1 - \frac{T_o}{T_1}\right) + \dot{Q}_2 \left(1 - \frac{T_o}{T_2}\right) - \dot{W}_1 - \dot{W}_2 \geq 0. \tag{1.19}$$

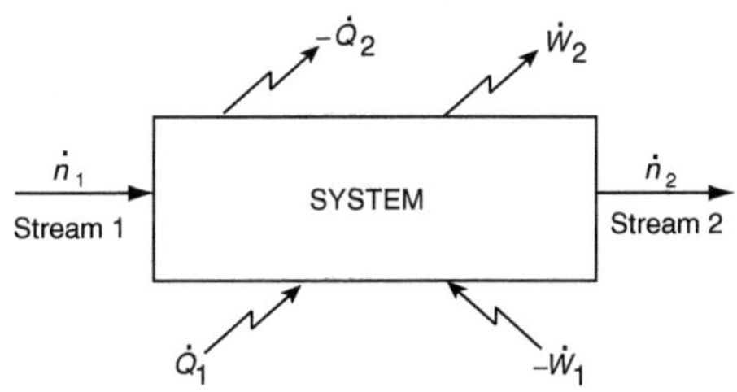

Fig. 1.10. General thermal system.

The above expression can also be expressed in terms of the exergy of the streams as follows:

$$\dot{n}_1\, \mathrm{ex}_1 - \dot{n}_2\, \mathrm{ex}_2 + \dot{Q}_1 \left(1 - \frac{T_o}{T_1}\right) + \dot{Q}_2 \left(1 - \frac{T_o}{T_2}\right) - \dot{W}_1 - \dot{W}_2 \geq 0, \qquad (1.20)$$

which can be generalized as follows:

$$\sum \dot{n}_{\mathrm{in}}\, \mathrm{ex}_{\mathrm{in}} - \sum \dot{n}_{\mathrm{out}}\, \mathrm{ex}_{\mathrm{out}} + \sum_i \dot{Q}_i \left(1 - \frac{T_o}{T_i}\right) + \sum_j -\dot{W}_j \geq 0, \qquad (1.21)$$

where first two terms represent the exergy of the streams entering and leaving the system, respectively. The third term in the above equation is the net power that can be obtained from the heat transfer rate to the system ($\dot{Q}_i$) using a reversible heat engine (see Fig. 1.8). The fourth term is the net power supplied to the system.[3] Equation (1.21) can be considered as the exergy balance equation for any system.

The left-hand side of Eq. (1.21) represents the difference between the input and output exergy, which is known as exergy loss, and can also be expressed as

$$\sum \text{exergy loss in each component} \geq 0. \qquad (1.22)$$

The sum of all exergy losses is zero if the system operates on reversible thermodynamic processes and is greater than zero if the system operates on irreversible thermodynamic processes. The exergy loss in different components of a cryogenic refrigeration or liquefaction system has been summarized in Table 1.1.

[3] $-\dot{W} > 0$.

Table 1.1. Exergy loss in different components

Equipment	Symbol	Exergy loss (kW)
Compressor	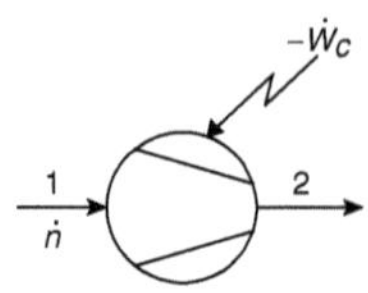	$\Delta \mathrm{ex}_{\mathrm{loss}} = \dot{n}(\mathrm{ex}_1 - \mathrm{ex}_2) - \dot{W}_c$
Expander	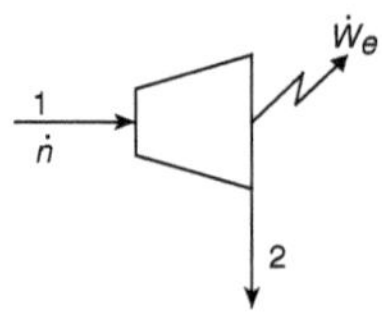	$\Delta \mathrm{ex}_{\mathrm{loss}} = \dot{n}(\mathrm{ex}_1 - \mathrm{ex}_2) - \dot{W}_e$
Throttle valve	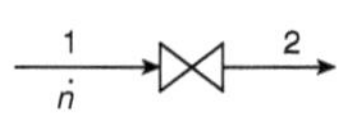	$\Delta \mathrm{ex}_{\mathrm{loss}} = \dot{n}(\mathrm{ex}_1 - \mathrm{ex}_2)$
Phase separator or stream splitter	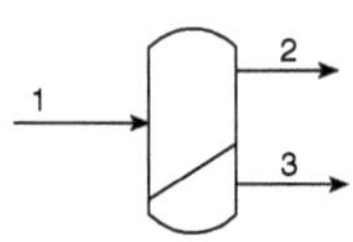	$\Delta \mathrm{ex}_{\mathrm{loss}} = \dot{n}_1 \mathrm{ex}_1 - \dot{n}_2 \mathrm{ex}_2 - \dot{n}_3 \mathrm{ex}_3$
Stream mixer	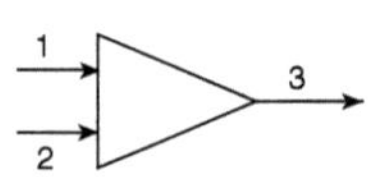	$\Delta \mathrm{ex}_{\mathrm{loss}} = \dot{n}_1 \mathrm{ex}_1 + \dot{n}_2 \mathrm{ex}_2 - \dot{n}_3 \mathrm{ex}_3$
Heat exchanger	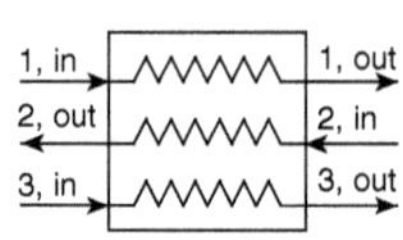	$\Delta \mathrm{ex}_{\mathrm{loss}} = \sum_{i=1}^{n} \dot{n}_i (\mathrm{ex}_{i,\mathrm{in}} - \mathrm{ex}_{i,\mathrm{out}})$
Condenser or aftercooler exchanging heat with ambient	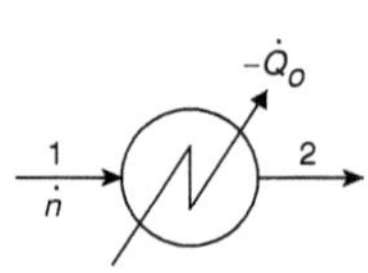	$\Delta \mathrm{ex}_{\mathrm{loss}} = \dot{n}(\mathrm{ex}_1 - \mathrm{ex}_2)$
Evaporator operating at low temperature	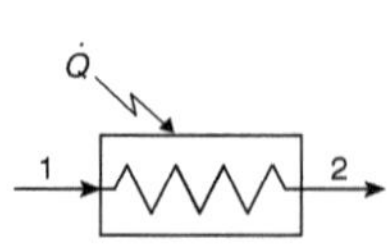	$\Delta \mathrm{ex}_{\mathrm{loss}} = \dot{n}(\mathrm{ex}_1 - \mathrm{ex}_2) + \dot{Q}(1 - T_o/T)$

$\dot{W}_c < 0, \dot{W}_e > 0, \dot{Q} > 0, \dot{Q}_o < 0.$

The exergy efficiency of any refrigeration or liquefaction system is defined as follows:

$$\eta_{\text{ex}} = \frac{\text{minimum power required by a reversible system}}{\text{actual power supplied}}. \tag{1.23}$$

or

$$\eta_{\text{ex}} = 1 - \frac{\sum \text{exergy loss in each component}}{\text{actual power supplied}}. \tag{1.24}$$

The concept of exergy efficiency can also be used in processes where there is no work transfer, for example, an open-cycle Linde–Hampson refrigerator. Similarly, the exergy efficiency can also be determined for part of a process (control volume). In all such cases, the actual power supplied is replaced by exergy expenditure as follows:

$$\eta_{\text{ex}} = 1 - \frac{\sum \text{exergy loss in each component}}{\text{exergy expenditure}}, \tag{1.25}$$

or

$$\eta_{ex} = 1 - \frac{\sum \dot{n}_{\text{in}}\,\text{ex}_{\text{in}} - \sum \dot{n}_{\text{out}}\,\text{ex}_{\text{out}} + \sum_i \dot{Q}_i \left(1 - \frac{T_o}{T_i}\right) + \sum_j -\dot{W}_j}{\text{exergy expenditure}}. \tag{1.26}$$

The exergy expenditure depends on the type of system. When a system receives heat and produces work as in a heat engine, the exergy expenditure of the system is $\dot{Q}(1 - T_o/T)$. On the other hand, when a system receives work and absorbs heat as in a refrigerator, the exergy expenditure of the system is $(-\dot{W})$. The expressions for input exergy, exergy loss, and exergy efficiency of different commonly used systems (Fig. 1.11) are summarized in Table 1.2.

Table 1.2. Expressions for input exergy, exergy loss, and exergy efficiency of some commonly used cryogenic process equipment [see Eq. (1.26)]

System	Figure	Input exergy (W)	Exergy loss (W)	Exergy efficiency
Refrigerator	Fig. 1.11(a)	$-\dot{W}$	$\dot{Q}(1 - T_o/T) - \dot{W}$	$\frac{\dot{Q}(T_o/T-1)}{-\dot{W}}$
Heat engine	Fig. 1.11(b)	$\dot{Q}(1 - T_o/T)$	$\dot{Q}(1 - T_o/T) - \dot{W}$	$\frac{\dot{W}}{\dot{Q}(1-T_o/T)}$
Gas cooler/ liquefier	Fig. 1.11(c)	$-\dot{W}$	$\dot{n}(\text{ex}_1 - \text{ex}_2) - \dot{W}$	$\frac{\dot{n}(\text{ex}_2-\text{ex}_1)}{-\dot{W}}$
Compressor	Fig. 1.11(d)	$-\dot{W}$	$\dot{n}(\text{ex}_1 - \text{ex}_2) - \dot{W}$	$\frac{\dot{n}(\text{ex}_2-\text{ex}_1)}{-\dot{W}}$
Turbine	Fig. 1.11(e)	$\dot{n}(\text{ex}_2 - \text{ex}_1)$	$\dot{n}(\text{ex}_1 - \text{ex}_2) - \dot{W}$	$\frac{\dot{W}}{\dot{n}(\text{ex}_1-\text{ex}_2)}$

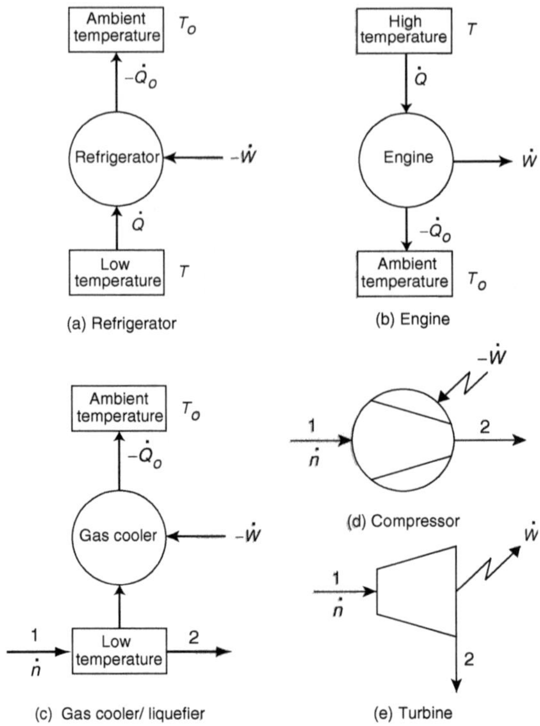

Fig. 1.11. Some commonly used systems.

1.6 Exergy efficiency of processes without any work interaction

The exergy efficiency was defined in the previous section with reference to the work interaction between a system and the surroundings when operating on irreversible and reversible processes.

Consider an open-cycle Linde–Hampson liquefier used in the cooling of infrared detectors on missiles (Fig. 1.12). High-pressure nitrogen is supplied from a gas bottle to the system (stream 1). There is no compressor or turbine in the process. The mole flow rate of the refrigerant through the system is $\dot{n}$.

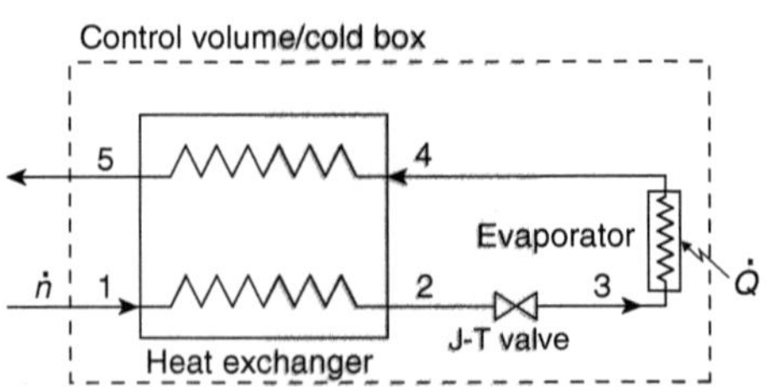

Fig. 1.12. open-cycle Linde–Hampson liquefier.

The work necessary for providing the required refrigeration is provided by the refrigerant stream. Therefore, the expenditure in this system is the difference between the exergy of the refrigerant entering and leaving the system $\dot{n}(\mathrm{ex}_1 - \mathrm{ex}_5)$. The exergy efficiency of the open-cycle Linde–Hampson liquefier can be obtained from Eq. (1.26) and Table 1.2 as follows:

$$\eta_{\mathrm{ex}} = 1 - \frac{\sum \text{exergy loss}}{\text{exergy expenditure}} = 1 - \frac{\dot{n}\,\mathrm{ex}_1 - \dot{n}\,\mathrm{ex}_5 + \dot{Q}\left(1 - \frac{T_O}{T}\right)}{\dot{n}\,\mathrm{ex}_1 - \dot{n}\,\mathrm{ex}_5} \tag{1.27}$$

or

$$\eta_{\mathrm{ex}} = \frac{\dot{Q}\left(\frac{T_O}{T} - 1\right)}{\dot{n}\left(\mathrm{ex}_1 - \mathrm{ex}_5\right)}. \tag{1.28}$$

Consider the Linde gas-cooling process shown in Fig. 1.13 for cooling or liquefying a gas. The compressor power input $(-\dot{W}_c)$ is the exergy expenditure when the entire process is considered. If only the cold box is considered, the exergy expenditure is the difference between the exergy of the refrigerant entering and leaving the cold box.

The overall exergy efficiency of the entire process can be expressed using Eq. (1.26) and Table 1.2 as follows:

$$\eta_{\mathrm{ex,\,o}} = 1 - \frac{\sum \text{exergy loss}}{\text{exergy expenditure}} = 1 - \frac{\dot{n}_7\,\mathrm{ex}_7 - \dot{n}_8\,\mathrm{ex}_8 - \dot{W}_c}{-\dot{W}_c} \tag{1.29}$$

or

$$\eta_{\mathrm{ex,\,o}} = \frac{\dot{n}_7(\mathrm{ex}_8 - \mathrm{ex}_7)}{-\dot{W}_c}. \tag{1.30}$$

Similarly, the exergy efficiency of the cold box of the process can be expressed using Eq. (1.26) and Table 1.2 as follows:

$$\eta_{\mathrm{ex,\,cb}} = 1 - \frac{\dot{n}_7\,\mathrm{ex}_7 - \dot{n}_8\,\mathrm{ex}_8 + \dot{n}_3\,\mathrm{ex}_3 - \dot{n}_6\,\mathrm{ex}_6}{\dot{n}_3\,\mathrm{ex}_3 - \dot{n}_6\,\mathrm{ex}_6} \tag{1.31}$$

or

$$\eta_{\mathrm{ex,\,cb}} = \frac{\dot{n}_7(\mathrm{ex}_8 - \mathrm{ex}_7)}{\dot{n}_3(\mathrm{ex}_3 - \mathrm{ex}_6)}. \tag{1.32}$$

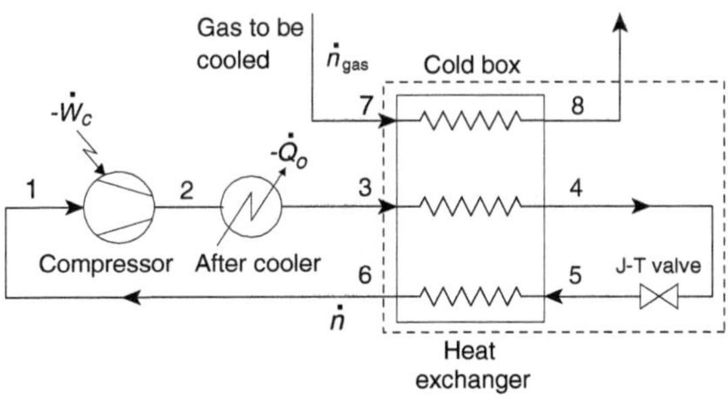

Fig. 1.13. Linde gas-cooling process.

1.7 Performance of an ideal gas cooler operating with a non-ideal expander

Consider the example of nitrogen to be cooled from 300 K to 125 K at a constant pressure of 10 bar using an ideal gas-cooling process shown in Fig. 1.5. The compression is assumed to occur at constant temperature, while the expansion process is assumed to be non-isentropic. The net power supplied to the system $-\dot{W}$ is the difference between that supplied to the compressor and that obtained from the expander ($-\dot{W} = -\dot{W}_c - \dot{W}_e$). Figure 1.14 shows the variation of compressor exit pressure and exergy efficiency of the process (Fig. 1.5) with the adiabatic efficiency of the expander. The exergy efficiency of the process is defined as the ratio of minimum work to actual work [Eq.(1.23)] as follows:

$$\eta_{\text{ex}} = \frac{-\dot{W}_{l,\text{rev}}}{-\dot{W}_c} = \frac{\dot{n}(\text{ex}_3 - \text{ex}_1)}{-\dot{W}_c}. \tag{1.33}$$

The adiabatic efficiency of the expander is defined as follows:

$$\eta_{\text{ad}} = \frac{\text{work obtained during non-isentropic process}}{\text{work obtained during isentropic expansion process}}. \tag{1.34}$$

An operating pressure above 450 bar is required when the expander adiabatic (isentropic) efficiency is 80%, which is typically the highest efficiency obtainable in practical expanders,[4] as shown in Fig. 1.14. Because of the high operating pressures, this process is only of theoretical interest.

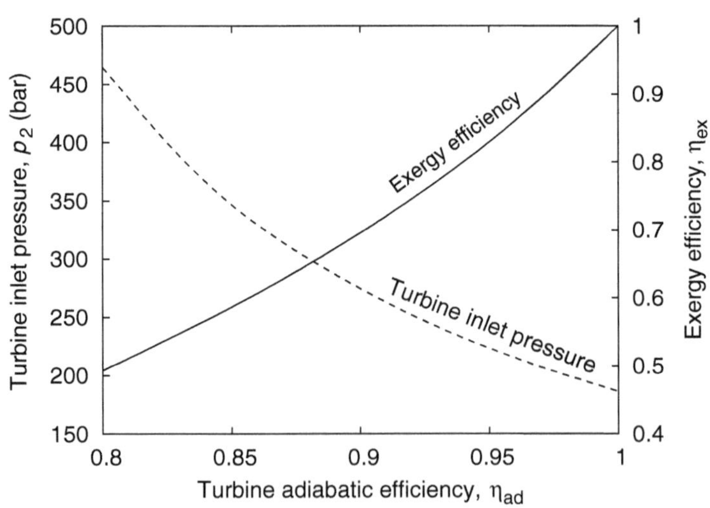

Fig. 1.14. Variation of exergy efficiency and compressor exit pressure of an ideal gas cooler shown in Fig. 1.5 to cool nitrogen from 10 bar, 300 K to 10 bar, 125 K using an isothermal compressor and a non-isentropic expander.

[4] The adiabatic efficiency of small expanders is typically less than 70%.

It can be seen that the exergy efficiency of an ideal gas-cooling system shown in Fig. 1.5 and operating with an ideal isothermal compressor and a turbine with an adiabatic efficiency (η_{ad}) of 80% is only about 50%. The net power input to the system is therefore double that of an ideal system operating with an isothermal compressor and an isentropic expander ($\eta_{ad} = 100\%$). The exergy efficiency of the process will be much lower than 50% when a non-isothermal compressor is used.

A high expander adiabatic efficiency is therefore a prerequisite for a high exergy efficiency of all turbine-based processes including the ideal gas-cooling process shown in Fig. 1.5. Exergy efficiencies greater than 70% can be obtained in most mixed refrigerant natural gas coolers and liquefiers operating with isothermal compressors. The exergy efficiency of practical mixed refrigerant natural gas liquefiers operating with non-isothermal compressors is typically greater than 50%.

1.8 Precooled ideal liquefaction process

The high pressure required in the ideal liquefaction process shown in Figs. 1.6 and 1.7 can be reduced by decreasing the temperature of the working fluid before expansion as shown in Fig. 1.15. Consider an ideal precooled nitrogen liquefaction process. Figure 1.16 shows the temperatures required at the entry of the expander to completely liquefy nitrogen at different operating pressures using the precooled process shown in Fig. 1.15. It can be observed that very low temperatures are required at the inlet of the expander for a reasonable liquid yield (the fraction of gas that gets liquefied on expansion). The working fluid should be partly condensed at the entry of the expander inlet at low operating pressures ($p < p_c$). The low-temperature precooling can be achieved using either an external or internal precooler. With an internal precooler, a part of the working fluid is diverted and expanded in an expander to provide the necessary refrigeration as in a Kapitza process.

Since only a part of the fluid gets liquefied even when very low temperature precooling is provided, the refrigeration available with the unliquefied vapor leaving the expander can be used to precool the high-pressure fluid entering the expander, as in the Solvay liquefaction process shown in Fig. 1.17. The Solvay liquefaction process

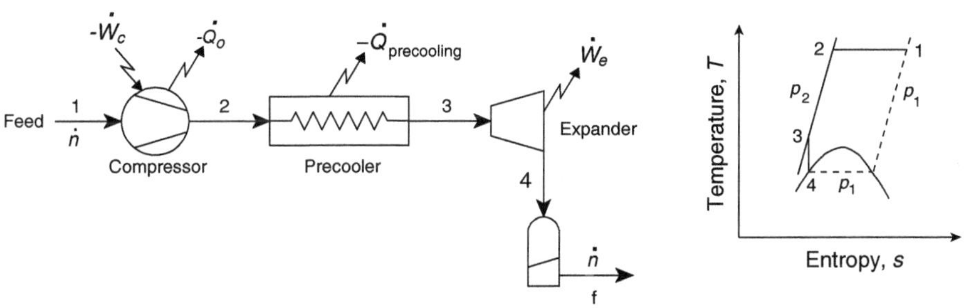

Fig. 1.15. Precooled ideal liquefaction process.

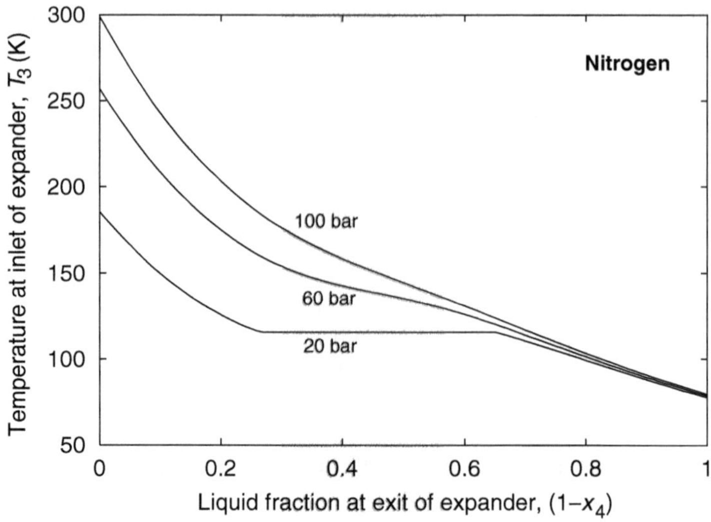

Fig. 1.16. Temperature at the inlet of the expander in an ideal precooled nitrogen liquefier operating with an isothermal compressor and an isentropic expander (Fig. 1.15).

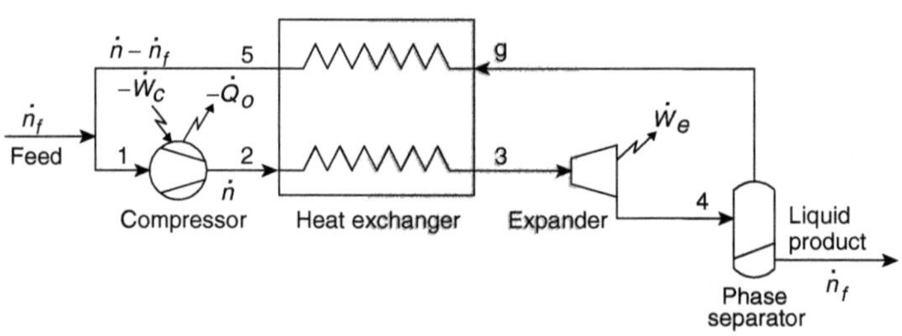

Fig. 1.17. Ideal Solvay liquefaction process.

has not been used in practice, essentially due to the problems associated with the expansion of two-phase fluids in expanders. Practical liquefaction systems have been built by replacing the isentropic expander with an isenthalpic throttle. A variant of the Solvay refrigerator known as the Gifford–McMahon (G-M) refrigerator is widely used for cryo vacuum pump applications. No liquefaction of the working gas occurs during the expansion process in the case of G-M refrigerators [13, 90].

1.9 Linde–Hampson refrigerators and liquefiers

Figure 1.18 shows an ideal Linde–Hampson liquefier, originally invented by Carl von Linde and William Hampson independently in 1895 to liquefy air. In an ideal liquefaction system, the gas to be liquefied is compressed isothermally at ambient

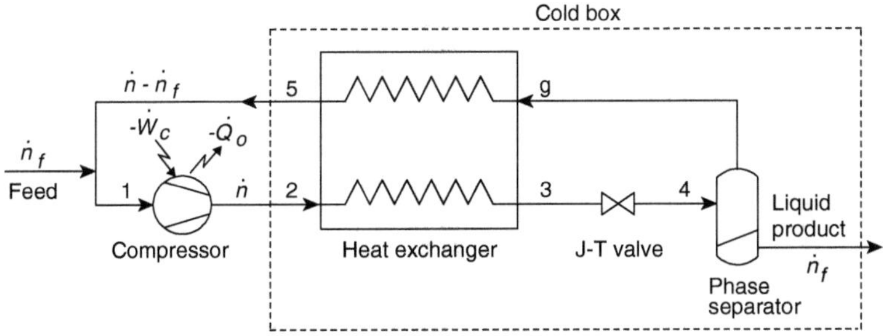

Fig. 1.18. Ideal Linde–Hampson liquefaction process.

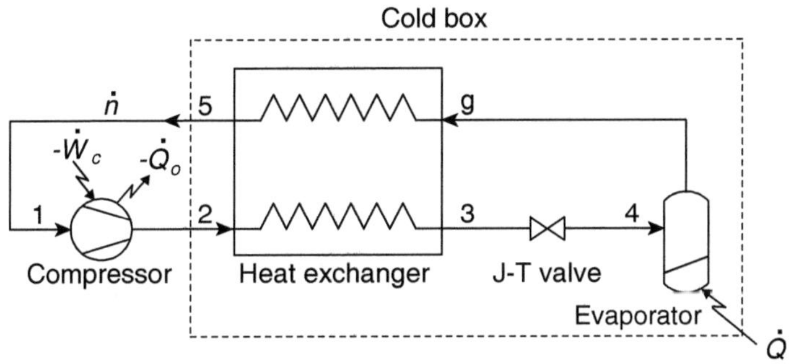

Fig. 1.19. Ideal Linde–Hampson refrigerator.

temperature.[5] The high-pressure gas is cooled in a heat exchanger by the return (low-pressure) stream and expanded to the desired pressure in a throttling (constant-enthalpy) device, more often called the J-T valve. The liquid and gaseous phases are separated in the phase separator, and the unliquefied gas is used to cool the high-pressure stream in the heat exchanger. With a minor change, the Linde–Hampson liquefier can be used to provide refrigeration at constant temperature, as shown in Fig. 1.19.

The Linde–Hampson refrigeration/liquefaction process can be represented on *T–s* and *T–h* planes as shown in Figs. 1.20 and 1.21.

A detailed study of the Linde–Hampson systems operating with pure fluids is a prerequisite to understanding the operation of the Linde–Hampson and other processes operating with gas mixtures.

An energy balance over a control volume that includes the heat exchanger, the expansion valve, and the phase separator (cold box) of a Linde–Hampson liquefier (Fig. 1.18) gives

$$\dot{n}h_2 = \dot{n}_f h_f + (\dot{n} - \dot{n}_f)h_5. \tag{1.35}$$

[5] In practical systems, the heat of compression is normally rejected to a coolant in an after-cooler.

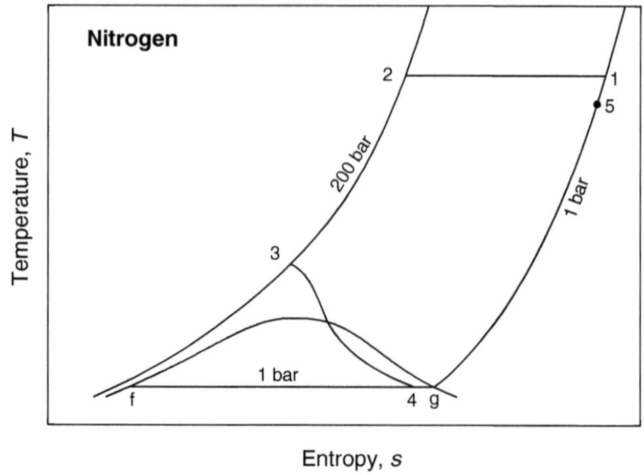

Fig. 1.20. Typical Linde–Hampson liquefaction/refrigeration system operating with nitrogen as the working fluid.

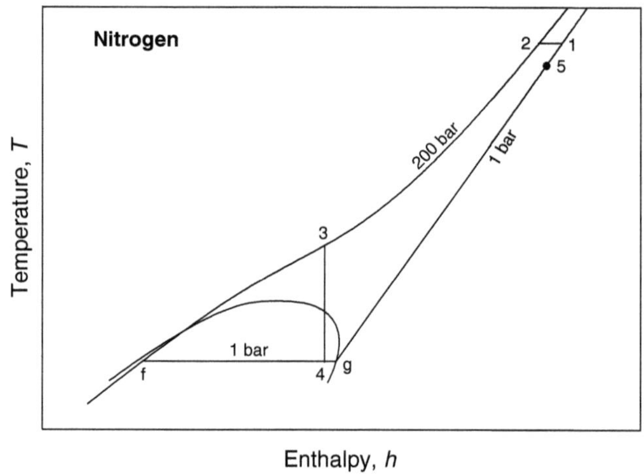

Fig. 1.21. Typical Linde–Hampson liquefaction/refrigeration system operating with nitrogen as the working fluid.

The fraction of the gas that gets liquefied on expansion is known as the liquid yield (Y) and can be expressed in terms of the enthalpy of the working fluid in the case of a Linde–Hampson liquefier using the above equation as follows:

$$\text{liquid yield, } Y = \frac{\dot{n}_f}{\dot{n}} = \frac{h_5 - h_2}{h_5 - h_f}. \tag{1.36}$$

The temperature of the low-pressure return gas at the exit of the heat exchanger (stream 5) depends on the effectiveness of the heat exchanger used. In the case of an ideal heat exchanger operating with pure fluids and a heat exchanger effectiveness $\varepsilon = 100\%$, $T_5 = T_2 = T_1$. It can be observed from Fig. 1.21 that the numerator in Eq. (1.36), the enthalpy difference between the low- and high-pressure fluids at the warm end of the heat exchanger $(h_5 - h_2)$, is much smaller than the denominator $(h_5 - h_f)$, which results in a very small fraction of liquid yield, typically less than 10%.

It can also be observed from Fig. 1.21 that the difference between the enthalpy of low- and high-pressure streams at any temperature $(h_1 - h_2)$ increases with a decrease in ambient temperature (T_1). The liquid yield therefore increases with a decrease in the ambient temperature.

The liquid yield will also increase when the temperature of the working gas entering the compressor (T_2) is decreased in a precooler before entry into the heat exchanger.

Figure 1.22 shows the variation of liquid yield for nitrogen, argon, and oxygen as a function of the temperature at the warm end of the heat exchanger. It can be seen that the maximum liquid yield, when the temperature at the warm end of the heat exchanger is 300 K, is less than 10% for a Linde–Hampson liquefier operating with nitrogen. On the other hand, the liquid yield for similar conditions will be much higher for oxygen and argon. In general, the higher the non-ideality of the gas at room temperature and operating pressures, the higher will be the liquid yield.

Argon is the most non-ideal among nitrogen, oxygen, and argon, at room temperature.[6] Hence the liquid yield is higher when argon is used as the working fluid in a Linde–Hampson liquefaction system, compared to that with other fluids, as shown in Fig. 1.22. The liquid yield will increase with a decrease in the temperature of the high- and low-pressure fluids at the warm end of the heat exchanger (T_2, T_5). The liquid yield is also a strong function of the operating pressure (p_2), as seen in Fig. 1.22. The liquid yield of a Linde–Hampson system can be increased by decreasing the temperature of the working fluid before it enters the heat exchanger (T_2) or by increasing the compressor exit pressure (p_2). The pressure at which the liquid yield is maximized at any given heat exchanger warm end temperatures $(T_2 = T_5)$ is given by the expression

$$\left(\frac{\partial Y}{\partial p_2}\right)_{T_2} = 0 \text{ or } \left(\frac{\partial h_2}{\partial p_2}\right)_{T_2} = 0 \text{ or } \mu_{\mathrm{J-T}}(p_2)\Big|_{T=T_2} = 0. \quad (1.37)$$

[6] The Joule–Thomson coefficient $(\mu_{\mathrm{J-T}})$ is 0.212, 0.265, and 0.361 K/bar respectively, for nitrogen, oxygen and argon at 300 K, 1 bar. The higher the value of $\mu_{\mathrm{J-T}}$, the higher is the non-ideality. $\mu_{\mathrm{J-T}}$ is zero for an ideal gas. In general, the larger the difference between ambient temperature and the critical point of a fluid, the smaller is the non-ideality at room temperature.

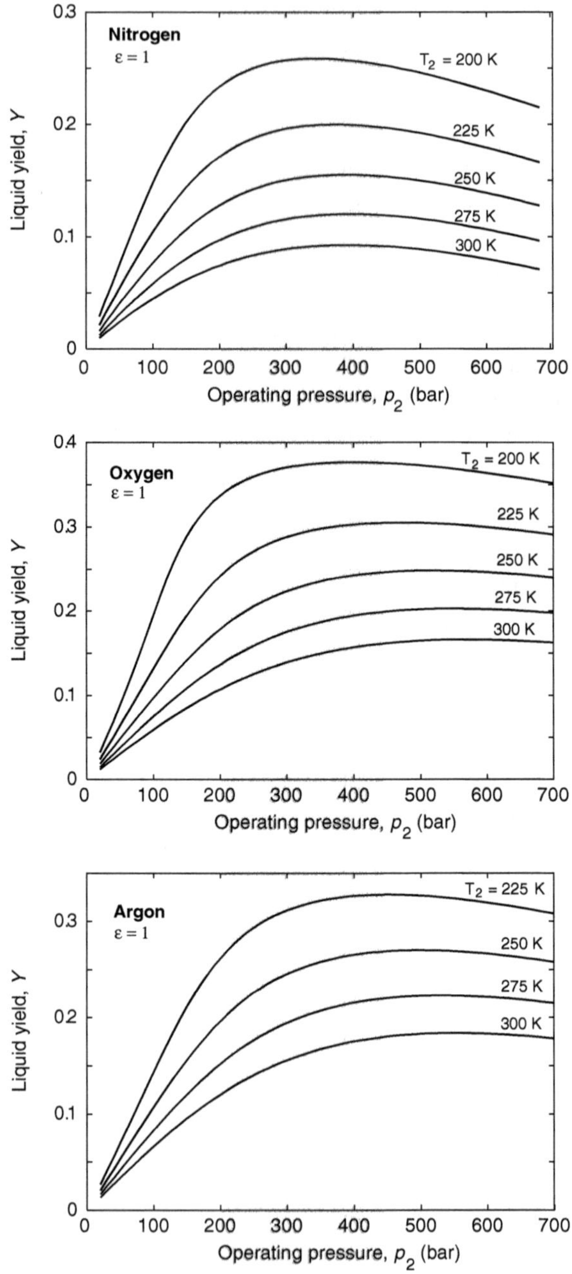

Fig. 1.22. Variation of liquid yield with temperature of the streams at the warm end of an ideal heat exchanger of a Linde–Hampson liquefier operating with different fluids ($p_1 = 1$ bar). The compression process has been assumed to be isothermal and the heat exchanger effectiveness $\varepsilon = 1$.

1.10 Joule–Thomson coefficient

Consider the temperature change a gas undergoes during an isenthalpic expansion or throttling process. James Prescott Joule and William Thomson (Lord Kelvin) studied the cooling of gases by isenthalpic (free) expansion of gases systematically. The effect has come to be known as the J-T effect named after them. Linde–Hampson refrigerators are also known as J-T refrigerators.

The Joule–Thomson coefficient ($\mu_{\mathrm{J-T}}$) of a gas is defined as follows:

$$\mu_{\mathrm{J-T}} = \left(\frac{\partial T}{\partial p}\right)_h . \tag{1.38}$$

Using the cyclic rule and the Maxwell relations,[7] Eq. (1.38) can be expressed in terms of the volume and temperature of the fluid as follows:

$$\mu_{\mathrm{J-T}} = -\left(\frac{\partial T}{\partial h}\right)_p \left(\frac{\partial h}{\partial p}\right)_T = \frac{1}{c_p}\left[T\left(\frac{\partial v}{\partial T}\right)_p - v\right]. \tag{1.39}$$

When a gas expands from a high pressure to a low pressure, the temperature of the gas will decrease when $\mu_{\mathrm{J-T}}$ is positive, will increase when $\mu_{\mathrm{J-T}}$ is negative, and will remain constant when $\mu_{\mathrm{J-T}}$ is zero.

Since $(\partial v/\partial T)_p = v/T$ for an ideal gas, the Joule–Thomson coefficient $\mu_{\mathrm{J-T}}$ is zero for ideal gases. Fortunately, all gases behave non-ideally at low temperatures (except at very high pressures), making it possible to cool gases using the Joule–Thomson (constant enthalpy) expansion process.

Figure 1.23 shows the variation of the J-T coefficient ($\mu_{\mathrm{J-T}}$) of helium with temperature at different pressures determined using the NIST 12 database [63]. The temperature where $\mu_{\mathrm{J-T}}$ is zero is called the inversion temperature and the line joining the inversion points is known as the inversion curve. It can be seen from Fig. 1.23 that the inversion temperature is a strong function of pressure. Since the specific heat is always positive, a negative Joule–Thomson coefficient ($\mu_{\mathrm{J-T}}$) indicates that the enthalpy of the gas increases with an increase in pressure at any given temperature T [Eq. (1.39)]. Consider an ideal Linde–Hampson liquefier. The liquid yield is proportional to the difference between the enthalpy of the gas at low and high pressures at ambient temperature [$(h_5 - h_2)$ in Eq. (1.36)]. The enthalpy of the low-pressure fluid is always greater than that of the high-pressure fluid at ambient temperature in the case of all gases other than helium, hydrogen, and neon (see Fig. 1.21). It can be seen from Fig.1.24 that $(h_5 - h_2) > 0$ is satisfied in helium only when the temperature is lower than the inversion temperature [$\mu_{\mathrm{J-T}} = 0$ or $(\partial h/\partial p)_T = 0$]. A similar behavior is observed in the case of neon and hydrogen. The Joule–Thomson coefficient is nearly independent of pressure at low pressures (see Fig. 1.23).

The maximum temperature at which the J-T coefficient is zero is known as the maximum inversion temperature. Since the maximum inversion temperature of helium, hydrogen and neon is lower than ambient temperature, only these gases need

[7] $\left(\frac{\partial A}{\partial B}\right)_C \left(\frac{\partial B}{\partial C}\right)_A \left(\frac{\partial C}{\partial A}\right)_B = -1$, $\left(\frac{\partial h}{\partial p}\right)_T = -T\left(\frac{\partial s}{\partial p}\right)_v + v$, and $\left(\frac{\partial s}{\partial p}\right)_v = \left(\frac{\partial v}{\partial T}\right)_p$.

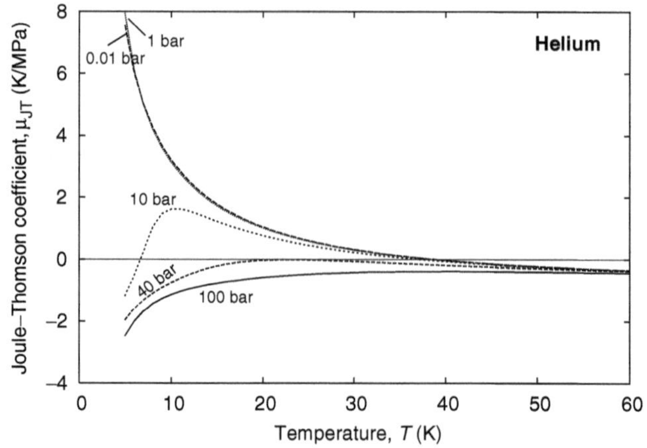

Fig. 1.23. Variation of J-T coefficient (μ_{J-T}) with temperature and pressure for helium.

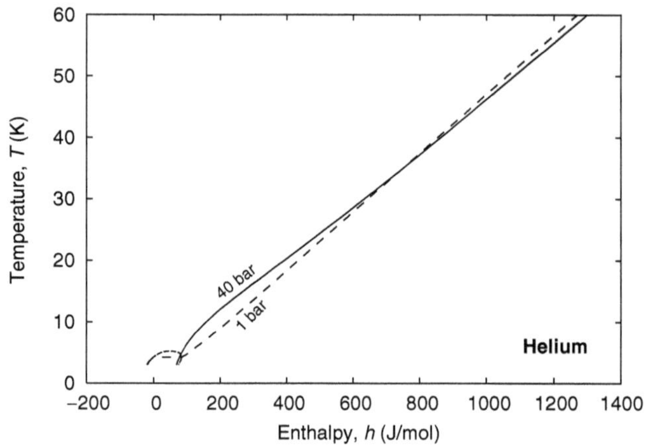

Fig. 1.24. Variation of enthalpy of helium with temperature at a pressure of 1 and 40 bar.

to be precooled to a temperature lower than their inversion temperature at the operating pressure before they can be liquefied using a Linde–Hampson process (Table 1.3).

Figure 1.25 shows a typical precooled Linde–Hampson liquefier that can be used for the liquefaction of hydrogen and neon. Since the inversion temperature of hydrogen and neon is greater than the normal boiling point of nitrogen (77.4K), nitrogen can be used as the precooling refrigerant for these fluids. The liquefaction of helium using the precooled Linde–Hampson liquefier requires the use of hydrogen as the precooling refrigerant. Practical helium liquefiers avoid the use of hydrogen as a precooling

Table 1.3. Maximum inversion temperature for different gases ($p \to 0$) estimated using the NIST 12 database [63]

Fluid	Maximum inversion temperature (K)
Helium	39
Hydrogen	195
Neon	220

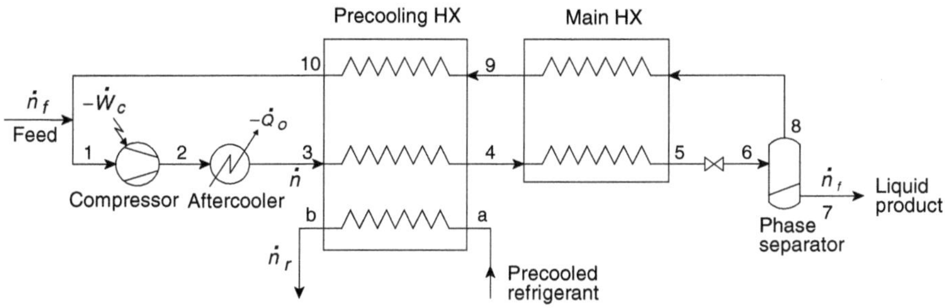

Fig. 1.25. Precooled Linde–Hampson liquefaction process.

refrigerant by using expanders to provide the refrigeration required to cool helium to less than its inversion temperature.[8]

A precooled Linde–Hampson process can also be used for the liquefaction of gases other than helium, hydrogen, and neon. Ammonia, hydroflurocarbon refrigerants such as R407C, R404A, R410A, as well as hydrocarbon refrigerants such as propane can also be used as the refrigerant in the precooler to liquefy gases other than helium, hydrogen, and neon. Precooled processes offer a higher liquid yield and a higher exergy efficiency.

An energy balance over a control volume that excludes the compressor and after-cooler of the precooled Linde–Hampson process (Fig. 1.25) gives

$$\text{liquid yield, } Y = \frac{\dot{n}_f}{\dot{n}} = \left(\frac{h_{10} - h_3}{h_{10} - h_7}\right) + \frac{\dot{n}_r}{\dot{n}}\left(\frac{h_b - h_a}{h_{10} - h_7}\right). \tag{1.40}$$

The first term on the right-hand side represents the liquid yield in a simple Linde–Hampson liquefier without any precooling, and the second term represents the additional yield due to precooling.

The exergy efficiency or the work required per unit mass of gas liquefied in the precooled Linde–Hampson liquefiers is also higher than that of simple Linde–Hampson liquefiers because of the higher liquid yield ($\dot{n}_f$). Precooled systems are widely used in the liquefaction of helium, hydrogen, and neon as well as in the liquefaction of natural gas.

[8] The temperature of all gases decreases on isentropic expansion to lower pressures.

1.11 Exergy efficiency of a Linde–Hampson liquefier

The performance of different liquefiers operating under different conditions, different processes, and different working fluids can be compared by comparing their performance with that of an ideal liquefier. In most liquefiers, only a part of the fluid that is compressed gets liquefied. All liquefiers, irrespective of the operating process, can be generalized as shown in Fig. 1.26. The exergy efficiency of any liquefier is defined as the ratio of minimum work of liquefaction [Eq. (1.8)] and the actual compressor work and can be expressed with respect to Fig. 1.26 as

$$\eta_{\text{ex},l} = \frac{\text{minimum power required for liquefaction}}{\text{compressor power input}} = \frac{-\dot{W}_{l,\text{rev}}}{-\dot{W}_c} = \frac{\dot{n}_f(\text{ex}_f - \text{ex}_1)}{-\dot{W}_c}. \tag{1.41}$$

The exergy expenditure in the process is the compressor power input. The exergy efficiency can also be derived using Eq. (1.26) as follows:[9]

$$\eta_{\text{ex},l} = 1 - \frac{\dot{n}_f\,\text{ex}_1 - \dot{n}_f\,\text{ex}_f - \dot{W}_c}{-\dot{W}_c} = \frac{\dot{n}_f(\text{ex}_f - \text{ex}_1)}{-\dot{W}_c}, \tag{1.42}$$

where $\dot{n}_f$ is the flow rate of the liquid product or feed (make-up gas), $-\dot{W}_c$ is the compressor power input, and ex is the exergy of the fluid (Fig. 1.26). Exergy efficiency is also known in the cryogenic literature as figure of merit (FOM) [13, 36, 90].

The exergy efficiency of the compressor section (compressor and aftercooler) shown in Fig. 1.18 can be expressed in terms of the exergy of the fluid at the entry of the compressor (state 1) and that at the exit of the aftercooler (state 2) as

$$\eta_{\text{ex,cs}} = \frac{\dot{n}(\text{ex}_2 - \text{ex}_1)}{-\dot{W}_c}. \tag{1.43}$$

The exergy efficiency of a Linde–Hampson liquefier can be expressed in terms of the exergy efficiency of the compressor as

$$\eta_{\text{ex},l} = \frac{-\dot{W}_{l,\text{rev}}}{-\dot{W}_c} = \frac{\dot{n}_f(\text{ex}_f - \text{ex}_1)}{\dot{n}(\text{ex}_2 - \text{ex}_1)/\eta_{\text{ex},c}} = \eta_{\text{ex,cb}}\,\eta_{\text{ex,cs}}, \tag{1.44}$$

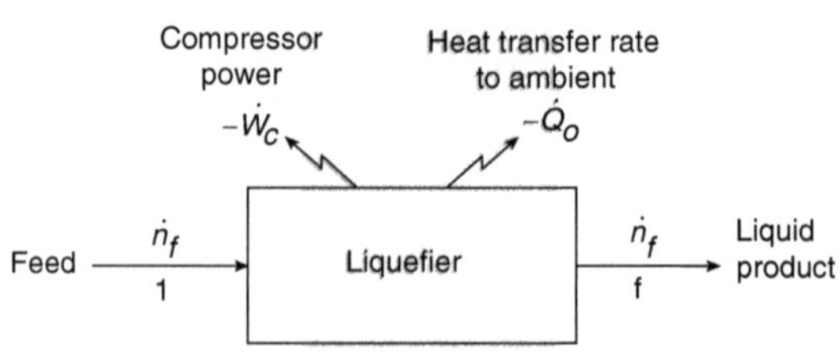

Fig. 1.26. Schematic of a generic liquefier.

[9] The exergy of heat rejected to ambient $\dot{Q}_o$ is zero.

where $\eta_{ex,cb}$ and $\eta_{ex,cs}$ refer to the exergy efficiency of the cold box and that of the compressor section (compressor and aftercooler), respectively.

The exergy efficiency of a refrigeration or liquefaction system can also be expressed in terms of exergy loss as

$$\eta_{ex} = 1 - \frac{\sum \text{exergy loss}}{\sum \text{compressor power input}}. \tag{1.45}$$

The expressions for exergy loss in different components of a cryogenic refrigeration/liquefaction system were presented in Table 1.1.

Consider an ideal Linde–Hampson liquefier operating with an isothermal compressor, an ideal heat exchanger (effectiveness $\varepsilon = 1$, and no pressure drop), and an ideal phase separator (no pressure drop, complete separation of phases). The exergy loss in such a system essentially occurs in the heat exchanger and the expansion device. The exergy efficiency of the system can be defined as follows:

$$\eta_{ex} = 1 - \frac{\text{exergy loss in heat exchanger} + \text{exergy loss in J-T valve}}{\text{compressor power input}}. \tag{1.46}$$

Figure 1.27 shows the variation of exergy losses in ideal Linde–Hampson liquefier operating with an ideal isothermal compressor, ideal heat exchanger ($\varepsilon = 100\%$), and ideal phase separator. It can be seen that the exergy loss in the valve is much higher than that in the heat exchanger at all operating pressures. The lowest total exergy loss or the highest exergy efficiency occurs at a operating pressure of about 365 bar. The exergy efficiency (η_{ex}) of the ideal Linde–Hampson nitrogen liquefier is less than 10% when the operating pressure is less than 100 bar.

1.11.1 Exergy losses in a non-ideal Linde–Hampson liquefier

Consider the exergy losses in a non-ideal Linde–Hampson nitrogen liquefier. Let the exergy efficiency of the compressor section (compressor and aftercooler) be 50%, and let the effectiveness of the heat exchanger be 92%. For the sake of simplicity, the pressure drop in the heat exchanger and phase separator is assumed to be zero. The compressor work is utilized in the liquefaction of nitrogen and in overcoming the exergy losses in the compressor, heat exchanger, and valve (Fig. 1.28). It can be seen that most of the exergy loss occurs in the throttle valve even when the heat exchanger effectiveness is 92%. When the exergy losses in the compressor are also taken into account [see Eq. (1.44)], the overall exergy efficiency of the system or the exergy utilization for liquefaction of nitrogen (useful effect) is only 3.35%.[10] Linde–Hampson liquefiers are not used in practice to liquefy gases due to the small exergy efficiency and the need for high operating pressures. Small Linde–Hampson systems operating with pressurized nitrogen/argon stored in gas bottles are, however, used in military applications such as the cooling of infrared detectors in missiles [72].

[10] $6.7\% * 0.5 = 3.35\%$.

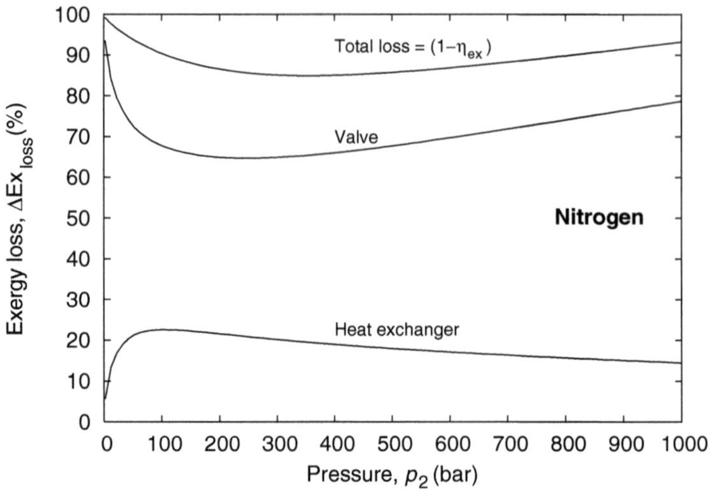

Fig. 1.27. Variation of exergy loss in an ideal Linde–Hampson nitrogen liquefier operating with an ideal isothermal compressor, ideal heat exchanger ($\varepsilon = 100\%$), and ideal phase separator. $p_1 = 1$ bar, $T_1 = T_2 = 300$ K.

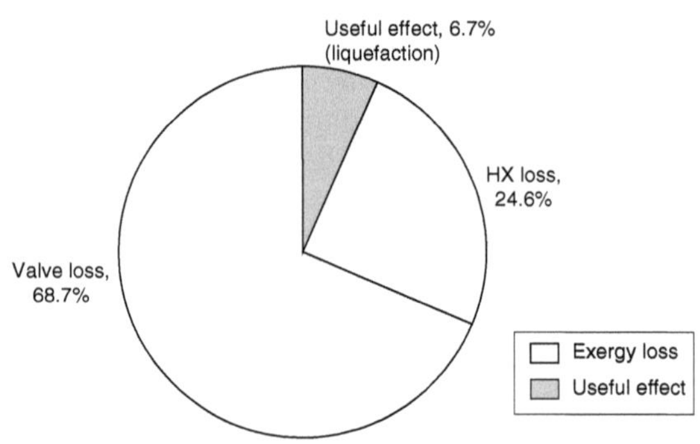

Fig. 1.28. Utilization of input exergy in the cold box of a Linde–Hampson nitrogen liquefier. The heat exchanger effectiveness $\varepsilon = 0.92$, $p_2 = 200$ bar, $p_1 = 1$ bar.

1.12 Temperature profiles in heat exchangers operating with single phase fluids

Appreciating the variation of temperature of the high- and low-pressure streams along the length of the heat exchanger is the first step in understanding the reason for the low exergy efficiency of the cryogenic refrigerators and liquefiers operating with pure fluids. Broadly, three different types of temperature profiles can exist in any countercurrent (counter flow) heat exchanger operating with single-phase fluids whose specific heat does not vary with temperature. These are shown in Fig. 1.29.

The minimum temperature difference between the streams occurs at the warm end ($\Lambda = 0$) or the cold end of the heat exchanger ($\Lambda = 1$). The minimum temperature

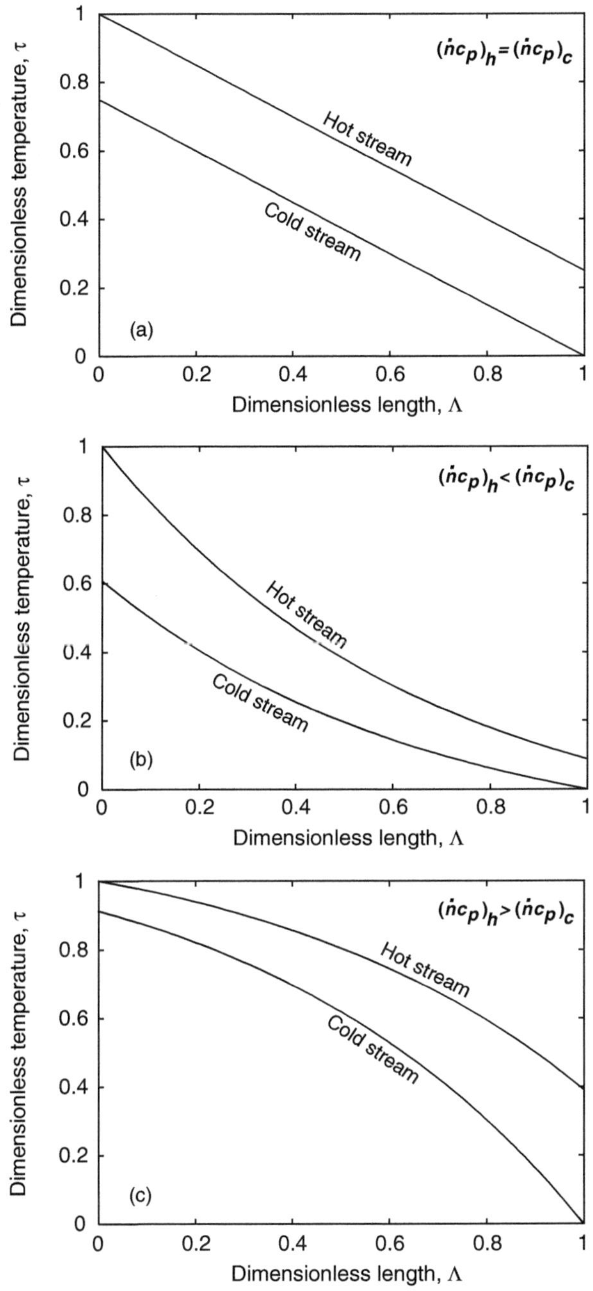

Fig. 1.29. Different types of temperature profiles that can exist in heat exchangers operating with single-phase fluids. The subscripts h and c refer to the hot and cold streams, respectively.

difference between the streams is also known as the minimum temperature approach. The temperatures of the hot and cold streams vary linearly with length when the heat capacity rate ($\dot{n}c_p$) of the two streams is the same. The minimum temperature approach occurs at the cold end of the heat exchanger ($\Lambda = 1$) when the heat capacity rate ($\dot{n}c_p$) of the hot fluid is less than that of the cold fluid [Fig. 1.29(b)], and at the warm end ($\Lambda = 0$) when the heat capacity rate ($\dot{n}c_p$) of the hot fluid is greater than that of the cold fluid [Fig. 1.29(c)].

In the case of a Solvay or Linde–Hampson refrigerator, the flow rate of both the hot (h) and cold (c) streams is the same. On the other hand, the flow rate of the hot stream ($\dot{n}$) is higher than that of the cold stream ($\dot{n} - \dot{n}_f$) in the case of a Linde–Hampson liquefier (see Figs. 1.18 and 1.19). However, the specific heat of the low-pressure stream is always lower than that of the high-pressure stream (see Figs. 1.21 and 1.32) for all pure fluids, at any given temperature. Therefore, the heat capacity rate ($\dot{n}c_p$) of the hot (high-pressure) fluid is always greater than that of the cold (low-pressure) fluid in the case of a Linde–Hampson liquefier or refrigerator operating with pure fluids. Therefore, the temperature profiles of type (c) shown in Fig. 1.29 occur in a Linde–Hampson refrigerator/liquefier operating with pure fluids.

In the first case [Fig. 1.29(a)], when the heat capacity rates ($\dot{n}c_p$) of the hot and cold fluid are equal, the temperature profiles of the hot and cold fluid streams approach each other when the heat exchanger effectiveness is 100%. The temperature difference between the hot and cold fluid streams decreases to zero at all locations of the heat exchanger for this case. When the heat capacity rates ($\dot{n}c_p$) of the hot and cold fluid are not equal, the temperature approach between the streams becomes zero at the warm or the cold end of the heat exchanger.

In the case of a Linde–Hampson liquefier operating with an ideal heat exchanger ($\varepsilon = 100\%$), the temperature approach between the streams becomes zero at the warm end of the heat exchanger. The temperature approach at the cold end, on the other hand, is likely to be on the order of 50 to 70 K (depending on the type of fluid used, operating pressure, etc.) (see Figs. 1.30 and 1.31).

Figure 1.32 shows the variation of the specific heat at constant pressure (c_p) with temperature for nitrogen at 200 bar and 1 bar, typical operating pressures in a Linde–Hampson liquefaction system. It can be seen that the specific heat at constant pressure (c_p) at 200 bar is more than that at 1 bar at all temperatures. While the variation of c_p with temperature is small when the pressure is 1 bar, the variation of c_p with temperature is large when the pressure is 200 bar. The constant specific heat assumptions made in Fig. 1.29 are therefore not applicable in a Linde–Hampson liquefaction system.

The large variation of c_p at low temperatures (Fig. 1.32) contributes to the large temperature approach at the cold end of the heat exchanger. In the case of Linde–Hampson systems operating with pure fluids, the temperature approach between the streams at the cold end will be identical to the temperature drop of the working fluid in the expansion valve. The large temperature drop results in a large irreversibility during the isenthalpic expansion process (see Figs. 1.20 and 1.27).

Temperature profiles similar to that shown in Fig. 1.29(a) occur in regenerative heat exchangers used in Stirling, Gifford–McMahon, and pulse tube refrigerators

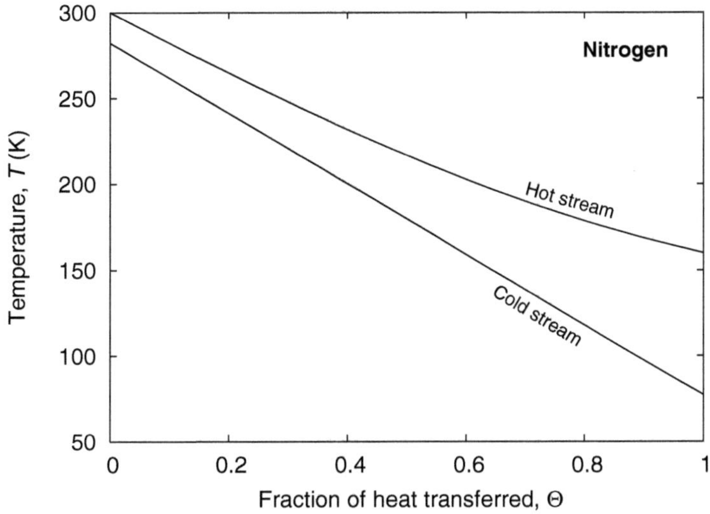

Fig. 1.30. Temperature profiles of the hot and cold streams of a Linde–Hampson nitrogen liquefier. $p_2 = 100$ bar, $p_1 = 1$ bar, $\varepsilon = 0.92$.

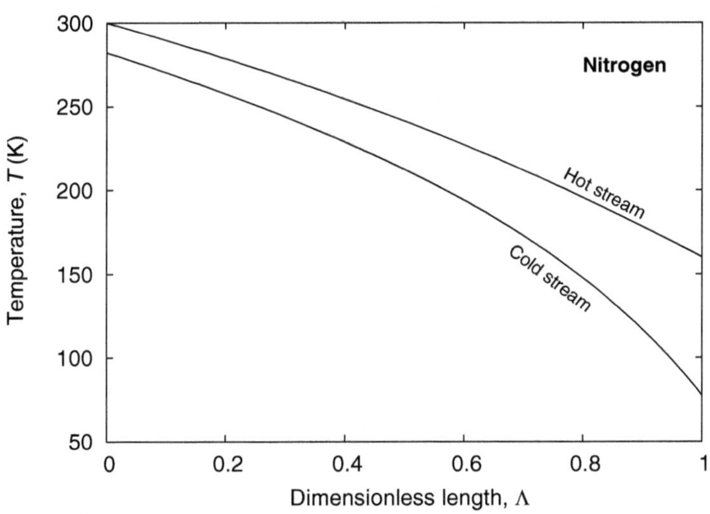

Fig. 1.31. Temperature profiles of hot and cold streams along the length of the heat exchanger of a typical Linde–Hampson nitrogen liquefier. $p_2 = 100$ bar, $p_1 = 1$ bar, $\varepsilon = 0.92$.

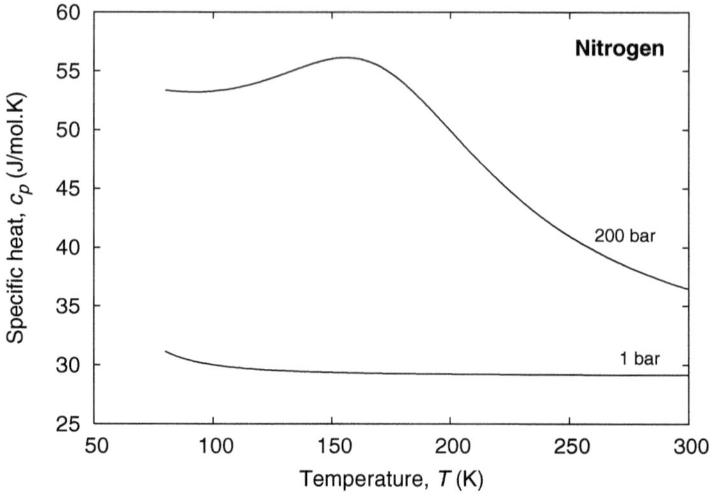

Fig. 1.32. Variation of specific heat (c_p) of nitrogen with temperature.

operating between 300 and 80 K with helium as the refrigerant. The temperature profiles shown in Fig. 1.29(b) can occur near the warm end of the second heat exchanger of a Kapitza nitrogen liquefier.

1.13 Heat exchanger effectiveness

The effectiveness of a heat exchanger (ε) is defined as follows:

$$\varepsilon = \frac{\text{actual heat transfer rate}}{\text{maximum possible heat transfer rate}}. \tag{1.47}$$

The effectiveness of a counter-flow heat exchanger (Fig. 1.33) can be defined in terms of the temperatures at the inlet (in) and exit (out) of the heat exchanger as follows:

$$\varepsilon = \frac{\left(\dot{n}c_p\right)_h \left(T_{h,\text{in}} - T_{h,\text{out}}\right)}{\left(\dot{n}c_p\right)_{\min} \left(T_{h,\text{in}} - T_{c,\text{in}}\right)} = \frac{\left(\dot{n}c_p\right)_c \left(T_{c,\text{out}} - T_{c,\text{in}}\right)}{\left(\dot{n}c_p\right)_{\min} \left(T_{h,\text{in}} - T_{c,\text{in}}\right)}. \tag{1.48}$$

Since the specific heat at constant pressure[11] of all cryogenic fluids changes with temperature (see Figs. 1.21 and 1.32), the above expression for effectiveness, which assumes that the specific heat of the fluid does not change over the length of the heat exchanger, cannot be used. Instead, the following expressions based on the enthalpy are used:

$$\varepsilon = \frac{h_{h,\text{in}} - h_{h,\text{out}}}{h_{h,\text{in}} - h'_{c,\text{in}}} \qquad \text{for } \left(\dot{n}c_p\right)_h < \left(\dot{n}c_p\right)_c, \tag{1.49}$$

[11] Specific heat at constant pressure $c_p = (\partial h/\partial T)_p$ for single phase fluids.

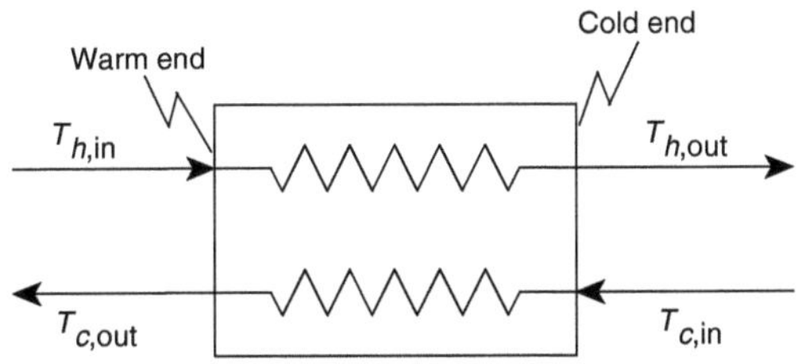

Fig. 1.33. Schematic of a counter-flow heat exchanger.

$$\varepsilon = \frac{h_{c,\,\text{out}} - h_{c,\,\text{in}}}{h''_{h,\,\text{in}} - h_{c,\,\text{in}}} \qquad \text{for } \left(\dot{n}c_p\right)_h > \left(\dot{n}c_p\right)_c, \tag{1.50}$$

where $h'_{c,\,\text{in}}$ is the enthalpy of the hot stream at temperature $T_{c,\,\text{in}}$ and pressure $p_{h,\,\text{out}}$. Similarly, $h''_{h,\,\text{in}}$ is the enthalpy of the stream at temperature $T_{h,\,\text{in}}$ and pressure $p_{c,\,\text{out}}$.

In the case of the Solvay or Linde–Hampson liquefier operating with a pure fluid (Figs. 1.17 and 1.18), the flow rate of the hot stream of the heat exchanger ($\dot{n}$) is greater than that of the cold stream ($\dot{n} - \dot{n}_f$). Similarly, the specific heat (c_p) of the hot (high-pressure) stream is greater than that of the cold (low-pressure) stream. Consequently, the effectiveness of the heat exchanger can be expressed as follows:

$$\varepsilon = \frac{h_5 - h_g}{h_1 - h_g}. \tag{1.51}$$

The liquid yield of a Solvay liquefaction system is a strong function of the effectiveness of the heat exchanger, as shown in Fig. 1.34. For example, in the case of a Solvay nitrogen liquefier, the operating pressure required for a liquid yield of 20% is 32 bar when the heat exchanger effectiveness (ε) is 100%, 49.5 bar when $\varepsilon = 90\%$ and 95 bar when $\varepsilon = 70\%$. No liquefaction occurs below a certain pressure at any given heat exchanger effectiveness ($\varepsilon < 100\%$). The minimum operating pressure (p_2) increases with a decrease in the heat exchanger effectiveness (ε), as seen in Fig. 1.34.

The performance of a Linde–Hampson liquefier or a Linde–Hampson refrigerator is closely related to the effectiveness of the heat exchanger. The liquid yield in a Linde–Hampson liquefier [Eq. (1.36)] can be expressed in terms of the heat exchanger effectiveness using Eq. (1.51) as follows:

$$\text{liquid yield, } Y = \frac{h_5 - h_2}{h_5 - h_f} = \frac{\varepsilon(h_1 - h_g) + (h_g - h_2)}{\varepsilon(h_1 - h_g) + (h_g - h_f)}. \tag{1.52}$$

Figure 1.35 shows the variation of liquid yield in a Linde–Hampson nitrogen liquefier with heat exchanger effectiveness (ε). The liquid yield is only 3% at a heat exchanger effectiveness of 90% and slightly less than 6% at a heat exchanger effectiveness of 95% when the operating pressure (p_2) is 200 bar. A high operating

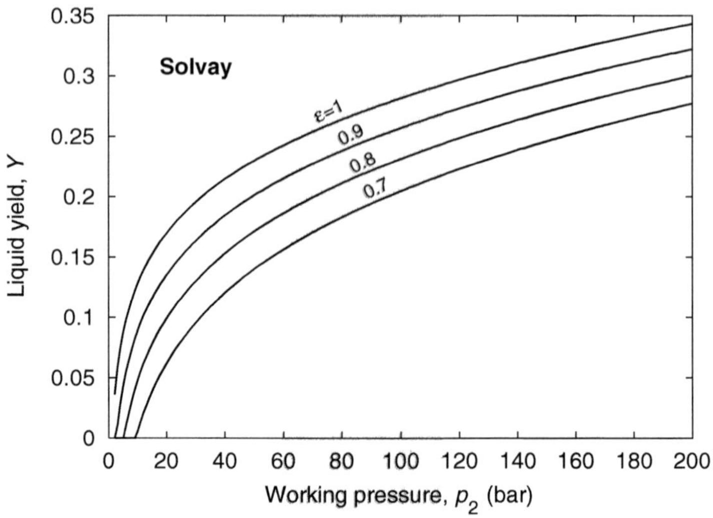

Fig. 1.34. Liquid yield in an ideal Solvay nitrogen liquefaction system at different heat exchanger effectiveness (ε). $p_1 = 1$ bar.

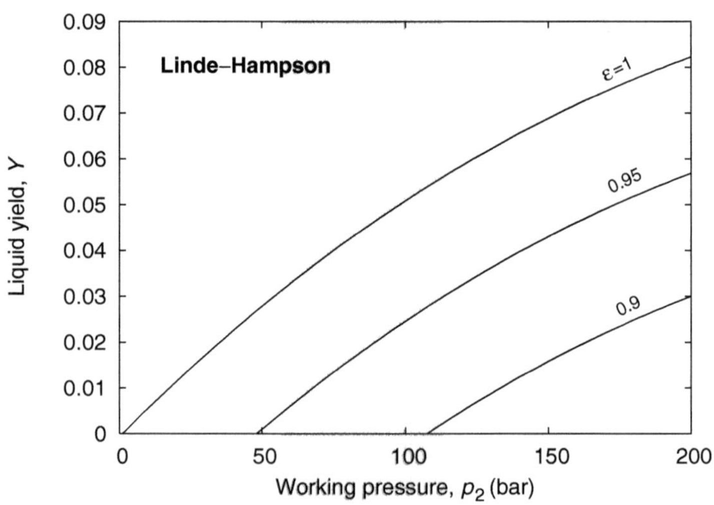

Fig. 1.35. Liquid yield in an ideal Linde–Hampson nitrogen liquefaction system at different heat exchanger effectiveness (ε). $p_1 = 1$ bar.

pressure, typically 125 to 200 bar, and a heat exchanger effectiveness greater than 95% are therefore required for a reasonable liquid yield in a Linde–Hampson nitrogen liquefier. The small liquid yield is essentially due to the small variation of enthalpy of nitrogen with pressure at room temperature (Fig. 1.21).

The liquid yield will become zero when $h_5 = h_2$. The heat exchanger effectiveness at which the liquid yield (Y) becomes zero can be obtained from Eq. (1.52) as follows:

$$\varepsilon_{min} = \frac{h_2 - h_g}{h_1 - h_g}. \tag{1.53}$$

The effectiveness of the heat exchanger (ε) should be greater than the minimum heat exchanger effectiveness (ε_{min}) for a Linde–Hampson refrigerator/liquefier to function.

Figure 1.36 shows the minimum heat exchanger effectiveness (ε_{min}) necessary for a Linde–Hampson liquefier or Linde–Hampson refrigerator to operate at different operating pressures. It can be seen that the lower the operating pressure, the higher the minimum effectiveness required. Since argon and oxygen are more non-ideal at room temperature compared to nitrogen, the minimum effectiveness required for a Linde–Hampson system to function is lower when argon or oxygen is used as the operating fluid, compared to nitrogen. The relationship between the minimum effectiveness required for a Linde–Hampson liquefier or refrigerator to function and the warm end temperature of the heat exchanger (T_2) has been estimated for nitrogen as the operating fluid at different operating pressures (Fig. 1.37). It can be seen that the minimum heat exchanger effectiveness required for the system to function decreases with a decrease in the temperature of the high-pressure stream at the warm end of the heat exchanger (T_2). The exergy efficiency of the cold box of a Linde–Hampson nitrogen liquefier is strongly dependent on the effectiveness of the heat exchanger used (Fig. 1.38). The exergy efficiency of the system decreases from 13.5% to 6.7%

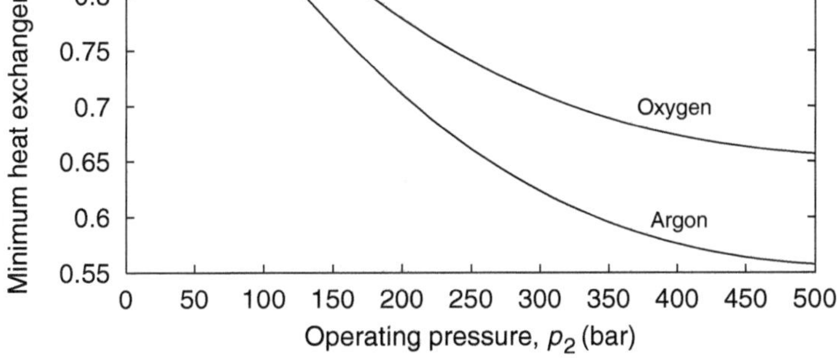

Fig. 1.36. Variation of minimum heat exchanger effectiveness (ε_{min}) required for a Linde–Hampson liquefier to function with operating pressure (p_2).

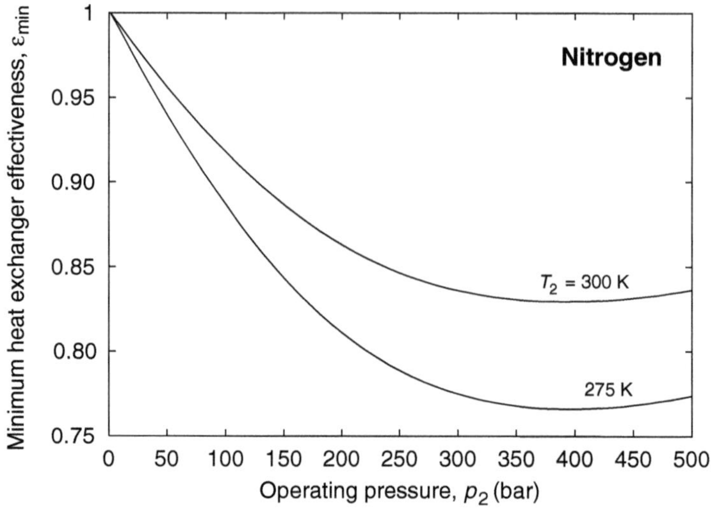

Fig. 1.37. Variation of minimum heat exchanger effectiveness (ε_{min}) required for a Linde–Hampson liquefier to function with temperature T_2.

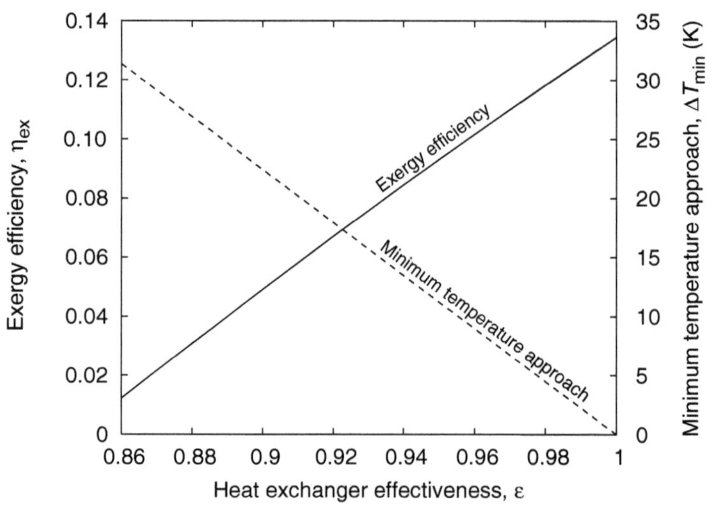

Fig. 1.38. Variation of exergy efficiency of a Linde–Hampson nitrogen liquefier with heat exchanger effectiveness (p_2 = 200 bar, p_1 = 1 bar, $T_1 = T_2$ = 300 K). The compression process is assumed to be isothermal.

when the effectiveness of the heat exchanger decreases from 100% to 92%. Since an effectiveness greater than 94% is difficult to realize in large commercial heat exchangers due to longitudinal heat conduction through the heat exchanger walls, the theoretical maximum exergy efficiency of the system at an operating pressure of 200 bar is limited to about 8%, with isothermal compression. When non-isothermal compression is used, the exergy efficiency of a Linde–Hampson liquefier reduces to less than about 4%. The exergy efficiency of most practical nitrogen liquefiers operating on the Linde–Hampson process varies normally between 1 and 3%.

It can also be seen from Fig. 1.38 that a small temperature approach or a high heat exchanger effectiveness is required for a reasonable system exergy efficiency. A minimum temperature approach of less than 3–5 K is normally used in the heat exchangers of most small cryogenic systems operating at 77 K. Even smaller heat exchanger temperature approaches are required at still lower operating temperatures such as those encountered in helium liquefiers.

1.14 Exergy efficiency of the Solvay and Linde–Hampson liquefaction processes

The only difference between the Solvay (Fig. 1.17) and Linde–Hampson liquefiers (Fig. 1.18) is in the process used for expansion of the fluid from high to low pressure. The use of an isenthalpic process in the place of an isentropic process results in a lower liquid yield and lower exergy efficiency in the case of any Linde–Hampson nitrogen liquefier, as shown in Fig. 1.39. The liquid yield, however, remains well below 100% even when the operating pressure is 1,000 bar in the case of a Solvay nitrogen liquefier, as shown in Fig. 1.40.

In the case of an ideal Solvay liquefier, exergy loss occurs only in the heat exchanger, whereas in the case of an ideal Linde–Hampson liquefier, exergy loss occurs both in the heat exchanger and in the expansion valve. The exergy loss in the heat exchanger even when the effectiveness is 100% is due to the finite temperature difference between the streams at all locations except the warm end of the heat exchanger. It can be seen from Fig. 1.39 that this exergy loss can be up to 30%, even at pressures of 100 bar in the case of nitrogen.

It is evident from Fig. 1.39 that the key to improving the exergy efficiency of the Linde–Hampson liquefier lies in decreasing the exergy losses in the heat exchanger and the expansion device. The exergy loss in the heat exchanger is large due to the large temperature difference between the streams at the cold end of the heat exchanger and the heat capacity rates ($\dot{n}c_p$) of the two streams being unequal. The exergy loss during expansion of the operating fluid is proportional to the ratio of pressures at the entry and exit of the throttle. The liquid yield is very small unless very high operating pressures are used. But high operating pressures result in high exergy loss across the throttle. The Linde–Hampson liquefaction process has been replaced by other processes that overcome this difficulty.

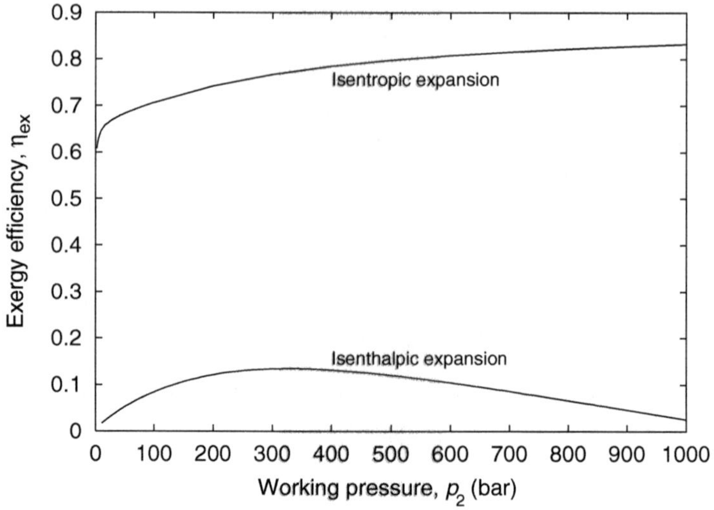

Fig. 1.39. Comparison of exergy efficiency of an ideal Linde–Hampson liquefier operating with isenthalpic and isentropic expansion (Solvay liquefier), and isothermal compression. $p_1 = 1$ bar, $T_1 = T_2 = T_5 = 300$ K.

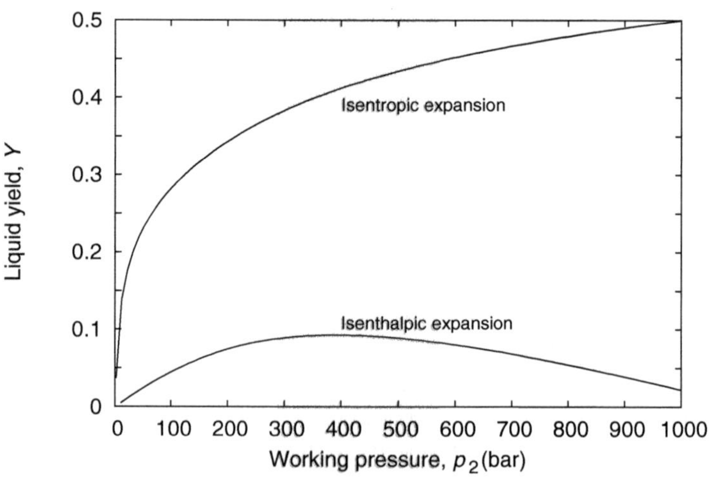

Fig. 1.40. Comparison of liquid yield of a Linde–Hampson liquefier operating with isenthalpic, isentropic expansion (Solvay liquefier), and isothermal compression. $p_1 = 1$ bar, $T_1 = T_2 = T_5 = 300$ K.

1.15 The Kapitza liquefaction process and its variants

The large variation of specific heat at constant pressure (c_p) of working fluids with pressure, particularly at low temperatures (see Fig. 1.32), results in a large temperature approach between the streams in the case of a Linde–Hampson liquefier operating with pure fluids such as nitrogen (see Fig. 1.31). The temperature change across the throttle (J-T valve) is identical to the temperature approach at the cold end of the heat exchanger. A large temperature change across the throttle results in a large exergy loss. Consider the thermal resistance (conduction/convection) shown in Fig. 1.41. For the sake of simplicity, let $T_1 > T_2 > T_o$, where T_o is the ambient temperature. The exergy loss during heat transfer across a thermal resistance can be expressed as follows:

$$\Delta\text{ex}_{\text{loss}} = \dot{Q}\left(1 - \frac{T_o}{T_1}\right) - \dot{Q}\left(1 - \frac{T_o}{T_2}\right). \tag{1.54}$$

The heat transfer rate across the resistance can be expressed as

$$\dot{Q} = UA(T_1 - T_2), \tag{1.55}$$

where UA refers to the thermal conductance. The exergy loss during heat transfer across the thermal resistance is given by the expression

$$\Delta\text{ex}_{\text{loss}} = UAT_o\frac{(T_1 - T_2)^2}{T_1T_2} \approx UAT_o\left(\frac{\Delta T}{T}\right)^2, \tag{1.56}$$

where ΔT is the temperature difference between and high and low temperatures ($T_1 - T_2$). Since exergy loss increases with a decrease in the operating temperature, the temperature approach between streams in the heat exchanger should decrease with a decrease in temperature of an ideal heat exchanger. In other words, a temperature profile similar to that in Fig. 1.29(b) is desirable to minimize the exergy loss in the heat exchanger. Such a temperature profile requires that the flow rate of the cold (low-pressure) stream be many times more than that of the hot (high-pressure) stream.

This concept is used in the Kapitza liquefaction process as shown in Fig. 1.42. In this process, a large part of the high-pressure fluid is diverted to an expander (stream 9). This satisfies the condition required for obtaining the temperature profile in Fig. 1.29(b). The gas passing through the expander undergoes a large temperature change. This cold gas is used to cool and condense the high-pressure fluid in a second heat exchanger (HX-2). The refrigeration obtained in the expander also helps in precooling the high-pressure nitrogen in the first heat exchanger (HX-1) before it

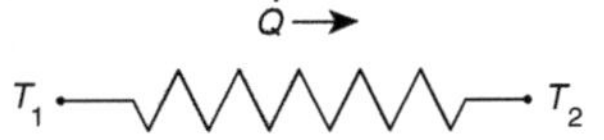

Fig. 1.41. Heat transfer across a conduction/convection resistance. $T_1 > T_2 > T_o$ (ambient temperature).

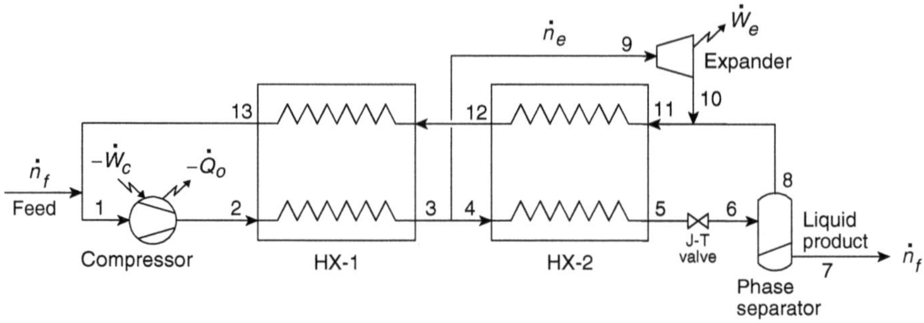

Fig. 1.42. Ideal Kapitza liquefaction process.

enters the expander and the second heat exchanger (HX-2). The high-pressure fluid is in a subcooled liquid state at the entry of the throttle valve when the operating pressure p_2 is lower than the critical pressure (p_c). The temperature of the high-pressure gas at the entry of the throttle valve is much lower than that in a Linde–Hampson system when the operating pressure (p_2) is greater than the critical pressure (p_c).

The expression for liquid yield (Y) in a Kapitza liquefier is obtained by an energy balance over a control volume that excludes the compressor as follows:

$$\text{liquid yield, } Y = \frac{\dot{n}_f}{\dot{n}} = \left(\frac{h_{13} - h_2}{h_{13} - h_7}\right) + \left(\frac{\dot{n}_e}{\dot{n}}\right)\left(\frac{h_9 - h_{10}}{h_{13} - h_7}\right). \tag{1.57}$$

The first term on the right-hand side of Eq. (1.57) is essentially the liquid yield we get when a simple Linde–Hampson liquefier is used. The second term represents the additional yield due to precooling using the expander.

The power extracted from the turbine can be used to reduce the compressor power in a Kapitza liquefier. The net power $-\dot{W}$ required in an ideal Kapitza liquefier is given by the expression

$$-\dot{W} = -\dot{W}_c - \dot{W}_e = \dot{n}(\text{ex}_2 - \text{ex}_1) - \dot{n}_e(\text{ex}_9 - \text{ex}_{10}). \tag{1.58}$$

The exergy efficiency of a Kapitza liquefier can be expressed as follows:

$$\eta_{\text{ex}} = \frac{\text{minimum power for liquefaction}}{\text{compressor power input}} = \frac{-\dot{W}_{\text{rev},l}}{-\dot{W}} = \frac{\dot{n}_f(\text{ex}_7 - \text{ex}_1)}{-\dot{W}}. \tag{1.59}$$

Figure 1.43 shows a Kapitza nitrogen liquefaction process on a *T–h* plane. The operating pressure was assumed to be 40 bar.[12] The turbine adiabatic efficiency is assumed to be 80% and the minimum temperature approach to be 10 K in the first heat exchangers and 20 K in the second heat exchanger.

[12] The critical pressure of nitrogen is 33.96 bar.

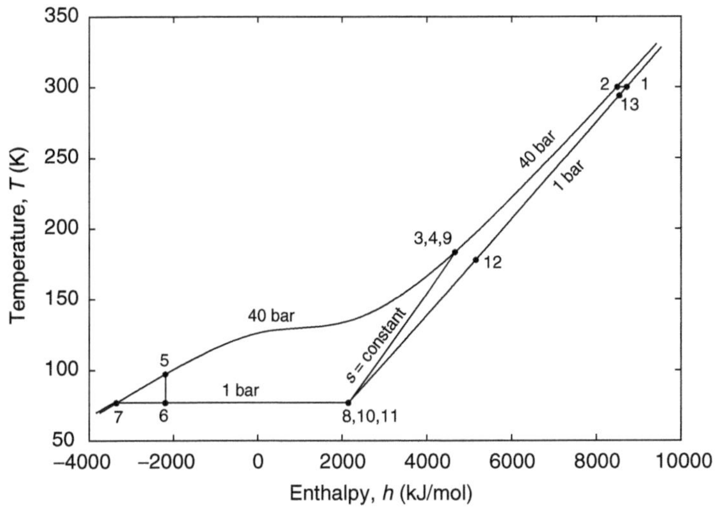

Fig. 1.43. Kapitza nitrogen liquefier operating with a minimum temperature approach of 10 K in both the heat exchangers and operating pressures of 40/1 bar, isothermal compressor, and isentropic expander.

Table 1.4 shows the temperature, pressure, vapor fraction, and flow rate at different streams of the Kapitza nitrogen liquefier. Figure 1.44 shows the temperature profiles in the two heat exchangers. It can be seen from Fig. 1.44 that the temperature approach between the streams increases continuously with a decrease in stream temperature in the first heat exchanger (HX-1), whereas it first decreases and thereafter remains about the same with a decrease in the stream temperatures in the second heat exchanger (HX-2). The peculiar shape of the stream temperature profiles in the second heat exchanger is due to the large variation of heat capacity (c_p) of nitrogen at temperatures close to the critical temperature.

It can be observed from Table 1.4 and Fig. 1.44 that the temperature drop in the J-T valve is very small compared to that in any Linde–Hampson liquefier, leading to a small exergy loss during the throttling process.

Figure 1.45 shows the utilization of input power (input exergy) in a Kapitza nitrogen liquefier operating at 40/1 bar and a minimum temperature approach of 10 K in the first heat exchanger (HX-I) and 20 K in the second heat exchanger. The adiabatic efficiency of the turbine was assumed to be 80%. The compression process was assumed to be isothermal. It can be seen that the exergy loss in the throttling valve is only 2.7% in this case, compared to about 60 to 70% in most Linde–Hampson liquefiers. The smaller exergy loss in the throttling valve is due to the expansion of a subcooled liquid with a temperature change of about 20 K, compared to the expansion of a superheated vapor with a temperature change of 60 to 80 K in most Linde–Hampson liquefiers. The exergy loss in the two heat exchangers and that in the turbine are very close (23.5% and 25.8%, respectively). The high exergy efficiency

Table 1.4. Temperature, pressure, and vapor fraction of different streams of a Kapitza nitrogen liquefier (Fig. 1.42)

	Stream						
	1	2	3	4	5	6	7
Temperature, K	300	300	300	184.2	97.4	77.4	77.4
Pressure, bar	1	40	40	40	40	1	1
Vapor fraction	1	1	1	1	0	0.213	0
Flow rate, mol/s	1	1	1	0.205	0.205	0.205	0.161

	Stream					
	8	9	10	11	12	13
Temperature, K	77.4	184.2	77.4	77.4	133.5	290
Pressure, bar	1	40	1	1	1	1
Vapor fraction	1	1	1	1	1	1
Flow rate, mol/s	0.044	0.795	0.795	0.839	0.839	0.839

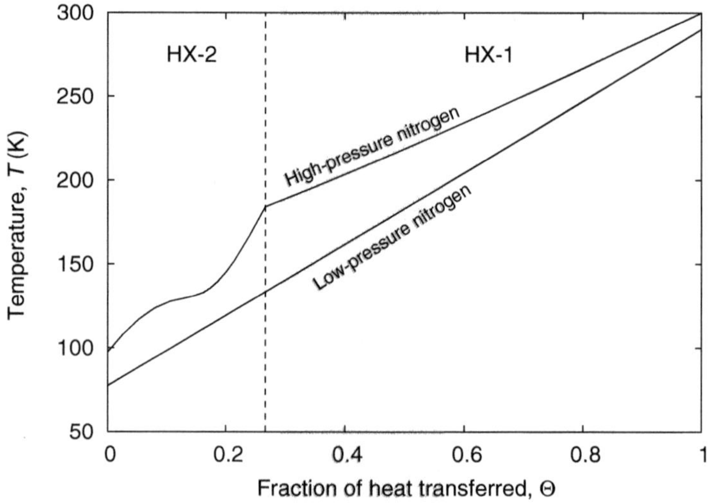

Fig. 1.44. Temperature profiles in an ideal Kapitza nitrogen liquefier operating at 40/1 bar with a minimum temperature approach of 10 K in HX-1 and 20 K in HX-2.

(48%) in this case is largely due to the reduction of exergy loss in the throttling valve compared to a Linde–Hampson liquefier (Fig. 1.28). The liquid yield (Y) is 20.5% in this case.

The entropy change during expansion of a subcooled liquid is much smaller than that during expansion of a superheated vapor. The exergy efficiency and liquid yield of a Kapitza liquefaction process are much greater than those of the Linde–Hampson liquefaction process essentially due to the expansion of a subcooled liquid instead of

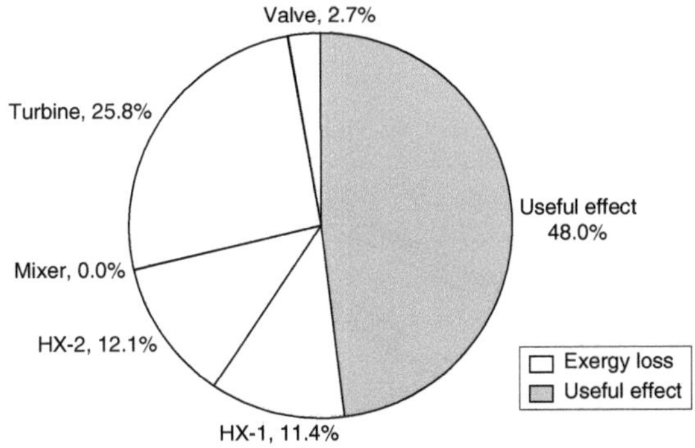

Fig. 1.45. Exergy utilization in an ideal Kapitza nitrogen liquefier operating at 40/1 bar with a minimum temperature approach of 10 K in HX-1 and 20 K in HX-2 and an isothermal compressor. Adiabatic efficiency of the expander η_{ad} is 80%.

a superheated vapor in the case of a Linde–Hampson process. The exergy loss in the heat exchangers is also lower in the case of a Kaptiza process since the temperature approach at the cold end is a fraction of that in a Linde–Hampson process. While the turbine provides most of the refrigeration necessary to liquefy and subcool the working fluid, the diversion of part of the working fluid to the turbine results in a small temperature approach between the streams at the cold end of the second heat exchanger, resulting in a smaller exergy loss in the heat exchanger.

Temperature profiles in the heat exchangers are slightly different from that shown in Fig. 1.44 when the operating pressures (p_2) is less than the critical pressure of nitrogen. Figure 1.46 shows the temperature profiles in the two heat exchangers of a Kapitza nitrogen liquefier operating at operating pressures of 20/1 bar. It can be observed that the high-pressure nitrogen condenses (horizontal portion in Fig. 1.46) and subcools in the second heat exchanger (HX-2), before expansion in the throttle valve. The minimum temperature approach in the second heat exchanger needs to be less than 12.5 K for the system to function. The exergy efficiency of the system (42%) is, however, lower than the previous case (40/1 bar) due to higher exergy loss in the heat exchangers (27.7%) and expansion valve (4.5%). The liquid yield in this case is 16%.

Table 1.5 shows the maximum exergy efficiency and liquid yield in a Kapitza nitrogen liquefier as a function of the minimum temperature approach between the streams in the first and second heat exchangers. It can be seen from cases 8 and 9 that the minimum temperature approach between the streams in the second heat exchanger ($\Delta T_{min,2}$) has more influence on the exergy efficiency of the system than that in the first heat exchanger ($\Delta T_{min,1}$).

The maximum exergy efficiency and liquid yield are obtained in case 6 corresponding to a zero temperature approach between the streams in both heat exchangers. An infinite heat exchanger surface area needs to be provided when a zero temperature approach is used in the heat exchangers and is therefore never used in practice. It is evident from Table 1.5 that a large temperature approach in the first heat exchanger

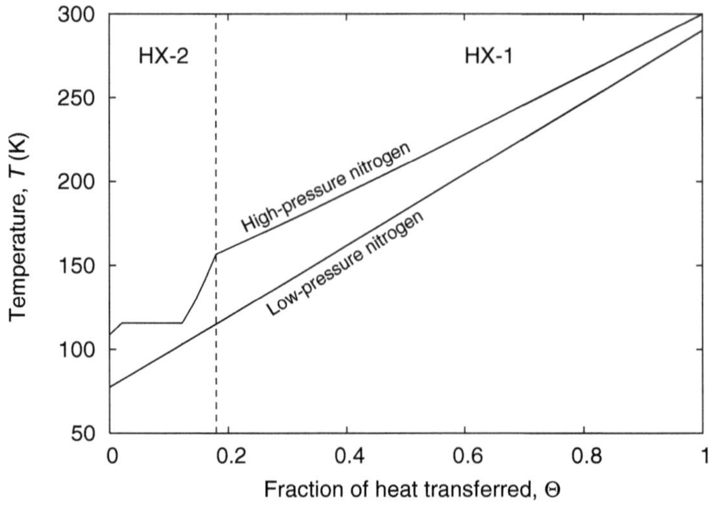

Fig. 1.46. Temperature profiles in an ideal Kapitza nitrogen liquefier operating at 20/1 bar with a minimum temperature approach of 10 K in HX-1 and 10 K in HX-2, and an expander adiabatic efficiency of 80%.

Table 1.5. Variation of the maximum exergy efficiency (η_{ex}) and liquid yield (Y) of a Kapitza nitrogen liquefier with the minimum temperature approach in the two heat exchangers ($p_2 = 40$ bar, $p_1 = 1$ bar, $T_1 = T_2 = 300K$, and turbine adiabatic efficiency $\eta_{ad} = 0.80$)

Case	Minimum temperature approach in heat exchangers		Exergy efficiency, η_{ex}	Liquid yield, Y
	$\Delta T_{min,1}$	$\Delta T_{min,2}$		
1	10	20	0.480	0.161
2	0	20	0.488	0.168
3	0	10	0.540	0.182
4	0	5	0.547	0.183
5	0	3	0.590	0.194
6	0	0	0.601	0.197
7	5	3	0.577	0.189
8	10	3	0.563	0.183
9	20	3	0.535	0.171

(case 8) does not lead to a significant drop in exergy efficiency and liquid yield compared to the best case (case 6) with zero temperature approach in both heat exchangers.

The exergy efficiency of the system decreases considerably when a non-isothermal compressor is used, and the turbine adiabatic efficiency is lower than 80%. The Kapitza liquefaction process is a minor variant of the Claude liquefaction process (Fig. 1.47). The heat transferred in the third heat exchanger of a Claude process is very small when a turbine expander is used, and can therefore be eliminated. The Kapitza process is widely used in the large-scale liquefaction and separation of air, with an operating pressure of 6.5 to 7 bar.

The Collins liquefaction process shown in Fig. 1.48 uses two expansion turbines and is a variant of the Kapitza and Claude processes. The Collins liquefaction process is widely used to liquefy helium. An operating pressure of 15 to 20 bar is normally used in the these systems.

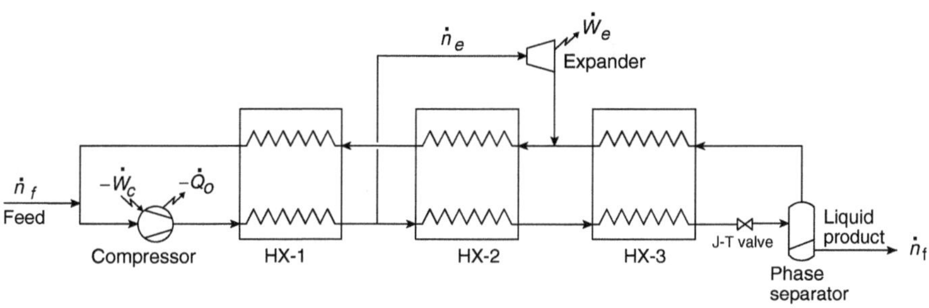

Fig. 1.47. Claude liquefaction process.

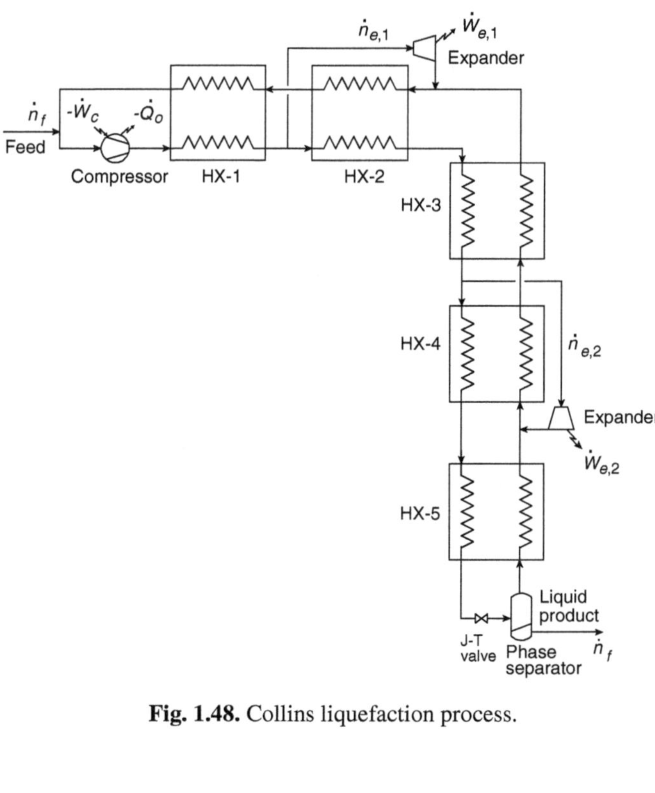

Fig. 1.48. Collins liquefaction process.

1.16 Pinch points

Figures 1.29(b) and (c) show the typical temperature profiles of the hot and cold fluid streams in a heat exchanger in which the heat capacity rate ($\dot{n}c_p$) of the hot and cold fluid streams is not the same. The minimum temperature approach between the streams occurs at the warm or cold end of the heat exchanger in Figs. 1.29(b) and (c). In some cases, the specific heat of the fluids[13] varies along the length of the heat exchanger such that the minimum heat capacity rate fluid $[(\dot{n}c_P)_{\min}]$ at one end of the heat exchanger becomes the maximum heat capacity rate fluid $[(\dot{n}c_P)_{\max}]$ at the other end, as shown in Figs. 1.49(a) and (b). The minimum temperature approach between the hot and cold fluid streams of the heat exchanger occurs between the two ends in the first case [Fig. 1.49(a)], while the temperature profiles get pinched at both the warm and cold ends of the heat exchanger in the second case [Fig. 1.49(b)]. The location where the temperature approach between the streams is minimum is called the pinch point. Normally, the first case [Fig. 1.49(a)] is considered an example of a heat exchanger with a pinch point, whereas that in the second case [Fig. 1.49(b)] is also a special case of a heat exchanger showing a pinch point.

Figures 1.50 and 1.51 show the temperature profile in the second heat exchanger (HX-2) in a Kapitza nitrogen liquefaction process (Fig. 1.42) with a minimum temperature approach of 20 K in the first heat exchanger and 5 K in the second heat exchanger and operating pressures of 40/1 bar. The heat exchanger length was determined using Aspen MUSE, a commercial program for the design of plate-fin heat exchangers. Pinch points occur in these two cases because of the large variation of the specific heat of nitrogen at 40 bar at temperatures close to the critical temperature of nitrogen (Fig. 1.52).

A large variation of specific heat at constant pressure (c_p) with temperature is observed in all real fluids at temperatures close to the critical point when the pressure p is greater than the critical pressure. The specific heat at constant pressure (c_p) tends to infinity at critical pressure and critical temperature for all fluids.[14]

The flow rate of the low-pressure (cold) fluid is 3.89 times that of the high-pressure (hot) fluid in the second heat exchanger of a Kapitza liquefier operating at 40/1 bar considered in this example. The specific heat (c_p) of nitrogen at 40 bar is 17.5 times that at 1 bar at a temperature of 130 K. The cold stream is therefore the maximum heat capacity rate fluid $[(\dot{n}c_p)_{\max}]$ (because of higher flow rate) above 130 K, while the hot stream is the maximum heat capacity rate fluid (because of higher specific heat) below 130 K. The switching of a minimum heat capacity rate fluid $[(\dot{n}c_p)_{\min}]$ to a maximum heat capacity rate fluid $[(\dot{n}c_p)_{\max}]$ and vice versa results in the pinching of temperature profiles. Pinch points are common in heat exchangers operating with refrigerant mixtures [93, 94]. The minimum temperature approach between the streams occurs in between the warm and cold ends of the second heat exchanger in this case. The heat exchanger effectiveness cannot be defined easily as a closed-form expression when pinch points occur. Hence, it is customary to specify

---

13 $(\partial h/\partial T)_p$ in the case of zeotropic mixtures undergoing phase change.

14 $(\partial p/\partial v) = 0$ at $T = T_c$ and $p = p_c$.

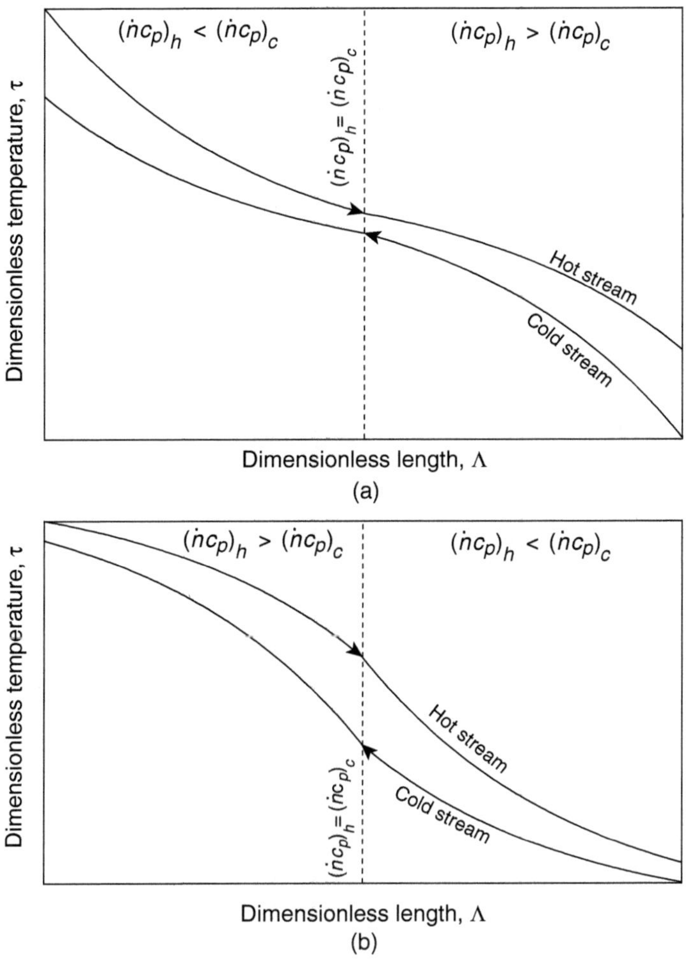

Fig. 1.49. Temperature profiles in heat exchanger showing occurrence of pinch points.

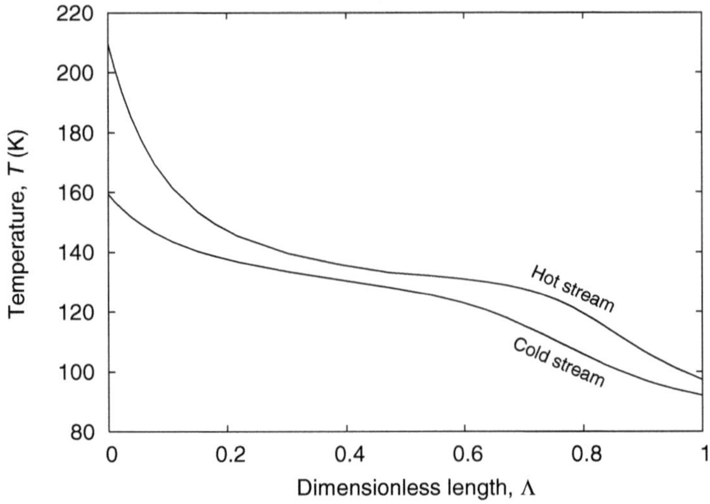

Fig. 1.50. Temperature profiles of the hot and cold streams in the second heat exchanger (HX-2) of a Kapitza nitrogen liquefier ($p_2 = 40$ bar, $p_1 = 1$ bar, $\eta_{ad} = 80\%$).

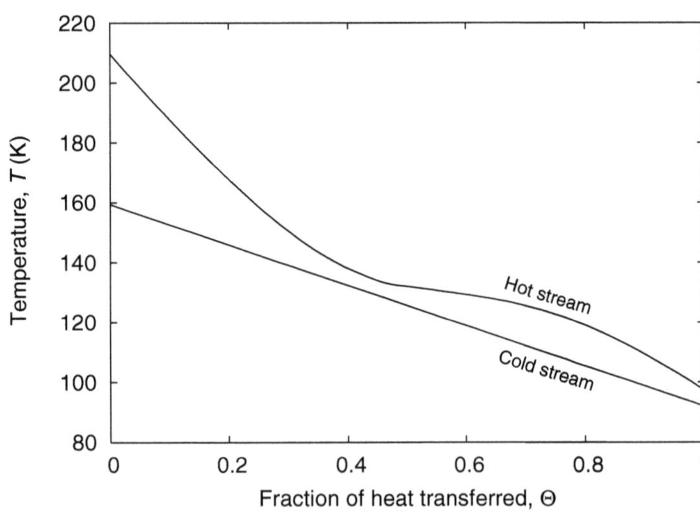

Fig. 1.51. Variation of the temperature of hot and cold streams in the second heat exchanger (HX-2) of a Kapitza nitrogen liquefier with heat transferred.

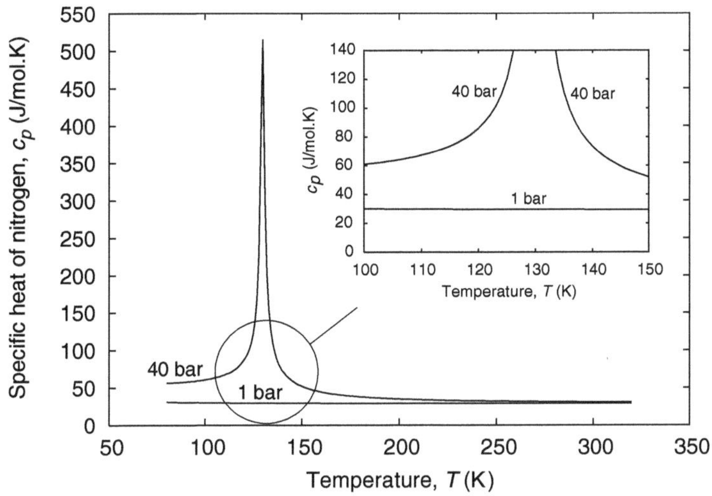

Fig. 1.52. Variation of specific heat of nitrogen with temperature and pressure.

the minimum temperature approach between the streams (at the pinch point) instead of a heat exchanger effectiveness in mixed refrigerant processes.

1.17 Types of refrigerant mixtures

Refrigerant mixtures can be broadly classified into two groups based on the temperature change during the condensation or evaporation process (also known in the refrigeration literature as "glide") as (1) zeotropic mixtures and (2) azeotropic mixtures. Glide is zero in the case of azeotropic mixtures (at the azeotropic mixture composition) and nonzero in the case of zeotropic mixtures. Zeotropic refrigerant mixtures are also known as non-azeotropic refrigerant blends (NARBs).

Figure 1.53 shows the relationship between the dew and bubble point temperatures of a typical zeotropic mixture of nitrogen and methane at a pressure of 1 bar. Consider four different states a, b, c, d at a nitrogen concentration of 0.5. The mixture is in a superheated vapor state at a, saturated vapor state at b, saturated liquid state at c, and subcooled liquid state at d. The temperature of the mixture at saturated vapor state b and at saturated liquid state c is called the dew and bubble point temperature, respectively. The line passing through the dew points is called the dew line, and that through the bubble points is called the bubble line.

The equilibrium composition of vapor and liquid will be different in the two-phase region. For example, the concentration of vapor in equilibrium with liquid at state c will be greater than 0.5 (state f), whereas the concentration of liquid in equilibrium with vapor at state b would be less than 0.5 (state e). Zeotropic mixtures can therefore be defined as mixtures in which the concentration of the coexisting phases is not the same.

Figure 1.54 shows the typical variation of bubble and dew point temperatures of an azeotropic mixture of refrigerants R23 (CHF_3) and R13 (CF_3Cl). The refrigerant glide becomes zero when the concentration of R23 in the mixture is 0.4997. The concentration of the vapor and liquid phases is also the same at this concentration. Azeotropic mixtures are widely used for constant-temperature refrigeration. However, they are not suitable for the processes described in this monograph (except as precooling refrigerant in some cases). Unless stated otherwise, a mixture refers to a zeotropic mixture for the rest of the monograph.

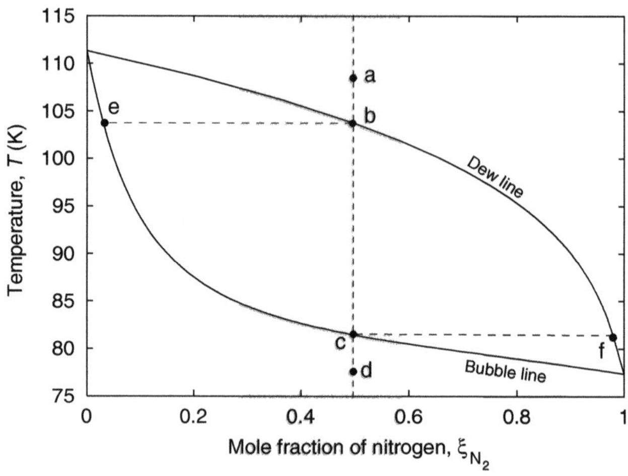

Fig. 1.53. Zeotropic mixture of nitrogen and methane at 1 bar.

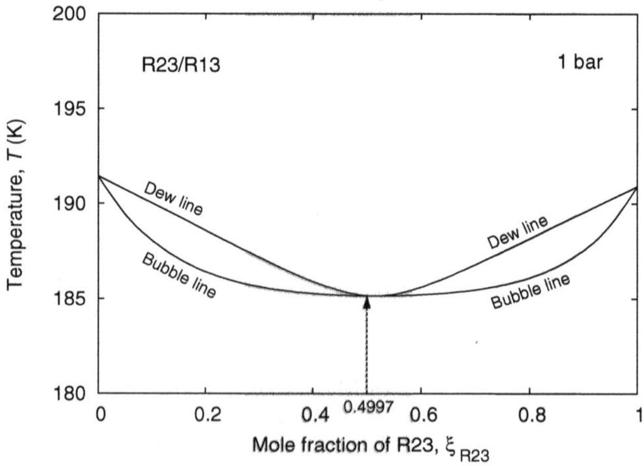

Fig. 1.54. Mixture of R23 (CHF_3) and R13 (CF_3Cl) at 1 bar exhibiting an azeotropic behavior. The bubble and dew point temperatures of the mixture are the same when the concentration of R23 in the mixture is 0.4997.

2

Simulation of cryogenic processes

Process simulation is a widely used technique in the design, analysis, and optimization of process plants. Simulators are computer programs that simulate the behavior of the process plants using appropriate mathematical models. Simulators are used for a variety of purposes:

- to perform material balance and energy balance of processes,
- to determine the detailed specifications of all units of a process,
- to troubleshoot startup and shut-down operations,
- to determine performance under off-design conditions,
- to design and troubleshoot control strategies.

Simulators are also extremely useful teaching tools to understand the behavior of individual units as well as interconnected units, namely a complete plant.

Cryogenic processes differ somewhat from general chemical processes. Some of the features special to cryogenic processes include multistream heat exchangers, the possibility of internal pinch points in heat exchangers, large variation in thermophysical properties across the heat exchanger length, operation close to critical temperatures, and double distillation columns. General-purpose commercial process simulators such as Aspen Plus or Hysim have features that accommodate many of these special requirements and therefore can be used to simulate most cryogenic liquefaction and refrigeration processes, including the mixed refrigerant processes studied in this monograph. Cubic equations of state such as the Peng–Robinson or Soave–Redlich–Kwong used in general-purpose simulators, however, do not predict properties of helium, hydrogen, or neon accurately at low temperatures. Higher-order equations of state such as the modified Benedict–Webb–Rubin equation of state need to be used for the simulation of helium, hydrogen, and neon.

Process simulators can be divided into three types based on their architecture, namely

- sequential modular,
- equation-oriented,
- simultaneous modular.

G. Venkatarathnam, *Cryogenic Mixed Refrigerant Processes*,
DOI: 10.1007/978-0-387-78514-1_2, © Springer Science+Business Media, LLC 2008

2.1 Sequential modular simulators

Sequential modular simulators are the most widely used simulators in the industry. The mathematical models representing individual units are coded separately as subroutines in these simulators. The mathematical models are developed so that the output stream data including pressure, temperature, enthalpy, entropy, etc. can be calculated for given input stream data and equipment operating parameters such as pressure ratio, outlet pressure, efficiency of the equipment, etc. While simulating the performance of a process plant, the subroutines representing different units are called in succession, with the output of one unit serving as the input of the next. The computation proceeds unit by unit from the feed to the product streams. When there are recycle loops in the process, the recycle loops are torn at suitable points and estimated values are assigned to these streams. Recycle loops are sequentially solved until the assumed values of the tear streams match the computed stream information.

2.1.1 Example: Open-cycle Linde–Hampson nitrogen liquefier

The sequential modular approach can be better understood by considering a simple example of an open-cycle Linde–Hampson nitrogen liquefier (Fig. 2.1). The design conditions for different equipment of the liquefier are shown in Table 2.1. The simulation starts with the heat exchanger. The conditions of outlet streams 3 and 5 can be determined from that of inlet streams 2 and g and the operating characteristic of the heat exchanger given the effectiveness or temperature approach between the streams at the warm or cold end of the heat exchanger. The temperature and flow rate of one of the two input streams of the heat exchanger (stream 2) are known. The temperature and flow rate of the second input stream, stream g in Fig. 2.1, are, however, not known. The temperature of streams 3 and 5 can be determined only when that of stream g is known. The flow rate of stream g, however, is dependent on the temperature of stream 3 entering the J-T valve. Thus, the conditions of stream 3 and g are interdependent. Streams that are interdependent are known as recycle streams. In such cases, the recycle is broken by tearing a recycle stream as shown in Fig. 2.2. The flow rate, pressure, and temperature of stream g' in Fig. 2.2, are assumed. With the enthalpy and flow rate of the two input streams known, the temperature (enthalpy) of the two outlet streams of the heat exchanger (3, 5) can be calculated from the specified heat

Table 2.1. Design specifications for a Linde–Hampson liquefier shown in Fig. 2.1

Compression process	Isothermal
Compressor outlet pressure	200 bar
Heat exchanger effectiveness	95%
Pressure drop in heat exchanger	0 bar
Valve outlet pressure	1 bar
Pressure drop in phase separator	0 bar

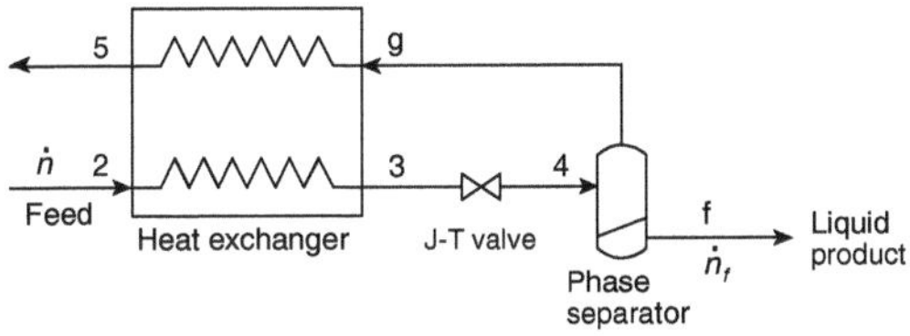

Fig. 2.1. Open-cycle Linde–Hampson liquefier.

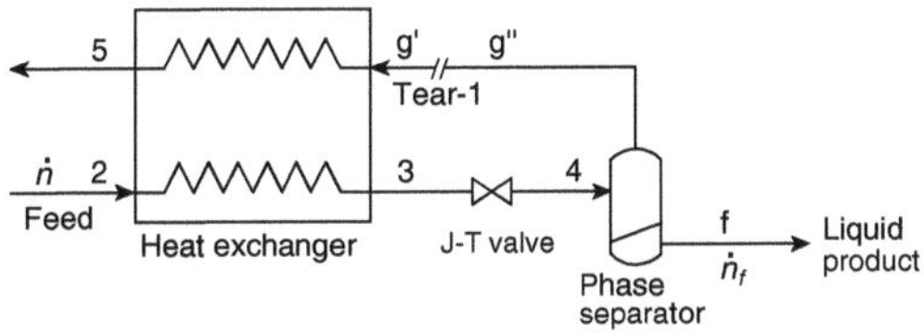

Fig. 2.2. Sequential modular simulation model of an open-cycle Linde–Hampson liquefier with tear stream g.

exchanger effectiveness or temperature approach between the streams at the warm or the cold end of the heat exchanger [see Eq. (1.50)].

The relationship between the mole flow rate of the tear stream g′ and that calculated after solving the governing equations of the heat exchanger, J-T valve, and the phase separator (stream g″) can be mathematically expressed as follows:

$$\dot{n}_{g''} = f(\dot{n}_{g'}). \tag{2.1}$$

The assumed mass flow rate $\dot{n}_{g'}$ will be the same as that calculated ($\dot{n}_{g''}$) when only the tear streams g″ and g′ are converged. The Wegstein method or a gradient method can be used to revise $\dot{n}_{g'}$ so that $\dot{n}_{g'} = \dot{n}_{g''}$. Convergence of any tear stream is assumed to have occurred when the enthalpy, flow rate, and concentration of the different components of the tear streams are about the same.

The temperature and vapor fraction at the exit of the J-T valve can be determined from the enthalpy of the inlet stream ($h_3 = h_4$). The determination of flow rates of the liquid (stream f) and vapor outlet stream (stream g″) of the phase separator is straightforward in the case of pure fluids. An isothermal flash calculation needs to be performed to determine the composition and flow rates of the liquid and vapor outlet streams in the case of mixtures (see refs. [67, 95] for the possible methods). The assumed stream information (in this case, the mass flow rate of g′) is altered at the end of each computation cycle to achieve convergence between assumed stream g′ and computed streams g″. The simulation procedure is shown in Fig. 2.3.

Table 2.2 shows the input file for the simulation of the open-cycle Linde–Hampson nitrogen liquefaction process shown in Fig. 2.2 using CRYOSIM, a process simulator developed by the author and his students [53, 74, 85]. When the converged stream g′ does not match stream g″, the computation shifts to the heat exchanger block *hx* with a new (revised) g′. The stream numbers need to be preceded by any letter/word (s in the present example) with CRYOSIM.

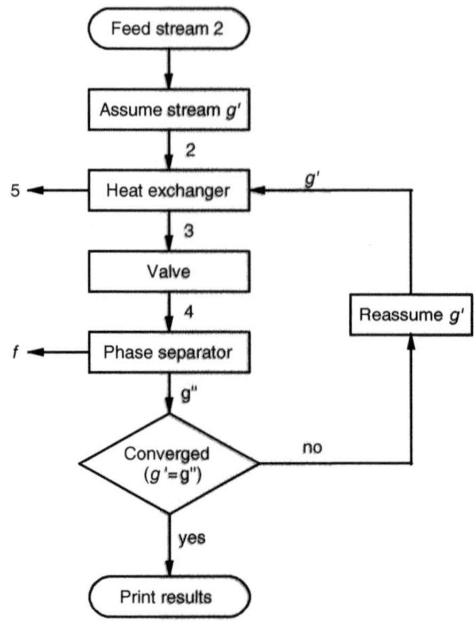

Fig. 2.3. Sequential modular simulation model for an open-cycle Linde–Hampson liquefier shown in Fig. 2.2.

Table 2.2. Input file for the simulation of a Linde–Hampson liquefier shown in Fig. 2.2 using the CRYOSIM process simulator.

```
Linde_Hampson_liquefier flowsheet {
        prop_pack = thermopack
        eos_type=peng_robinson
        components=n2 }
s2 = stream {
        flowrate = 100 kmol/hr
        pressure = 200 bar
        temperature = 300 K
        composition = (n2=1) }
sg' = stream {
        flowrate = 95 kmol/h
        pressure = 1 bar
        vapor_fraction = 1.0
        composition = (n2 = 1) }
hx = heatex2(s2:in, sg':in, s3:out, s5:out) {
        effectiveness = 1 }
jt = jt_valve(s3:in,s4:out) {
        pressure = 1 bar }
ps = phase_sep(s4:in, sf:liquid, sg':vapor)
c1 = converge(sg'':computed, sg':assumed) {
        go_to = hx
        convergence_method = wegstein }
```

Both graphical user interface and text-based inputs such as that shown in Table 2.2 can be used in commercial process simulators such as Aspen Plus. Consider an open-cycle Kapitza nitrogen liquefier (Fig. 2.4). There are two heat exchangers in the process. Hence, there must be two tear streams. The cold streams entering the heat exchanger (11, 12) are torn as shown in Fig. 2.4. Figure 2.5 shows the procedure to be followed for the simulation of the Kapitza open cycle using the sequential modular approach. The two tear stream convergence loops are nested, resulting in a large number of iterations when the estimated values of the tear streams are far from the converged values. The maximum number of nested convergence loops among the processes discussed in Chapter 1 occurs in the case of the Collins helium liquefaction process due to the presence of five heat exchangers.

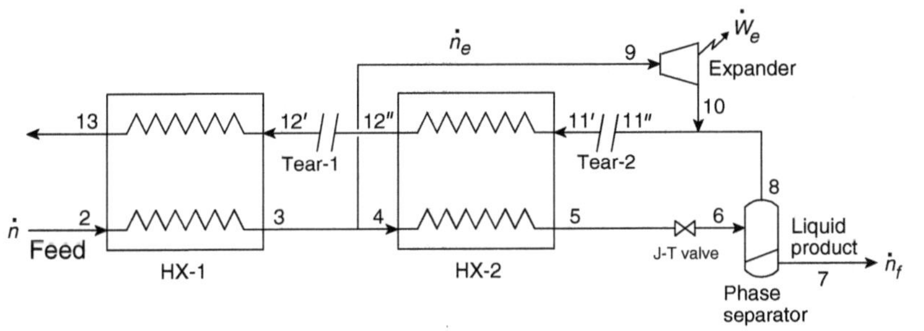

Fig. 2.4. Schematic of a Kapitza liquefaction process.

Advantages

The sequential modular method offers the following advantages:

- Process computations follow the material flow through the process. It is therefore easy to debug convergence failures.
- The mathematical models of different units (multistream heat exchangers, distillation columns, etc.) can be developed and coded separately with different solution procedures for different equipment modules.
- New types of equipment modules can be easily added. The only criterion that needs to be kept in mind while developing the new module is that the output stream information should be calculated for given input stream information and operating conditions of the equipment.
- The overall solution procedure is not affected by the complexities incorporated in each module.

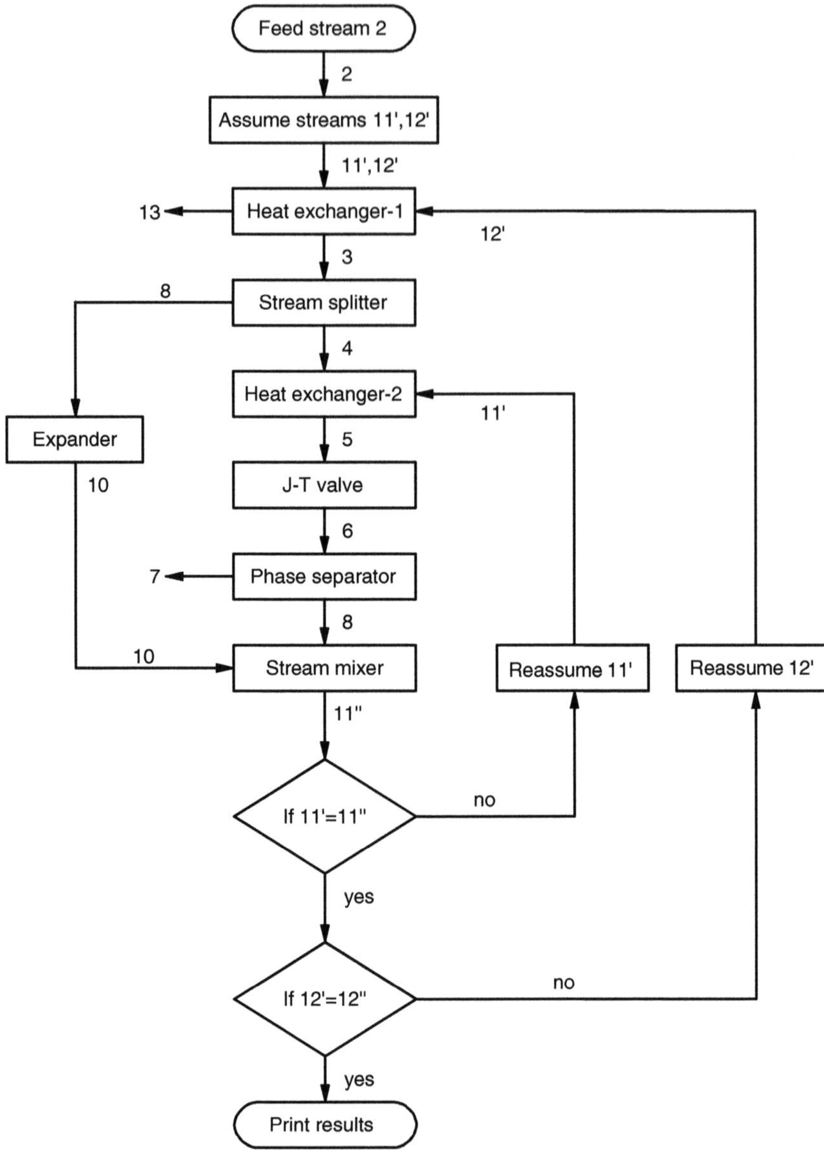

Fig. 2.5. Sequential modular simulation model for an open-cycle Kapitza liquefier (Fig. 2.4).

Disadvantages

The sequential modular approach, however, has two disadvantages:

- Computations can sometimes fail to converge when the process is strongly interconnected or when the number of tear streams is large.
- The computation time is high when the number of tear streams is large.

2.1.2 Tearing of recycle streams

Heat exchangers in which the cold stream is derived from the hot stream are sometimes known as regenerative heat exchangers. These are different from heat exchangers known as regenerators, in which heat transfer takes place between a solid and only one fluid at any point of time. One recycle stream needs to be torn for each regenerative heat exchanger. There is one recycle stream in the open-cycle Linde–Hampson liquefier shown in Fig. 2.2, and two in the closed-cycle Linde–Hampson liquefier (Fig. 1.18): one due to recycling of heat in the heat exchanger, and another due to recycling of mass (stream 5). There are two recycle streams for an open-cycle Kapitza liquefier (Fig. 2.4) and six for a closed-cycle Collins liquefier (Fig. 1.48).

Recycle streams need to be torn at appropriate locations in sequential modular simulators. The theory for automatic tearing of complex flowsheets is well developed (see Ref. [91]). Most modern process simulators tear the recycle streams automatically without the need for any user intervention. However, convergence problems do arise in some cases, and it is helpful to specify the tear streams manually. Tearing of the cold streams entering each regenerative heat exchanger has usually been found to be effective in our studies [74, 85]. If more than one cold stream exists in any heat exchanger, only one cold stream needs to be torn. The stream leaving the condenser (or aftercooler) can be torn in the case of closed-cycle systems.

2.2 Equation-oriented simulators

The governing equations of each process unit is solved one at a time, sequentially, in the case of a sequential modular approach, while the governing equations of all the units are solved together simultaneously in an equation-oriented approach. In the equation-oriented approach, each equipment module contributes the governing equations to be solved. Tearing of streams is not necessary in the equation-oriented approach since all the governing equations are solved simultaneously.

The equation-oriented approach offers the following advantages over the sequential modular method:

- As all equations are solved simultaneously, there is no need for nested iteration loops, which makes it suitable for simulation of strongly interconnected processes with many recycle loops.
- It is best suited for design optimization and dynamic simulation studies where the process needs to be simulated thousands of times.

The equation-oriented approach, however, has the following demerits:

- Good initial estimates are required for all variables for convergence.
- It is hard to handle errors when there is inconsistency in specifications.
- The addition of new equipment modules is not simple.
- A general-purpose, robust, nonlinear equation solver is required.
- Inequality constraints involving design variables are harder to implement in design optimization studies compared to sequential and simultaneous modular approaches.

While the sequential modular approach continues to be the workhorse of the industry, process simulators that have both sequential modular and equation-oriented capabilities such as the Aspen Plus are now commercially available. A few iterations in the sequential modular mode are normally needed to estimate the initial values of all variables, before the equation-oriented method can be used to reach convergence.

2.3 Simultaneous modular simulators

Consider a process with multiple tear streams. In a sequential modular approach, all tear stream convergence loops are nested one inside the other, as shown in Fig. 2.5. Only one tear stream is revised in each of these loops. This leads to a large number of iterations for convergence of all the tear streams in a sequential modular approach. In the simultaneous modular method, also known as the two-tier method, the sequential modular method is used as the basic technique for the solution of process flowsheets, with simultaneous reassumption of all tear streams as in the equation-oriented method (during convergence). The simultaneous modular approach thus combines the good features of both the sequential modular and equation-based approaches. Simultaneous modular approach is also favored for design optimization studies. In this approach, the tear stream convergence becomes an additional equality constraint in the optimization problem. The design variables and tear stream variables are varied at the end of each iteration simultaneously to reach the maxima (or minima) of the objective function while converging the tear streams simultaneously. A combination of simultaneous and sequential modular approaches is sometimes preferred. The first few iterations of the optimization problem are solved to converge the tear streams in a sequential modular approach, and the final convergence is achieved using the simultaneous modular approach. The simultaneous modular approach is also used in CRYOSIM [74] for optimization studies.

Most cryogenic processes discussed in this monograph have fewer than 10 recycle streams, and convergence is fairly rapid with the sequential modular approach. Hence, both the sequential modular and equation-oriented approach can be used to simulate mixed refrigerant processes. The simultaneous modular approach is recommended for design optimization studies.

2.4 Simulation of heat exchangers with pinch points

Pinch points form in cryogenic heat exchangers due to the large variation of specific heat or enthalpy with temperature as discussed in Section 1.16. Consider a Kapitza nitrogen liquefaction process (Fig. 1.42). The cold box that excludes the compressor and aftercooler can be simulated separately, as shown in Fig. 2.4. The flowsheet needs to be torn at two places (streams 11 and 12). While the design effectiveness or the temperature approach at the warm end can be specified for the first heat exchanger, the heat exchanger effectiveness cannot be specified for the second heat exchanger because of the possibility of an occurrence of a pinch point (Figs. 1.50 and 1.51). It is customary to specify the minimum temperature approach between the streams in the heat exchangers with pinch points. In this section, the methods to simulate the performance of a heat exchanger with a pinch point are described.

Consider a heat exchanger shown in Fig. 2.6. The temperatures of the hot and cold streams $T_{h,o}$, $T_{c,o}$, and $T_{c,n}$ are specified, along with the mass flow rates of the two streams. The temperature of the hot stream leaving the heat exchanger $T_{h,n}$ can be determined from an energy balance across the entire heat exchanger. The subscripts o refers to the warm end and n refers to the cold end.

The heat exchanger is divided into n parts such that the heat transferred across each part is the same, or the temperature change of the maximum heat capacity rate fluid stream $(\dot{n}c_p)_{\max}$ across each part is the same. The heat transfer rate across any ith part $(d\dot{Q}_i)$ will be known when either of these methods is adopted.

Consider any ith part of the heat exchanger (Fig. 2.7). The temperatures of the hot and cold streams entering the heat exchanger are designated as $T_{h,i-1}$ and $T_{c,i}$, respectively. The temperatures of the hot and cold fluid streams leaving the heat exchanger are designated as $T_{h,i}$ and $T_{c,i-1}$, respectively, as shown in Fig. 2.7. The simulation of a heat exchanger with pinch begins at the warm end of the heat exchanger (part 1). The temperatures of the streams at the warm end of part 1 ($T_{h,o}$ and $T_{c,o}$) are known. Since the heat transferred across part 1 $(d\dot{Q}_1)$ is also known, the temperature of the hot and fluid streams at the cold end of part 1 ($T_{h,1}$ and $T_{c,1}$) can be estimated from a simple energy balance across part 1. With the temperature of the hot and cold fluid streams at the warm end of part 2 (cold end of part 1) now known, the procedure is repeated from parts 2 to n. The smallest of the temperature approaches between the streams $(T_{h,i} - T_{c,i})$ gives the minimum temperature approach between the

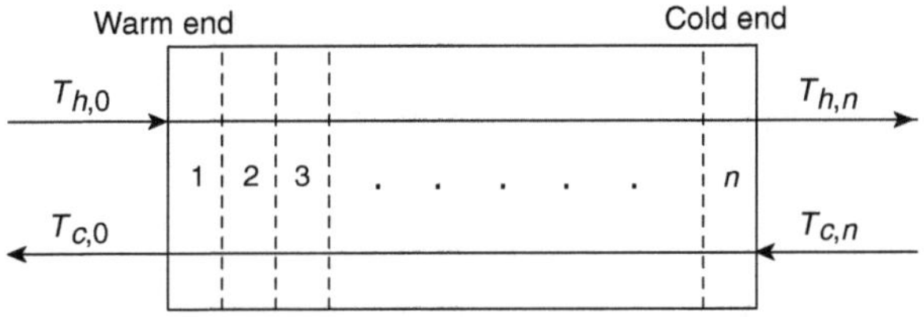

Fig. 2.6. Model of a two stream heat exchanger divided into n parts.

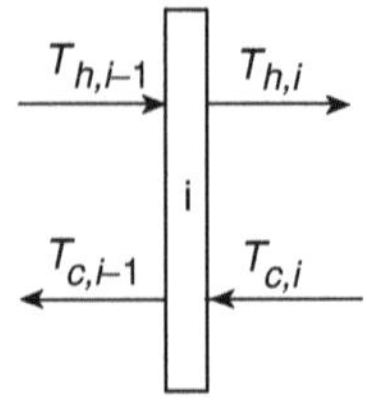

Fig. 2.7. Temperatures at the entry and exit of the ith part of a two-stream heat exchanger divided into n parts.

streams. If there are points of inflection in the temperature approach-heat load curves, the relevant parts are subdivided to determine the minimum temperature approach. CRYOSIM uses the Brent method [26] to subdivide the parts and determine the minimum temperature approach.

The above method can also be extended to multistream heat exchangers. All but one stream temperature needs to be specified to determine the minimum temperature approach in a multistream heat exchanger. Similar methods are used in the MHeatX module of the Aspen Plus process simulator, the heatexm module of the CRYOSIM process simulator, and the LNG heat exchanger module of some other simulators.

Consider the second heat exchanger of the open-cycle Kapitza nitrogen liquefaction process shown in Fig. 2.4. When only the minimum temperature approach between the streams is specified as the operating condition, the outlet temperature of the hot stream (T_5 in Fig. 2.4) is changed iteratively to meet the desired minimum temperature approach. For the second heat exchanger in Fig. 2.4, this relationship between the temperature of the hot stream leaving the heat exchanger (T_5) and the minimum temperature approach between the streams is shown in Fig. 2.8.

A minimum temperature approach of 5 K can be obtained when T_5 is 105.1 K. The desired pinch point temperature can thus be obtained by varying T_5. The minimum temperature approach condition is solved using the following expression:

$$f(T_5) = \Delta T_{\text{min}} - \Delta T_{\text{min, specified}} = 0. \tag{2.2}$$

The Newton method can be used to converge the function $f(T_5) = 0$. Most commercial process simulators have provisions for similar functions to be specified as design specifications.[1] Alternately, one can converge $f(T_5)$ and the tear streams simultaneously in the case of the simultaneous modular approach. $f(T_5)$ becomes an additional equation to be solved in the case of the equation-oriented approach.

Figure 2.9 shows the pinch point in the second heat exchanger of a Kapitza nitrogen liquefier operating at 30/1 bar for different values of temperature of the high-pressure fluid leaving the second heat exchanger (T_5). It can be seen that one or more temperature crossovers is observed at low values of T_5. Pinch point occurs in this case due to the specific heat becoming infinite during the constant temperature condensation process. The large variation of the specific heat (c_p) at temperatures close to the

[1] Design Spec in Aspen Plus.

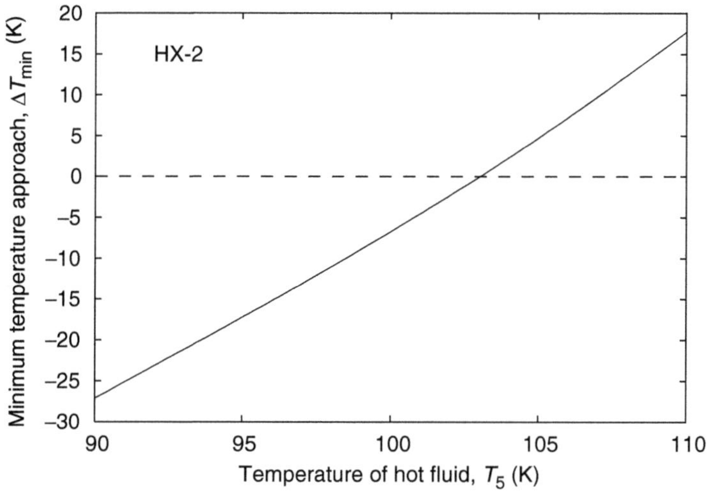

Fig. 2.8. Variation of the minimum temperature approach between the streams with temperature of the hot stream leaving the second heat exchanger (T_5) of the Kapitza nitrogen liquefaction process shown in Fig. 2.4.

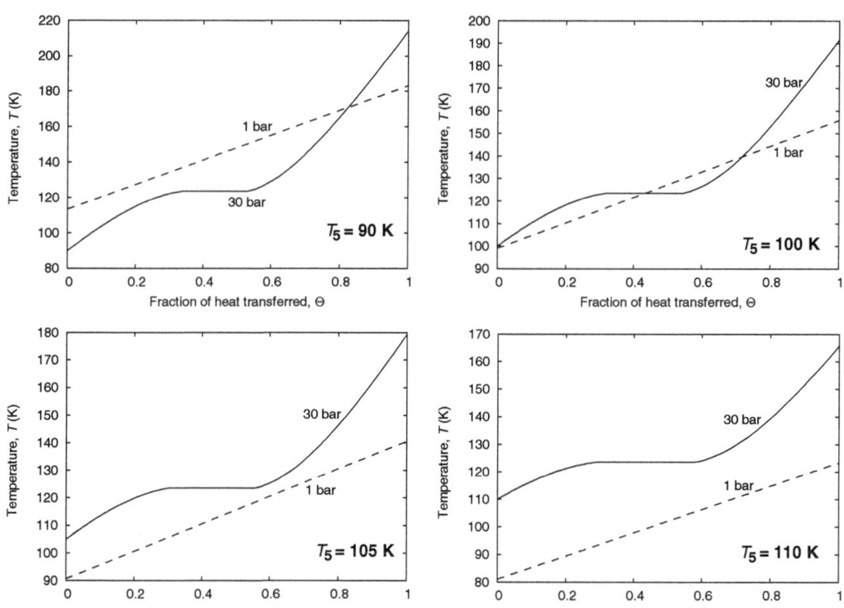

Fig. 2.9. Variation of the temperature of the hot (30 bar) and cold (1 bar) streams in the second heat exchanger of a Kapitza nitrogen liquefier at different temperatures of the hot stream leaving the heat exchanger (T_5).

critical temperature results in pinch points when the operating pressure is above the critical pressure, as discussed in Section 1.16.

2.5 Optimization of a Kapitza nitrogen liquefier

Consider the open-cycle Kapitza nitrogen liquefaction process shown in Fig. 2.4. The exergy efficiency of the process is strongly dependent on the minimum temperature approach between the streams in the two heat exchangers and the flow rate through the turbine. The simultaneous modular approach can also be used effectively for optimization studies. Table 2.3 shows the approach to be used to optimize an open-cycle Kapitza nitrogen liquefier (Fig. 2.4) operating at operating pressures of 40/1 bar using the simultaneous modular approach. The adiabatic efficiency of the turbine was assumed as 80%. The temperature approach between the streams was assumed to be 5 K in the first heat exchanger and 3 K in the second heat exchanger. The flow rate through the turbine ($\dot{n}_9$), the temperature of the cold stream leaving the first heat exchanger (T_{13}), and the temperature of the hot stream leaving the second heat exchanger (T_5) are assumed as the design variables of the optimization study. The minimum temperature approach between the streams of the heat exchanger is constrained to be greater than or equal to the specified values. The temperature at the exit of the turbine expander is constrained to prevent occurrence of two-phase flow conditions in the turbine.

Both sequential modular and simultaneous modular approaches can be used to optimize the performance of the Kapitza process. The tear streams are converged separately, as a part of each simulation run with the sequential modular approach,

Table 2.3. Method for the optimization of a Kapitza nitrogen liquefier.

Objective function: Maximize exergy efficiency, η_{ex}

Subject to constraints:
- Minimum temperature approach between the streams in the first heat exchanger, $\Delta T_{min,\,hx-1} \geq \Delta T_{min,\,hx-I,\,specified}$
- Minimum temperature approach between the streams in the second heat exchanger, $\Delta T_{min,\,hx-2} \geq \Delta T_{min,\,hx-II,\,specified}$
- No two-phase condition in the turbine or $T_{10} > T_{dew,\,10}$
- Convergence of tear stream 11, $\dot{n}_{11'} - \dot{n}_{11''} = 0, h_{11'} - h_{11''} = 0$
- Convergence of tear stream 12, $\dot{n}_{12'} - \dot{n}_{12''} = 0, h_{12'} - h_{12''} = 0$

Design variables:
- Flow through the turbine, $\dot{n}_9$
- Temperature of the cold stream leaving the first heat exchanger, T_{13}
- Temperature of the hot stream leaving the second heat exchanger, T_5

Constants:
- Specified pressure drop of the high- and low-pressure streams of the first heat exchanger
- Specified pressure drop of the high- and low-pressure streams of the second heat exchanger
- Specified adiabatic efficiency of the turbine

while the tear streams are converged together with other constraints in the case of optimization using the simultaneous modular approach.

In the simultaneous modular approach, the flow rate and enthalpy of the two tear streams are also considered equality constraints, as shown in Table 2.3.

Figure 2.10 shows the convergence of the design variables during the optimization process using the Aspen Plus process simulator. The Sequential Quadratic Programming (SQP) method was used to optimize the exergy efficiency of the process.

The optimization of the performance of other cryogenic processes operating with pure fluids is similar to that described above for the Kapitza liquefaction process. The optimization of mixed refrigerant processes additionally involves a choice of refrigerants and their composition, as well as the operating pressures. The methods for the optimization of mixed refrigerant processes are discussed in Chapter 5 with practical examples.

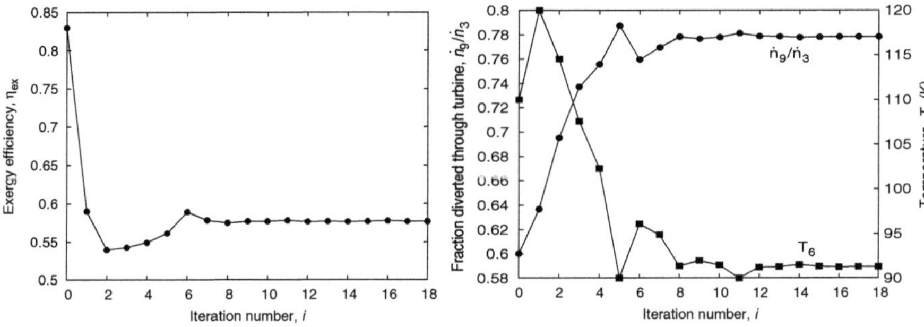

Fig. 2.10. Convergence of the design variables and objective function during the optimization of a Kapitza nitrogen liquefaction process operating at 40/1 bar, and a turbine adiabatic efficiency of 80%. The temperature approach between the streams was assumed to be 5 K in the first heat exchanger and 3 K in the second heat exchanger.

3

Need for refrigerant mixtures

Cryogenic refrigerators and liquefiers were operated with pure fluids for much of the last century. The importance of using a mixture of refrigerants appears to have been first proposed in 1936 by Podbielniak [69], but was implemented in practice in the 1960s in large-scale natural gas liquefaction after the pioneering work of Kleemenko [50]. The use of mixtures in refrigeration applications was pioneered in the former Soviet Union [4, 17]. In this chapter, the methods for overcoming the low exergy efficiency of refrigeration and liquefaction systems operating with pure fluids using refrigerant mixtures are discussed.

3.1 Refrigeration systems

The advantages in using refrigerant mixtures in place of pure fluids can be well understood with reference to any process. The Linde–Hampson refrigeration process (Fig. 3.1) has been chosen to compare its performance with pure fluids and zeotropic mixtures because of its simplicity.

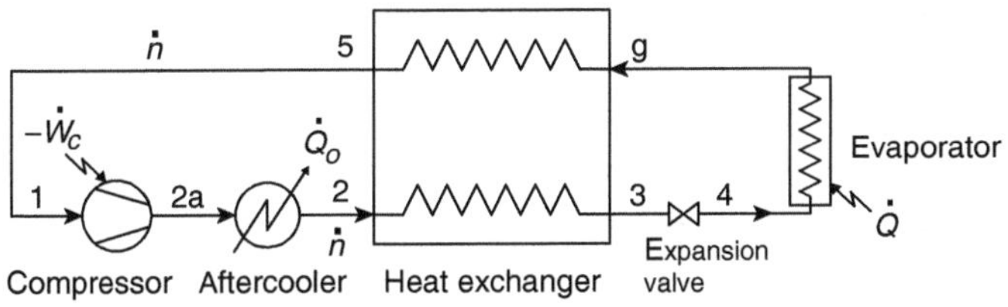

Fig. 3.1. Schematic of a Linde–Hampson refrigerator.

An energy balance around the heat exchanger, expansion valve, and evaporator of a Linde–Hampson refrigerator (Fig. 3.1) gives the specific refrigeration effect ($\dot{Q}/\dot{n}$) as follows:

G. Venkatarathnam, *Cryogenic Mixed Refrigerant Processes*,
DOI: 10.1007/978-0-387-78514-1_3, © Springer Science+Business Media, LLC 2008

$$\frac{\dot{Q}}{\dot{n}} = h_5 - h_2 = \Delta h_{\min}, \tag{3.1}$$

where $\dot{Q}$ is the refrigeration load at temperature T, and $\dot{n}$ is the flow rate through the process. h_5 and h_2 refer to the enthalpy of the low- and high- pressure fluids at the warm end of the heat exchanger, respectively. $\Delta h_{\min}$ is the minimum enthalpy difference between the cold and hot streams at any location in the heat exchanger. In the case of pure fluids, $\Delta h_{\min}$ always occurs at the warm end of the heat exchanger (see Fig. 1.21). The enthalpy of a zeotropic mixture can vary nonlinearly with temperature in the two-phase region, with points of inflection on the enthalpy-temperature curves [93, 94]. The minimum temperature difference between the streams can therefore occur at any location in the heat exchanger.

Figure 3.2 shows the variation of the specific refrigerating effect $\Delta h_{\min}$ with operating pressure p_2 for a Linde–Hampson refrigerator operating with different pure fluid refrigerants. The heat exchanger effectiveness (ε) has been assumed to be 100%. Also, no condensation is assumed to occur in the aftercooler in all the cases. The operating pressure is limited to a small value in the case of propane and ethane to satisfy this condition. The pressure at the exit of the expansion valve (compressor suction) is fixed at 1.0 bar. It can be observed from Fig. 3.2 and Table 3.1 that the specific refrigeration effect ($\Delta h_{\min}$) obtained in a Linde–Hampson refrigerator increases with an increase in the normal boiling point of the refrigerant. The maximum specific refrigeration effect is limited to a value slightly more than 1000 J/mol in the case of nitrogen at very high operating pressures. Table 3.2 shows the pressure required for a specific refrigeration effect of 1000 J/mol with different working fluids. The

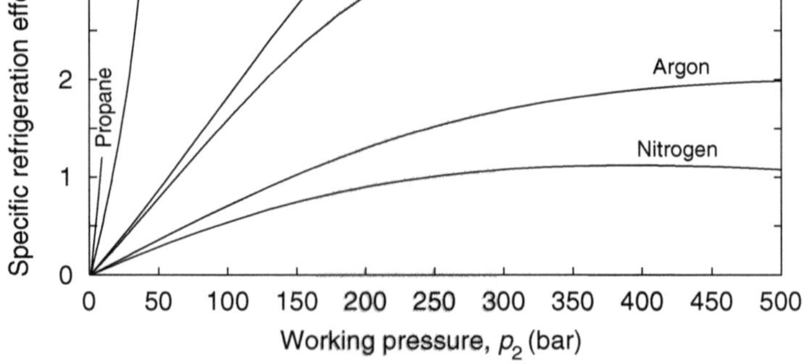

Fig. 3.2. Variation of specific refrigeration effect ($\dot{Q}/\dot{n} = h_5 - h_2 = \Delta h_{\min}$) with operating pressure for different working fluids of a Linde–Hampson refrigerator ($p_1 = 1$ bar, $T_1 = 300$ K, $\varepsilon = 100\%$, or $T_2 = T_5 = T_1$).

Table 3.1. Normal boiling point of different refrigerants studied [63].

Fluid	Normal boiling point (K)	Critical temperature (K)
Nitrogen	77.36	126.19
Argon	87.30	150.69
Methane	111.67	190.56
Krypton	119.78	209.46
Ethane	184.55	305.33
Propane	231.06	369.85

Table 3.2. Operating pressure (p_2) required for the specific refrigeration effect to be 1000 J/mol for different refrigerants of a Linde–Hampson refrigerator ($p_1 = 1$ bar, $T_1 = T_2 = 300$ K, $\varepsilon = 1$) estimated using the NIST 12 database [63].

Fluid	Operating pressure p_2 (bar)
Nitrogen	246.0
Argon	137.4
Methane	63.5
Krypton	57.0
Ethane	16.5
Propane	7.9

working pressures of ethane and propane have been limited to prevent condensation in the aftercooler. A specific refrigeration effect on the order of 600 to 1000 J/mol can be obtained with optimum mixtures of nitrogen, methane, ethane, and propane with operating pressures lower than 25 bar.

In the case of a Linde–Hampson refrigerator operating with an ideal heat exchanger ($T_5 = T_2 = T_1$) and pure fluid refrigerants, the specific refrigeration effect is related to the Joule–Thomson coefficient at the warm end of the heat exchanger as

$$h_5 - h_2 = -\int_{p_5}^{p_2} \left(\frac{\partial h}{\partial p}\right)_{T_1} dp = \int_{p_5}^{p_2} \left(\frac{\partial h}{\partial T}\right)_p \left(\frac{\partial T}{\partial p}\right)_h dp = \int_{p_5}^{p_2} c_p \mu_{\mathrm{J-T}} dp. \tag{3.2}$$

The value of the Joule–Thomson coefficient ($\mu_{\mathrm{J-T}}$) essentially represents the non-ideality of a fluid at a given pressure and temperature. The higher the value of the J-T coefficient, the higher is the non-ideality at the given temperature. The J-T coefficient of ideal fluids is zero. It is clear from Fig. 3.2 and Eq. (3.2) that, in general, the non-ideality at room temperature ($\mu_{\mathrm{J-T}}$) or the specific refrigeration effect (Δh_{min}) is higher for fluids whose critical temperature is closer to the ambient temperature such as propane than those whose critical temperature is much lower than the ambient temperature such as nitrogen.

With an appropriate mixture of high and low boiling point components, it is possible have a high specific refrigeration effect (high critical temperature) at a relatively low refrigeration temperature. The high critical temperature or a high specific refrigeration effect and a high dew point temperature result from the use of high boiling point components and a low bubble point or refrigeration temperature results from the use of low boiling point components in the mixture.

The bubble point of normal zeotropic mixtures will, however, be higher than that of the lowest boiling point fluid in the mixture. The difference between the boiling point of the lowest boiling point component or the most volatile fluid and the bubble point of the mixture depends on the composition of the mixture (for example, see Fig. 3.3 for a binary mixture of nitrogen and methane). Binary mixtures of hydrocarbons such as methane, ethane, propane, and butane show a behavior similar to that shown in Fig. 3.3. The addition of high boiling point components will increase the specific refrigeration effect, but it also increases the refrigerating temperature in such cases.

In some zeotropic mixtures, the liquid phase can separate into two or more immiscible liquid phases. Consider the example of a binary mixture of nitrogen and ethane. Figure 3.4 shows the temperature-concentration diagram of a nitrogen-ethane mixture at a pressure of 20 bar, estimated using the Peng–Robinson equation of state, with binary interaction parameter, acentric factor, critical pressure, and temperature data from Ref. [51] and a three-phase flash algorithm. It can be observed that the bubble point temperature remains nearly constant between a nitrogen concentration of 26.5% and 96.6%, where the liquid phase splits into two immiscible liquid phases at low temperatures (vapor liquid liquid equilibria or VLLE). The splitting of a nitrogen hydrocarbon mixture into immiscible liquid phases is observed typically

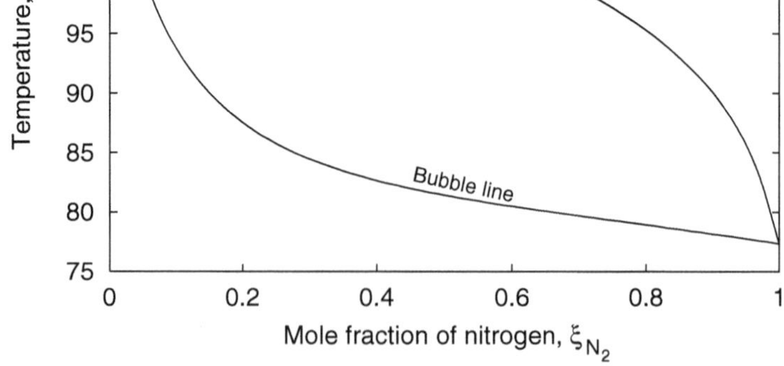

Fig. 3.3. Temperature concentration diagram of a binary mixture of nitrogen-methane at a pressure of 1 bar.

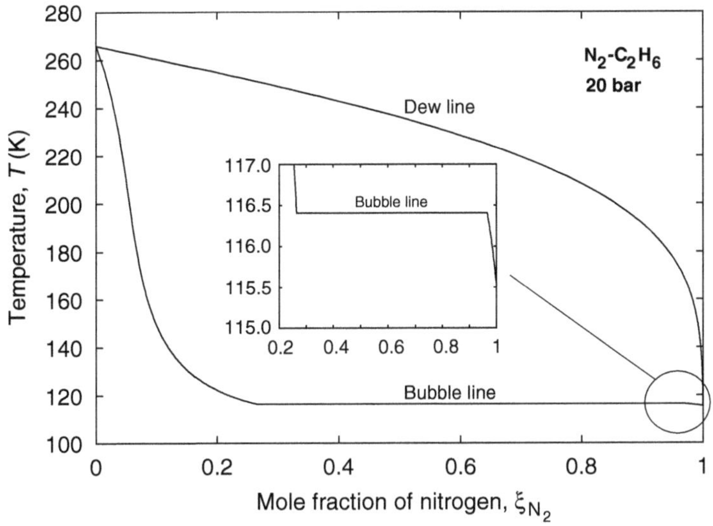

Fig. 3.4. Temperature concentration diagram for a binary mixture of nitrogen-ethane at a pressure of 20 bar. The inset shows the effect of the second liquid-phase occurrence on the bubble point temperature.

at temperatures lower than 130 K (for $p < 20$ bar). The phase split occurs only in some nitrogen-hydrocarbon mixtures and at certain mixture compositions. Similar phase splits can be observed with binary mixtures of perfluorocarbons and hydrocarbons such as $CF_4 - C_3H_8$ (R14-R290), $CF_4 - nC_4H_{10}$ (R14-R600), $CF_4 - iC_4H_{10}$ (R14-R600a), $C_2H_5 - C_4H_{10}$ (R125-R600), etc., in mixtures of perfluorocarbons and hydrofluorocarbons such as $CF_4 - C_2H_4F_2$ (R14-R152a), and in mixtures of perfluorocarbons and hydrochlorofluorocarbons such as $CF_4 - CHF_2Cl$ (R14-R22), etc. [34, 55].

It can be seen from the insert in Fig. 3.4 that the difference between the boiling point of the low boiler (nitrogen) and the bubble point of the mixture in the immiscible liquid phase region is extremely small (typically less than 1 K). Thus, the ethane content in the mixture increases the bubble point of the mixture in the immiscible liquid region only marginally. In other words, the addition of ethane to nitrogen increases the specific refrigeration effect of the refrigerant, but does not significantly change the refrigerating temperature in the immiscible liquid region.

Figure 3.5 shows the temperature profiles of the hot and cold fluid streams in the heat exchanger of a Linde–Hampson refrigerator operating with a mixture of nitrogen and ethane (46.9/53.1 mol%) and operating pressures of 20/3.4 bar. It can be seen that the temperature profiles get pinched at about 140 K. In general, pinch points occur between the two ends of the heat exchanger when binary mixtures are used. The maximum refrigeration effect is obtained when the pinch point occurs at the warm end of the heat exchanger. The refrigeration that can be obtained, and consequently the exergy efficiency of the process, is small when pinch points occur in between the

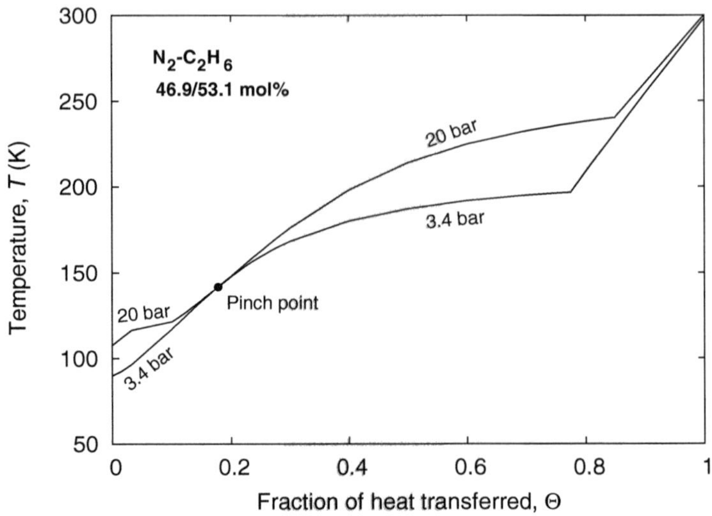

Fig. 3.5. Temperature profiles of hot and cold fluid streams of a Linde–Hampson refrigerator operating with a mixture of nitrogen and ethane (46.9/53.1 mol%) and operating pressures of 20/3.4 bar. The evaporator outlet temperature is 90 K.

two ends of the heat exchanger. This problem can be circumvented by using mixtures with at least three or more refrigerants.

Alfeev, Brodyanskii, and colleagues [7, 27] were probably the first to realize the importance of using a fluid in which the liquid-phase splits into two or more liquid-phases at low temperature to obtain a near-constant temperature in the evaporator. Many mixtures were suggested by Alfeev et al. in their patent.

Figure 3.6 shows the variation of the total enthalpy of a mixture of 30 mol% nitrogen, 30 mol% methane, 20 mol% ethane, and 20 mol% propane suggested by Alfeev et al. [7]. Since one of the immiscible liquid phases will be rich in the low boiling point fluid (in this case nitrogen), heat addition in the evaporator can take place at nearly constant temperature, close to the boiling point of nitrogen. The phase split of liquid into immiscible liquid phases allows us to obtain a constant-temperature refrigeration as in pure fluids. The refrigeration temperature is very close to that of the most volatile component (nitrogen), while the specific refrigeration effect ($\Delta h_{\min}$) is high due to the high critical temperature because of the presence of high boilers such as ethane and propane. The immiscible liquid region is normally limited to the evaporator or a tiny length of the heat exchanger at the cold end for high exergy efficiency.

Figure 3.7 shows the variation in the specific refrigerating effect with operating pressure for two patented mixtures (see Table 3.3) at pressures up to 25 bar, which are typical in a Linde–Hampson refrigerator operating with gas mixtures. It can be seen that with the two mixtures, it is possible to increase the refrigeration effect by nearly 4 to 5 times that of nitrogen or argon. The refrigerating temperature with mixture M1

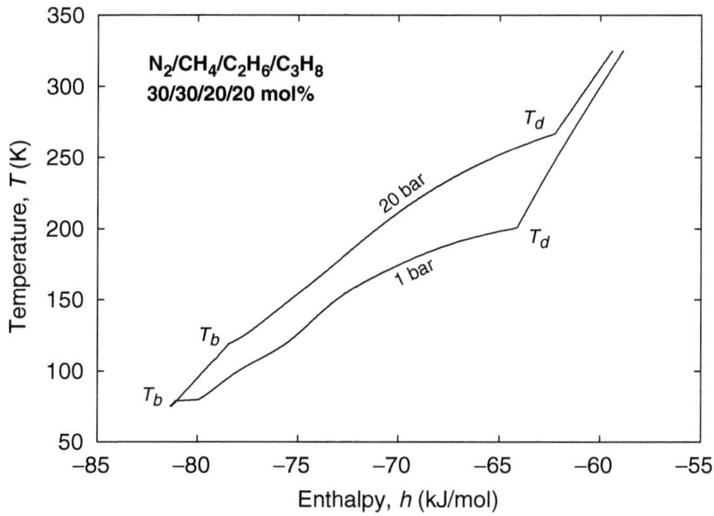

Fig. 3.6. Enthalpy temperature variation for a mixture of $N_2/CH_4/C_2H_6/C_3H_8$ (30/30/20/20 mol%) [7]. T_d and T_b are the dew point and bubble point temperatures, respectively. The near-horizontal portion on the 1 bar curve is due to the occurrence of two immiscible liquids.

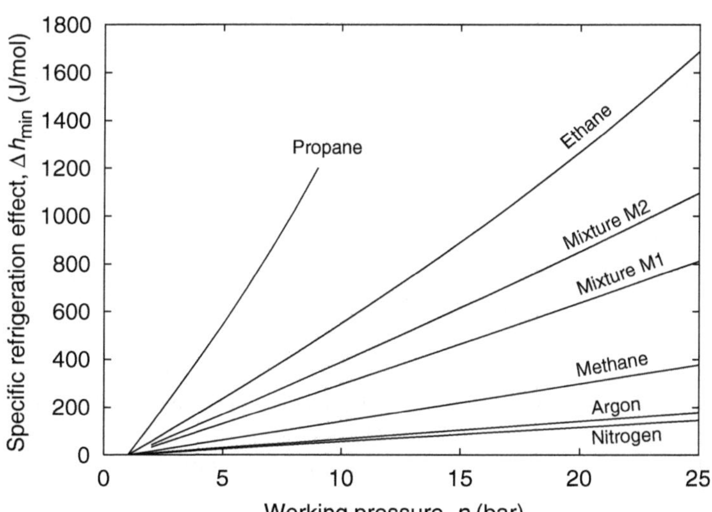

Fig. 3.7. Specific refrigeration effect with different refrigerants of a Linde–Hampson refrigerator. See Table 3.3 for the composition of mixtures M1 and M2.

Table 3.3. Two patented mixtures considered in this section.

Mixture	Fluid	Composition (mol%)	Source
M1	Nitrogen	30	[8]
	Methane	30	
	Ethane	20	
	Propane	20	
M2	Nitrogen	30	[20]
	Methane	15	
	Ethane	25	
	Propane	30	

and M2 would be 79.9 K and 78.9 K, respectively, compared to 77.24 K with nitrogen at an evaporating pressure of 1 bar.

It can be seen from Fig. 3.7 that a specific refrigeration effect of 1000 J/mol can be obtained at an operating pressure of about 23 bar with a nitrogen-hydrocarbon mixture, mixture M2 in Table 3.3, compared to about 246 bar required when nitrogen alone is used as the working fluid (see Table 3.2). The possibility of obtaining a large specific refrigeration effect at low operating pressures, typically less than 20 bar, has made it possible to develop closed-cycle Linde–Hampson refrigerators operating with refrigerant mixtures.

The specific refrigeration effect is high due to the large concentration of high boiling point components whose Joule–Thomson coefficient at room temperature [see Eq. (3.2)] is high, while the refrigeration temperature is low due to the large concentration of low boiling point components and the occurrence of two immiscible liquid phases at low temperatures.

3.2 Exergy efficiency of ideal Linde–Hampson refrigerators operating with refrigerant mixtures

The exergy efficiency of any refrigerator is the ratio of the work of compression of the refrigerator ($-\dot{W}_c$) and the work required by a reversible (ideal) refrigerator ($-\dot{W}_{r,\mathrm{rev}}$) to provide the same refrigeration load ($\dot{Q}$) at a specified temperature T, which is the same as the ratio of the COP of the refrigerator to the COP of an ideal refrigerator as shown below:

$$\eta_{\mathrm{ex}} = \frac{-\dot{W}_{r,\mathrm{rev}}}{-\dot{W}_c} = \frac{\dot{Q}}{-\dot{W}_c}\left(\frac{T_o}{T} - 1\right) = \frac{\mathrm{COP}}{\mathrm{COP}_{r,\mathrm{rev}}}. \tag{3.3}$$

Figure 3.8 shows the variation of theoretical exergy efficiency of an ideal Linde–Hampson refrigerator with operating pressure (p_2) for different pure fluids. The compression was assumed to be isothermal, the pressure drop in heat exchangers to be

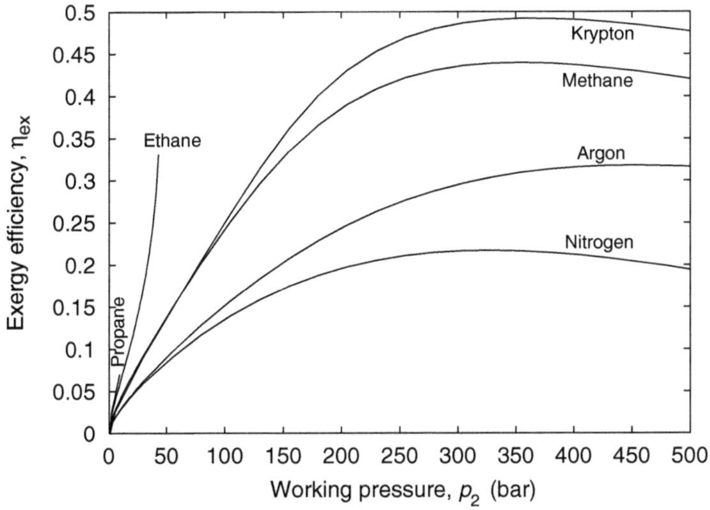

Fig. 3.8. Exergy efficiency of an ideal Linde–Hampson refrigerator operating with pure fluids and an isothermal compressor. $p_1 = 1$ bar, $T_1 = T_2 = T_5 = 300$ K, or $\varepsilon = 1.0$.

zero, and the effectiveness of the heat exchanger (ε) to be 100%. It can be seen from Fig. 3.8 and Table 3.1 that the exergy efficiency (η_{ex}) increases with an increase in refrigeration temperature, similar to the specific refrigeration effect (Δh_{min}). The exergy efficiency of practical systems is normally less than a fourth of the theoretical exergy efficiency predicted in Fig. 3.8 due to the exergy losses in the compressor and heat exchanger.

Figure 3.9 shows the variation of exergy efficiency with operating pressure for a Linde–Hampson refrigerator operating with nitrogen and the two nitrogen-hydrocarbon mixtures shown in Table 3.3. It can be observed from Figs. 3.8 and 3.9 that the exergy efficiency of an ideal Linde–Hampson refrigerator operating with gas mixtures at pressures below 20 bar is many times more than the maximum that can be achieved with nitrogen (at very high operating pressures). The reasons for this difference can be understood with reference to Fig. 3.10, which shows the thermodynamic processes followed in a Linde–Hampson refrigerator operating with a pure fluid and a refrigerant mixture.

The coefficient of performance of any refrigerator absorbing heat at a temperature (T) and rejecting heat at a temperature (T_o) is given by the expression

$$\mathrm{COP} = \frac{\dot{Q}}{-\dot{W}_c} = \frac{\dot{Q}}{\dot{Q}_o - \dot{Q}} = \frac{T_o\,|\Delta s|}{T_o|\Delta s_o| - T|\Delta s|}, \tag{3.4}$$

where Δs_o and Δs refer to the entropy change during the compression (heat rejection) and the heat addition processes, respectively. The entropy change during the heat addition and heat rejection processes is the same in all ideal refrigeration cycles such

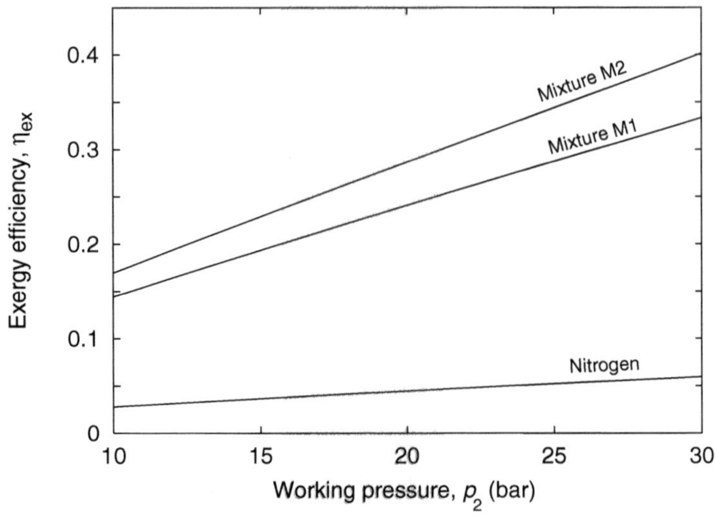

Fig. 3.9. Exergy efficiency of an ideal Linde–Hampson refrigerator operating with nitrogen and nitrogen-hydrocarbon mixtures and an isothermal compressor. $p_1 = 1$ bar, $T_1 = T_2 = T_5 = 300$ K, or $\varepsilon = 1$. The composition of the mixtures M1 and M2 is shown in Table 3.3.

as Carnot, Stirling and Erricson. On the other hand, it can be seen from Fig. 3.10 that the entropy change during the heat addition process of a Linde–Hampson refrigerator is only a very small fraction of that during the heat rejection process, resulting in a very small COP when pure fluids such as nitrogen are used as refrigerants compared to mixtures, leading to higher exergy efficiency in the case of mixtures. The entropy change during the heat addition process is small when compared to that during the heat rejection process when pure nitrogen is used as the refrigerant due to a large exergy loss or entropy gain during throttling, as well as that obtained in the heat exchanger.

In the case of mixtures, the high-pressure refrigerant is in a subcooled state at the entry of the throttle (stream 3 in Fig. 3.1). The refrigerant mixture undergoes a temperature change of less than 10 K in the throttle, versus the 70 to 80 K in the case of a pure fluid. Therefore, the exergy loss or entropy generation during expansion is only a very small part of the total exergy loss in the case of refrigerant mixtures. The exergy loss during heat transfer across a resistance with a temperature gradient ΔT is given by Eq. (1.56):

$$\Delta \mathrm{ex}_{\mathrm{loss}} \approx UAT_o \left(\frac{\Delta T}{T}\right)^2 . \tag{3.5}$$

The temperature approach between the streams in the heat exchanger increases with a decrease in the temperature of the high-pressure stream in the case of nitrogen [Fig. 3.11(a)]. However, it decreases with a decrease in the temperature of the high-pressure stream in the case of mixtures [Fig. 3.11(b)].

The temperature approach between the streams of the heat exchanger is as small as 5 K in many Linde–Hampson refrigerators operating at low temperatures with

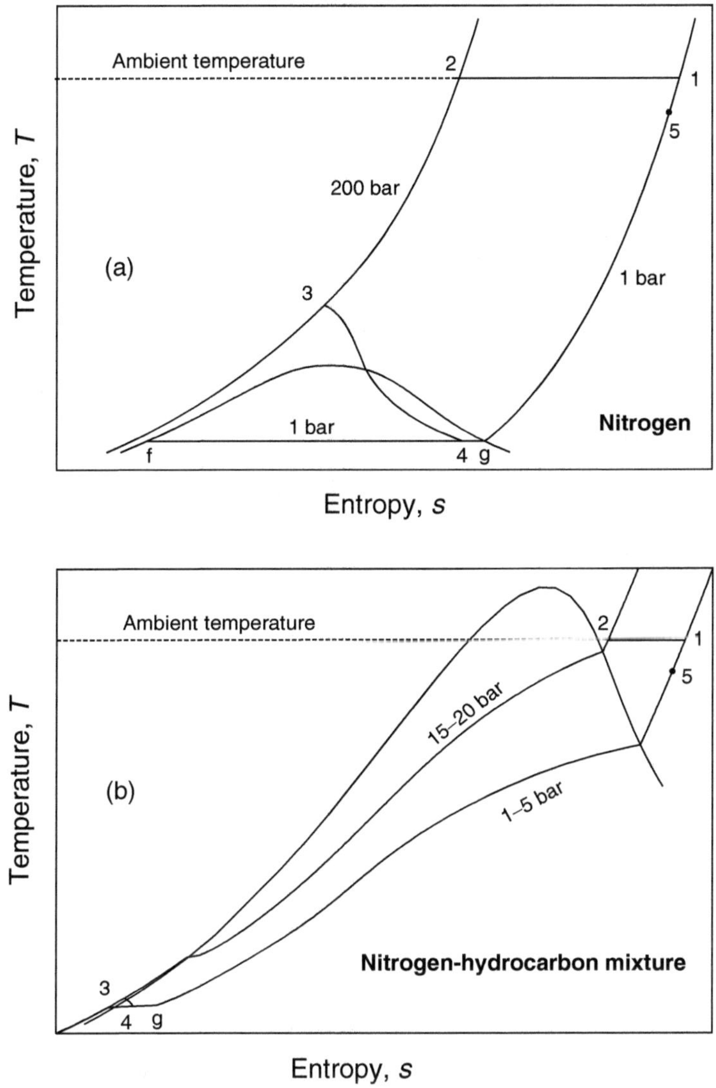

Fig. 3.10. Comparison of an ideal Linde–Hampson refrigerator operating with an isothermal compressor and with pure nitrogen and a typical nitrogen-hydrocarbon mixture as the working fluid.

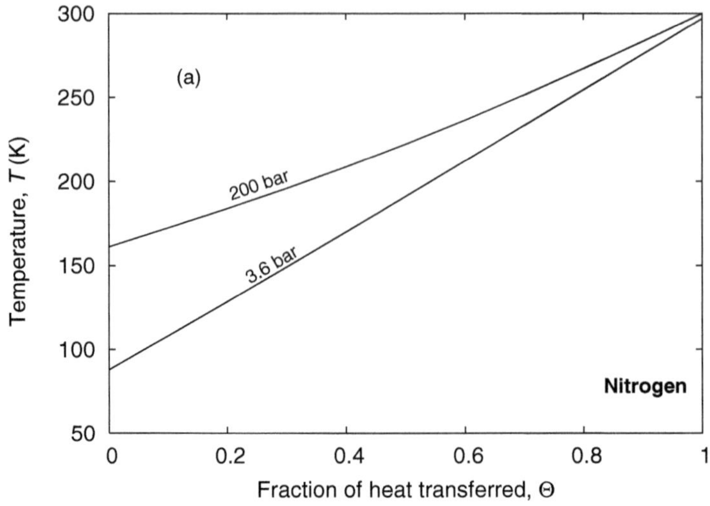

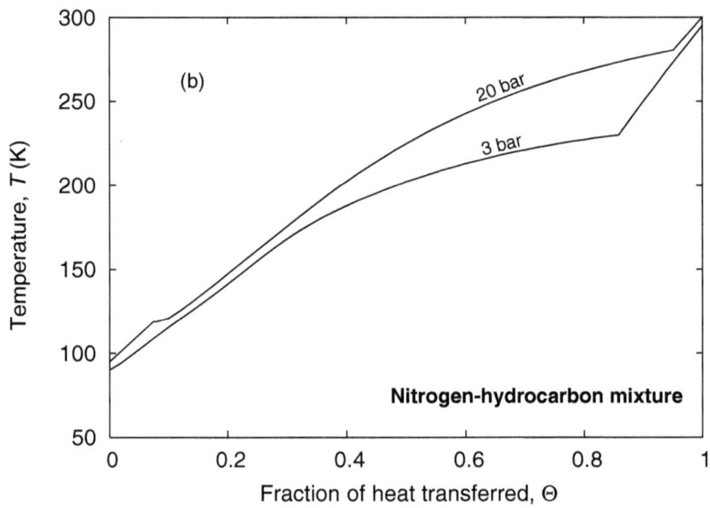

Fig. 3.11. Typical temperature profiles in the heat exchanger of a Linde–Hampson refrigerator operating at 90 K with pure nitrogen and a typical four-component nitrogen-hydrocarbon mixture.

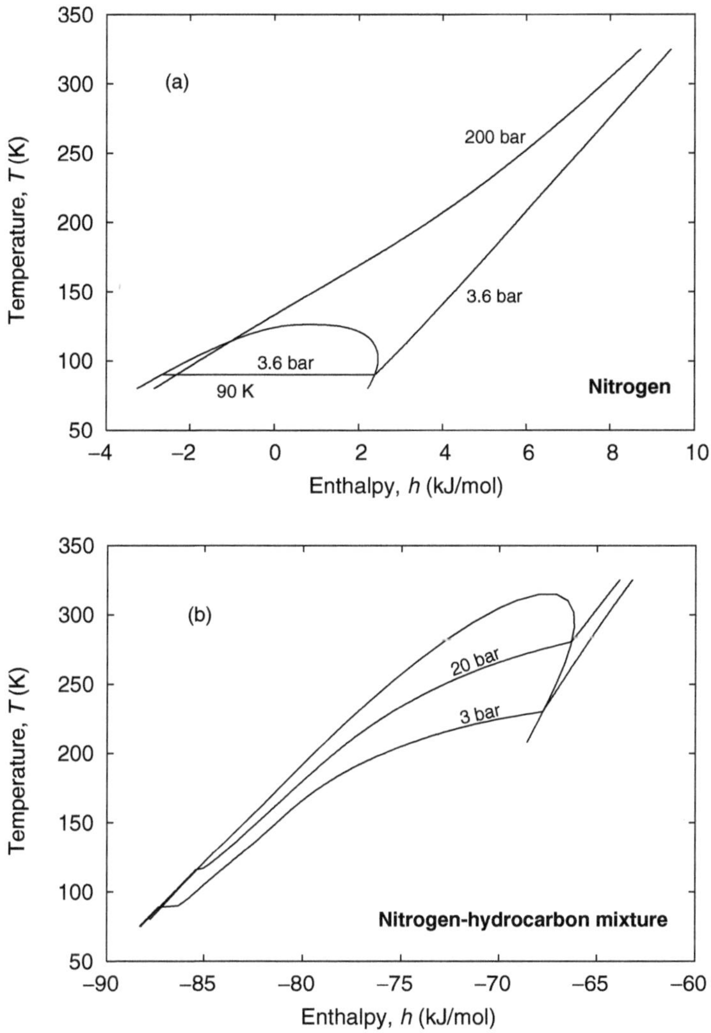

Fig. 3.12. Enthalpy-temperature diagram of (a) pure nitrogen and (b) a typical four component nitrogen-hydrocarbon mixture.

mixtures, compared to about 60 to 80 K in the case of pure fluids. This difference is due to the shape of the constant-pressure lines (at the compressor suction and discharge pressures) on a temperature-enthalpy diagram (Fig. 3.12).

Since the temperature profiles of the streams are nearly parallel in the heat exchanger at temperatures below 200 K, the exergy loss in the heat exchanger is much smaller when mixtures are used as refrigerants.

With the reduction of irreversibilities during the heat transfer and expansion process in Linde–Hampson refrigerators operating with mixtures, it is possible to reduce the operating pressures to about 20 to 25 bar and still obtain a relatively large amount of refrigeration compared to a Linde–Hampson refrigerator operating with pure fluids. While operating at pressures above 30 bar will ensure higher exergy efficiency,[1] commercially available R22/R410A refrigeration compressors can be used in Linde–Hampson refrigerators operating with gas mixtures when the operating pressures are typically less than 20 bar. The differences between a Linde–Hampson refrigerator operating with pure fluids and mixtures are summarized in Table 3.4.

Table 3.4. Comparison of Linde–Hampson refrigerators with pure fluids and typical nitrogen-hydrocarbon mixtures

		Refrigerant	
		Pure fluid	Mixture
1.	Typical operating pressures	100 - 200 bar	15-20 bar
2.	Refrigeration temperature at evaporator pressure of 1 bar	77.24 K	78–79 K
3.	Entropy change during heat addition process	Small	Large
4.	Heat transfer coefficients in heat exchanger	Small	Large
5.	Temperature approach at cold end of the heat exchanger	Large, typically 70–90 K	Small, typically 5–15 K
6.	Theoretical exergy efficiency	10–20%	30–40%
7.	Practical exergy efficiency	1–2 %	3–6 %

A Linde–Hampson refrigerator operating with refrigerant mixtures is also referred to as a throttle cycle refrigerator [3, 17, 28], auto refrigerating cascade (ARC) [60], auto cascade refrigeration system [9], or mixed refrigerant cascade (MRC) refrigerator [13] in the literature. Processes that operate with zeotropic refrigerant mixtures are commonly referred as mixed refrigerant processes. Commercial refrigeration compressors operate at high speeds, typically 48/57 Hz (corresponding to a supply

[1] It may be recalled that higher the pressure difference, the higher will be the refrigeration effect $(h_5 - h_2)$ in a Linde–Hampson process [see Eq. (3.2)].

frequency of 50/60 Hz), and the compression process is closer to an isentropic process than an isothermal process assumed in an ideal refrigerator (Fig. 1.19). An aftercooler therefore needs to be used in all MRC systems (Fig. 3.1).

An examination of Fig. 3.10 reveals that the schematic of a Linde–Hampson refrigerator operating with mixtures is somewhat similar to a traditional reverse Rankine vapor compression refrigerator (VCR) with an internal heat exchanger and operating with pure fluids (Fig. 3.13) or a Lorenz–Meutzner vapor compression refrigerator operating with zeotropic mixtures (Fig. 3.14). The reverse Rankine, Lorenz-Meutzner, and mixed refrigerant Linde–Hampson refrigerators belong to the broad category of vapor compression refrigerators. Linde–Hampson refrigerators operating with refrigerant mixtures, however, differ from reverse Rankine refrigerators in many respects. Some of them are outlined below:

- The difference between the dew and bubble point temperatures at the two operating pressures (refrigerant glide) is more than 100 to 120 K in the case of mixed refrigerant Linde–Hampson refrigerators, compared to 2 to 6 K in the case of vapor compression refrigerators operating with zeotropic refrigerant mixtures such as R407C.
- Similarly, the temperature drop of the liquid refrigerant in the internal heat exchanger is limited to a few degrees in a normal VCR, whereas in a mixed refrigerant Linde–Hampson refrigerator it can be as high as 150 to 200 K.
- Complete condensation of the refrigerant and a small amount of subcooling of the liquid (typically 2 to 5 K) are common in VCR condensers. On the other hand, the refrigerant either condenses partially or does not condense at all in the condenser/aftercooler in a mixed refrigerant Linde–Hampson refrigerator.
- Similarly, the refrigerant is evaporated completely in the evaporator of a normal VCR system, whereas only a very small fraction of the latent heat of the low-pressure refrigerant is used to cool the load in a mixed refrigerant Linde–Hampson refrigerator. Most of the latent heat is used in cooling and condensing the high-pressure refrigerant.
- The performance of a VCR is only marginally controlled by the effectiveness of the internal heat exchanger, whereas the performance of the MRC refrigerator is closely coupled to the high effectiveness of the heat exchanger as in the case of a Linde–Hampson refrigerator operating with pure fluids.

The mixed refrigerant Linde–Hampson refrigerator thus extends the range of operation of traditional vapor compression refrigeration systems to cryogenic temperatures. The use of commercially available domestic refrigerator/air conditioner compressors also makes it possible to build cryocoolers that reach liquid nitrogen temperature at a much smaller cost compared to cryocoolers such as the Gifford–McMahon or Stirling.

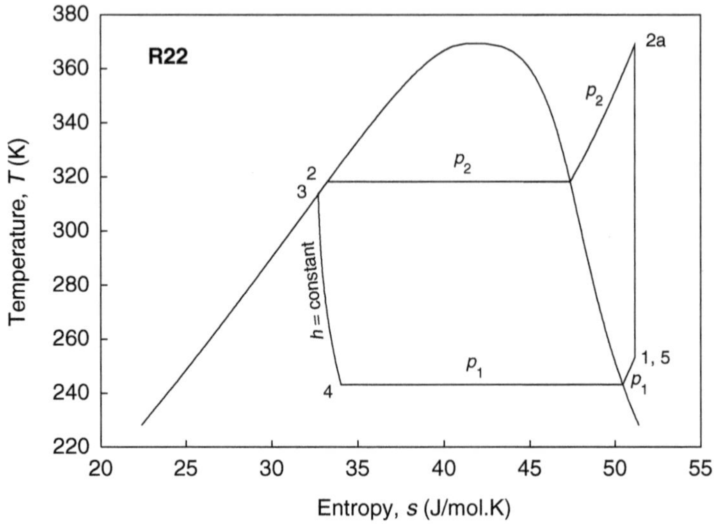

Fig. 3.13. Reverse Rankine vapor compression refrigerator process operating with pure fluid such as refrigerant R22.

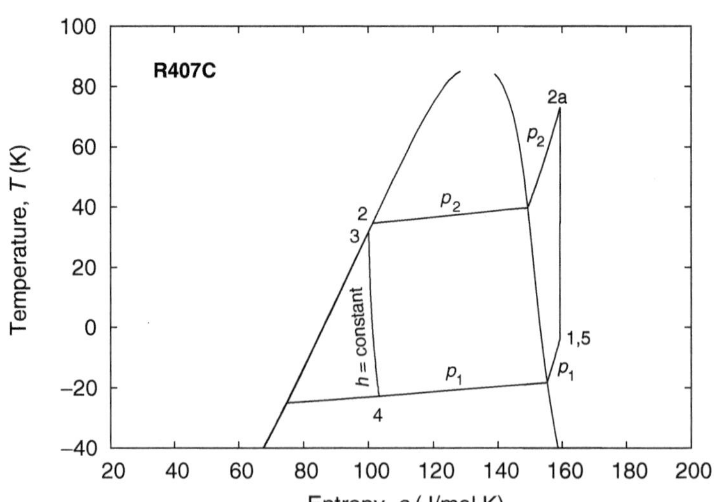

Fig. 3.14. Lorenz–Meutzner vapor compression refrigerator process operating with a zeotropic mixture such as refrigerant R407C.

3.3 Cooling of gases using mixed refrigerant processes

The ideal process for cooling a gas is shown in Fig. 1.5. The process is not used in practice because of the high operating pressures required. Figure 3.15 shows a process that can be used for the cooling and liquefaction of gases. The process is similar to the traditional Linde–Hampson process but provides refrigeration over a range of temperatures. Because of the similarity, the process may be termed the Linde gas-cooling (liquefaction) process.

Consider the possibility of cooling argon from 300 K to 173 K (−100 °C) using the Linde gas-cooling process (Fig. 3.15) with nitrogen as the refrigerant. The heat exchanger effectiveness is assumed to be 100% ($\Delta T_{\text{min}} = 0$). Table 3.5 shows the temperature, pressure, and flow rate of nitrogen and argon of different streams of the process. Both high-pressure nitrogen (refrigerant) and argon (process gas) are cooled to a temperature close to 173 K in the heat exchanger by the low-pressure nitrogen. The temperature drop across the expansion valve is quite large (96.1 K). Argon cannot be cooled to a temperature lower than 173.2 K in this process using nitrogen as the refrigerant. It can be seen that both nitrogen and argon are in the gaseous state in the entire process.

The exergy efficiency of the cold box of the Linde gas cooler shown in Fig. 3.15 is given by the expression

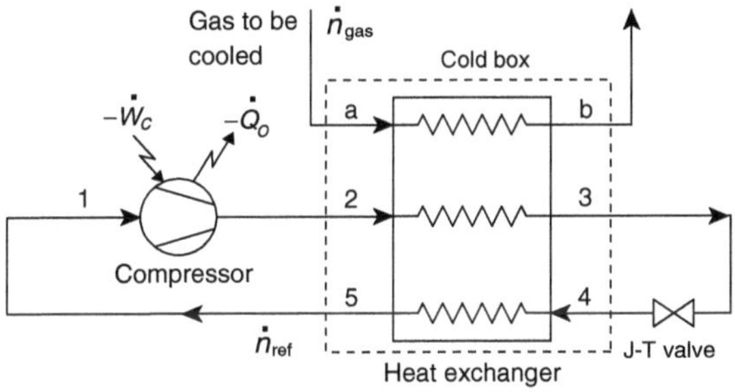

Fig. 3.15. Ideal Linde gas-cooling process operating with an isothermal compressor.

Table 3.5. Temperature, pressure, vapor fraction, and flow rate of the different streams of an ideal Linde argon cooler operating with nitrogen as the refrigerant (Fig. 3.15)

	Nitrogen				Argon	
	Stream 1	Stream 2	Stream 3	Stream 4	Stream a	Stream b
Temperature, K	300	300	173.5	77.4	300	173.2
Pressure, bar	1	200	200	1	10	10
Vapor fraction	1	1	1	1	1	1
Flow rate, mol/s	1	1	1	1	0.361	0.361

$$\eta_{\text{ex, cb}} = \frac{\dot{n}_{\text{gas}}(\text{ex}_b - \text{ex}_a)}{\dot{n}_{\text{ref}}(\text{ex}_2 - \text{ex}_5)}. \tag{3.6}$$

Figure 3.16 shows the exergy utilization in the cold box of the above process. The exergy efficiency is only 2.3% in this process, even when an ideal heat exchanger ($\varepsilon = 100\%$) is used. The large exergy loss in the expansion valve and the heat exchanger is due to the large temperature approach between the streams at the cold end of the heat exchanger (Fig. 3.17). The exergy loss in the expansion valve and the heat exchanger can be decreased only by decreasing the temperature approach between the streams at the cold end of the heat exchanger.

An energy balance over the heat exchanger gives

$$\dot{n}_{\text{ref}}[(h_5 - h_4) - (h_2 - h_3)] = \dot{n}_{\text{gas}}(h_a - h_b), \tag{3.7}$$

or

$$\dot{n}_{\text{ref}}(\Delta h_{\text{lp}} - \Delta h_{\text{hp}}) = \dot{n}\Delta h_{\text{gas}}. \tag{3.8}$$

Equation (3.8) can be expressed in terms of temperature change of the high-pressure (hp) and low-pressure (lp) refrigerants and the process gas being cooled (gas) as follows:

$$\dot{n}_{\text{ref}}\left[\int_4^5 \left(\frac{\partial h}{\partial T}\right)_{p,\text{lp}} dT - \int_3^2 \left(\frac{\partial h}{\partial T}\right)_{p,\text{hp}} dT\right] = \dot{n}_{\text{gas}} \int_b^a \left(\frac{\partial h}{\partial T}\right)_{p,\text{gas}} dT, \tag{3.9}$$

where the integration is to be carried out over the entire heat exchanger length. $\dot{n}_{\text{ref}}$ corresponds to the mole flow rate of the refrigerant of the Linde gas cooler, and $\dot{n}_{\text{gas}}$ that of the gas being cooled.

The exergy loss in the heat exchanger of the Linde gas cooler shown in Fig. 3.15 will be minimized when the temperature difference between the hot and cold streams remains constant throughout the heat exchanger. In such a case, Eq. (3.9) reduces to the following:

$$\left(\frac{\partial h}{\partial T}\right)_{p,\text{lp}} - \left(\frac{\partial h}{\partial T}\right)_{p,\text{hp}} = \frac{\dot{n}_{\text{gas}}}{\dot{n}_{\text{ref}}}\left(\frac{\partial h}{\partial T}\right)_{p,\text{gas}}. \tag{3.10}$$

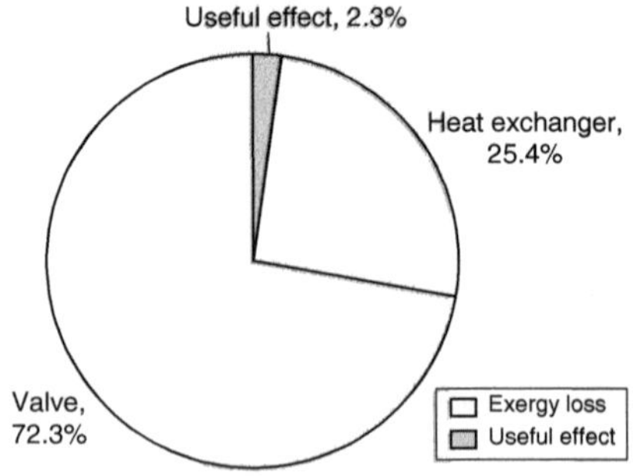

Fig. 3.16. Utilization of input exergy in the cold box of a Linde argon gas cooler operating with nitrogen as the refrigerant (see Table 3.5).

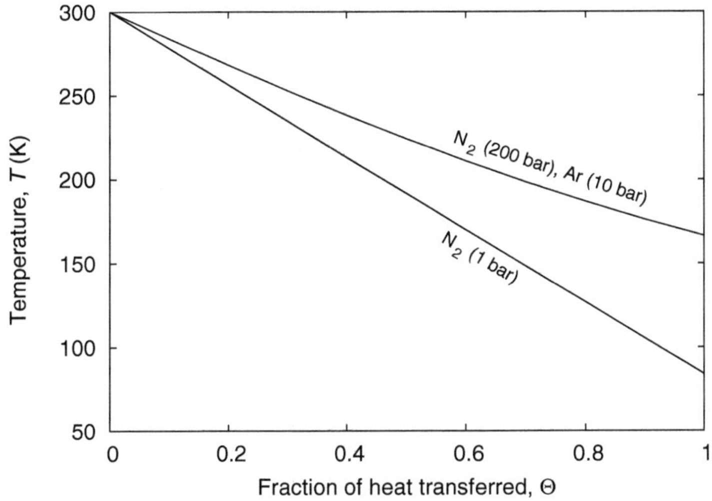

Fig. 3.17. Temperature profiles of hot and cold fluid streams in the heat exchanger of an ideal Linde argon gas cooler operating with nitrogen as the refrigerant (see Table 3.5).

When a pure fluid or a mixture of gases that does not undergo phase change in the heat exchanger is used as the working fluid (refrigerant) of the Linde gas-cooling process, then Eq. (3.10) reduces to the following expression:

$$c_{p,\,\mathrm{lp}} - c_{p,\,\mathrm{hp}} = \frac{\dot{n}_{\mathrm{gas}}}{\dot{n}} \left(\frac{\partial h}{\partial T} \right)_{p,\,\mathrm{gas}}. \tag{3.11}$$

Since the specific heat of the high-pressure refrigerant stream is always greater than that of the low-pressure refrigerant stream in the case of pure fluids or zeotropic mixtures in gaseous state (see Figs. 1.21 and 1.32), Eq. (3.11) is never satisfied in the case of pure fluid refrigerants. In other words, it is not possible to operate a Linde gas cooler with the same temperature approach between the hot and cold fluid streams throughout the heat exchanger when fluids that exist in the gaseous phase and do not undergo phase change in the heat exchanger are used as the refrigerant. High exergy efficiency therefore cannot be obtained in a Linde gas cooler operating with pure fluid refrigerants.

3.4 Linde gas cooler operating with mixtures

Consider a Linde gas cooler operating with mixtures. An overall energy balance around the three-stream heat exchanger in Fig. 3.15 since $h_3 = h_4$ gives

$$\dot{n}_{\mathrm{ref}}(h_5 - h_2) = \dot{n}_{\mathrm{gas}}(h_a - h_b). \tag{3.12}$$

Substituting Eq. (3.12) into Eq. (3.10) gives

$$\left(\frac{\partial h}{\partial T}\right)_{p,\mathrm{lp}} - \left(\frac{\partial h}{\partial T}\right)_{p,\mathrm{hp}} = \left(\frac{h_5 - h_2}{h_a - h_b}\right)\left(\frac{\partial h}{\partial T}\right)_{p,\mathrm{gas}}. \tag{3.13}$$

Consider a case where the process gas is a pure fluid at low-pressures ($p \ll p_c$). The enthalpy of such a gas can be assumed to vary nearly linearly with temperature (see Fig. 1.21), except at temperatures very close to the saturation temperature. In such cases, Eq. (3.13) reduces to the following expression:

$$\left(\frac{\partial h}{\partial T}\right)_{p,\mathrm{lp}} - \left(\frac{\partial h}{\partial T}\right)_{p,\mathrm{hp}} = \frac{h_5 - h_2}{T_a - T_b}. \tag{3.14}$$

Equation (3.14) is applicable at any heat exchanger location. The right-hand side of Eq. (3.14), however, is only dependent on the end states.

Let the variation of the enthalpy of the two streams with temperature be given by the expressions

$$h_{\mathrm{hp}}(T) = a_0 + \sum_{i=1}^{n} a_i\, T^i, \tag{3.15}$$

$$h_{\mathrm{lp}}(T) = b_0 + \sum_{i=1}^{n} b_i\, T^i. \tag{3.16}$$

Substituting Eqs. (3.15) and (3.16) in Eq. (3.14) gives

$$(b_1 - a_1) - \sum_{i=2}^{n} i\,(b_i - a_i)\, T^{i-1} = \frac{h_5 - h_2}{T_a - T_b}. \tag{3.17}$$

The right-hand side of Eq. (3.17) is only dependent on the end temperatures and is independent of temperature T. For the left-hand side of Eq. (3.17) to be independent of temperature, a_i and b_i should satisfy the following condition:

$$b_i = a_i \qquad \text{for } i = 2, n. \tag{3.18}$$

The difference between the slopes $(\partial h/\partial T)_p$ should be such that the enthalpy difference between the low- and high-pressure streams must decrease with an increase in temperature. Both a linear and nonlinear variation of enthalpy with temperature can satisfy equation Eq. (3.18). It is possible to satisfy Eq. (3.14) when a mixture of gases that undergo phase change is used as a refrigerant in the Linde gas cooler.

Consider a Linde nitrogen gas cooler in which nitrogen at a pressure of 10 bar is cooled from 300 K to 130 K. Figure 3.18 shows a mixture that satisfies the above requirements except at temperatures close to the dew point temperature of the low-pressure stream. A comparison of Fig. 3.18 with Fig. 1.21 shows that the constant-pressure lines converge at low temperatures in the case of the refrigerant mixture shown in Fig. 3.18, while the constant-pressure lines converge at high temperatures on a *T-h* plane in the case of nitrogen (Fig. 1.21).

The temperature profiles of the hot streams (high-pressure refrigerant and nitrogen) and the cold stream (low-pressure refrigerant) in the heat exchanger were determined

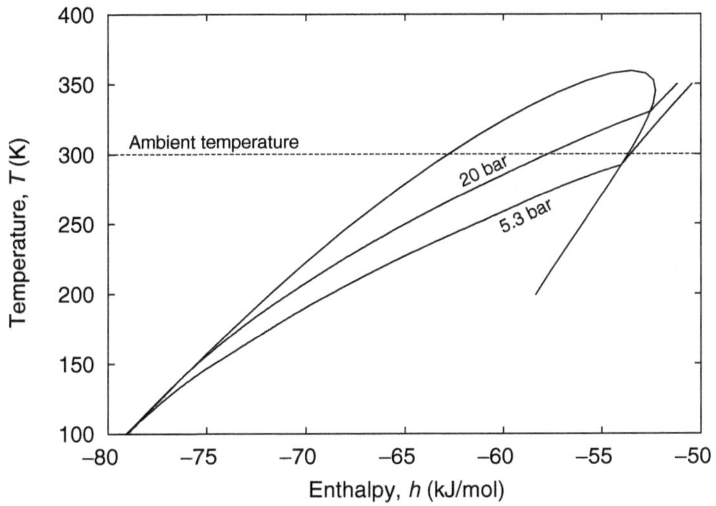

Fig. 3.18. Enthalpy-temperature diagram for a mixture used in a Linde gas cooler for cooling nitrogen from 300 K to 130 K. (N_2 : 8.6, CH_4 : 28.9, C_2H_4 : 25.5, C_3H_8 : 23.7, iC_5H_{12}: 13.2 mol%).

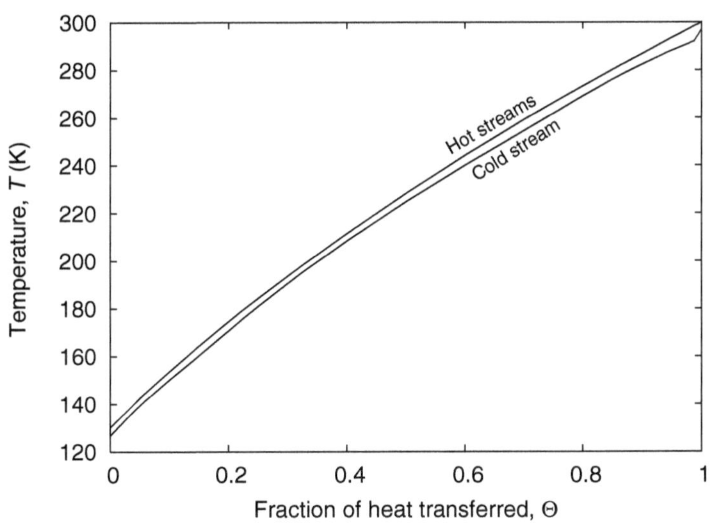

Fig. 3.19. Temperature profiles of the hot and cold streams of a Linde gas cooler for cooling nitrogen from 300 K to 130 K at a constant pressure of 10 bar and operating with a nitrogen-hydrocarbon mixture. (N_2 : 8.6, CH_4 : 28.9, C_2H_4 : 25.5, C_3H_8 : 23.7, iC_5H_{12}: 13.2 mol%, $p = 20/5.3$ bar), $p_{N_2} = 10$ bar.

for the mixture shown in Fig. 3.18. It can be seen from Fig. 3.19 that the temperature approach between the streams is small and nearly uniform over a large length of the heat exchanger when an optimum mixture shown in Fig. 3.18 is used.

The Linde gas cooler is a good example of a refrigerator that provides refrigeration over a range of temperatures efficiently. The exergy efficiency of a Linde gas cooler operating with refrigerant mixtures is normally higher than that of an ideal gas cooler (Fig. 1.5) operating with an expander whose adiabatic efficiency is about 70 to 80%.

In the case of an ideal mixed refrigerant Linde–Hampson refrigerator that provides refrigeration at a near-constant temperature, the relationship between the two-phase enthalpy temperature slopes for minimum exergy loss in the heat exchanger can be derived from Eq. (3.14) as follows:

$$\left(\frac{\partial h}{\partial T}\right)_{p,\,\mathrm{lp}} = \left(\frac{\partial h}{\partial T}\right)_{p,\,\mathrm{hp}}. \tag{3.19}$$

The above equation is satisfied over a large length of the heat exchanger in the case of a mixed refrigerant Linde–Hampson refrigerator operating in the LRS mode, and over a small length at low temperatures while operating in the GRS mode, as shown in Fig. 4.2.

3.5 Liquefaction of natural gas

Natural gas available from a well head contains many impurities such as water, carbondioxide, etc. These are removed in a pretreatment plant before natural gas is fed into the liquefier. The natural gas feed consists predominantly of methane, but also contains small amounts of nitrogen, ethane, and often high boilers, some of which are removed from the feed at a temperature of 225 to 245 K. The bubble point temperature of natural gas to be liquefied is normally much higher than that of pure methane because of the presence of ethane, propane, etc. A large amount of flash gas[2] will be generated when the high-pressure liquefied natural gas (LNG), typically at 40 to 70 bar, is expanded to a lower pressure, typically 1.0 to 1.05 bar. The quantity of flash gas produced can be reduced substantially by subcooling the LNG to a low temperature, typically 110 to 130 K.

The liquefaction of natural gas thus involves desuperheating or sensible cooling of a high-pressure natural gas feed at 40 to 70 bar and at ambient temperature to its dew point temperature (typically 210 to 240 K), condensation of natural gas to its bubble point temperature (typically 190 to 220 K), and subcooling it to a temperature of 110 to 130 K.

The specific heat of natural gas[3] is different in the above gaseous, two-phase, and liquid regions. Three different mixtures are therefore needed to satisfy Eq. (3.13) in these three regions. Many natural gas liquefaction processes have been invented to

[2] Flash gas is the vapor generated when a liquid is expanded to a lower pressure.

[3] $\left(\frac{\partial h}{\partial T}\right)_p$ during condensation.

meet this requirement. In some processes, a single mixture is split into three different mixtures for providing refrigeration in the three regions. In some other processes, two different mixtures are used, with one of the mixtures further split into two different mixtures. Three distinctly different mixtures are used in some other processes. The details of these and other processes including turbine-based processes are discussed in Chapter 6 for the liquefaction of natural gas, and in Chapter 7 for the liquefaction of nitrogen.

4

Constant-temperature refrigeration processes

Mixed refrigerant processes can be broadly classified into two groups: (1) those in which refrigeration is provided over a constant temperature or over a small range of temperatures, typically less than 1–5 K, and (2) those in which refrigeration is provided over a large range of temperatures, typically greater than 50 K, for example, in the cooling and liquefaction of gases. The processes belonging to the former category are described in this chapter.

The refrigeration processes operating with refrigerant mixtures can be classified (1) according to the state of the high-pressure refrigerant at the end of the aftercooler (or condenser) and (2) whether phase separators are used in the process, as shown in Fig. 4.1. They can also be classified into different groups based on the number of phase separators.

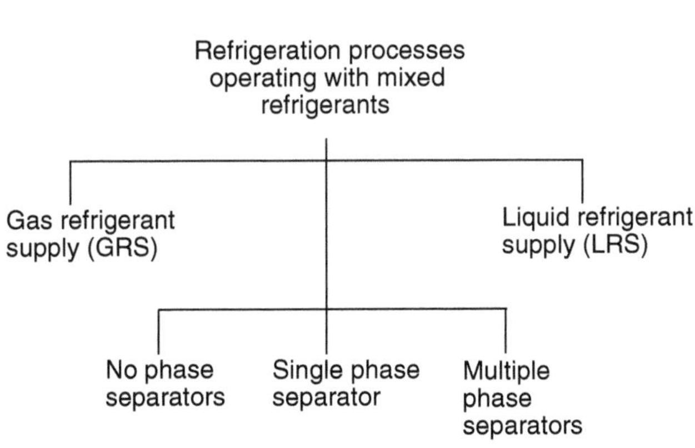

Fig. 4.1. Classification of refrigerators operating with mixed refrigerants.

G. Venkatarathnam, *Cryogenic Mixed Refrigerant Processes*,
DOI: 10.1007/978-0-387-78514-1_4, © Springer Science+Business Media, LLC 2008

4.1 Gas refrigerant supply and liquid refrigerant supply (GRS/LRS) processes

Refrigerators operating with refrigerant mixtures can be broadly classified into two types based on the state of the fluid leaving the aftercooler as follows: (1) gas refrigerant supply (GRS) systems in which no condensation occurs in the aftercooler, and (2) liquid refrigerant supply (LRS) systems in which partial condensation of the high-pressure refrigerant occurs in the aftercooler or condenser [19]. The dew point temperature of the refrigerant at both the compressor suction and discharge pressures is below the ambient temperature in the case of GRS systems. On the other hand, the dew point temperature of the high-pressure refrigerant is above the ambient temperature, and that of the low-pressure refrigerant is below the ambient in the case of LRS systems as shown in Fig. 4.2.

A larger amount of heat is rejected in the aftercooler when the refrigerant condenses partially. Consequently, the refrigeration obtained from an LRS system will be higher than that from a GRS system for the same power input. This can also be observed from the enthalpy difference between the high- and low-pressure refrigerants at ambient temperature, which is much larger in the case of LRS systems compared to GRS systems (Fig. 4.2). The constant-pressure lines on a *T-h* plane are nearly parallel (but not necessarily linear) over a large temperature range in an ideal mixture for an

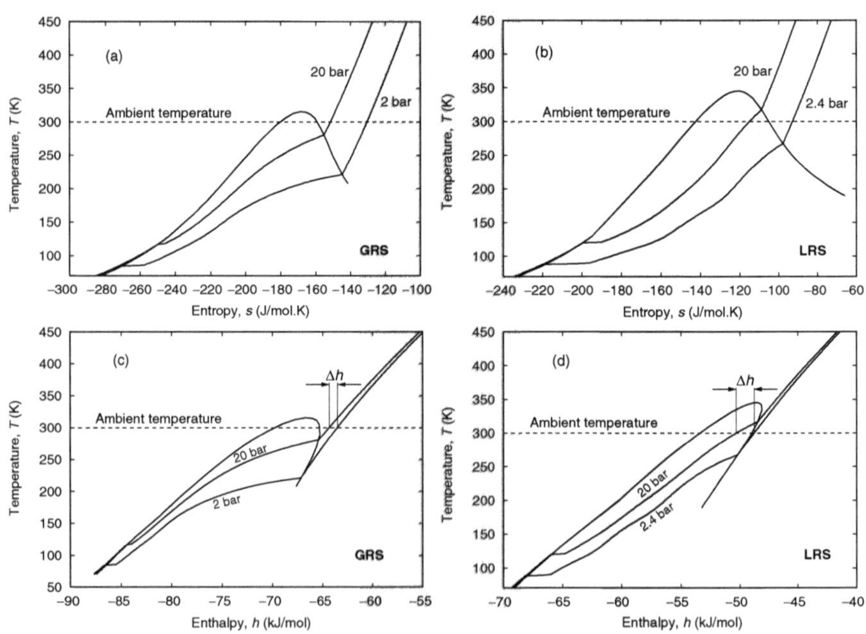

Fig. 4.2. Examples of mixtures for GRS and LRS refrigerators. The difference between GRS and LRS mixtures essentially is in the dew point temperature of the high-pressure refrigerant. Δh corresponds to the specific refrigeration effect.

LRS Linde–Hampson refrigerator, is as shown in Fig. 4.2. On the other hand, the high- and low-pressure lines on the *T-h* plane of an ideal mixture for a GRS Linde–Hampson refrigerators are parallel only at low temperatures, resulting in a large temperature approach between the streams close to the dew point temperatures. The exergy loss in the heat exchanger is therefore lower in the case of LRS systems compared to GRS systems. Similarly, the entropy change during the heat addition process is much larger in the case of the Linde–Hampson systems operating in the LRS mode, resulting in much higher exergy efficiency.

The maximum theoretical exergy efficiency that can be obtained with ideal GRS Linde–Hampson refrigerators operating at a operating pressure of 20 bar and temperature between 80 and 100 K varies between 35 and 40%, whereas the theoretical exergy efficiency of ideal LRS Linde–Hampson refrigerators varies between 60 and 75%. Practical exergy efficiencies are much smaller due to higher exergy loss in different components.

Most large systems are operated in the LRS mode due to its higher exergy efficiency and higher refrigeration capacity. The GRS mode is sometimes preferred in small cryocoolers that operate with flammable refrigerant mixtures such as nitrogen-methane-ethane-propane mixtures. The amount of flammable refrigerant that can be charged into a small cryocooler is sometimes limited by the allowable limits imposed by transporters, airlines, or other regulations, or the maximum operating pressure/work drawn by the compressor during the initial cool-down process.[1] In such cases, it is advantageous to use the GRS processes over the LRS processes.

4.2 Linde–Hampson refrigerators operating with refrigerant mixtures

The schematic of a Linde–Hampson refrigerator without any phase separator and operating with refrigerant mixtures is shown in Fig. 4.3. The compression process is closer to the isentropic process due to the high-speed compressors used in these systems. Most of the heat rejection occurs in the aftercooler in these systems.

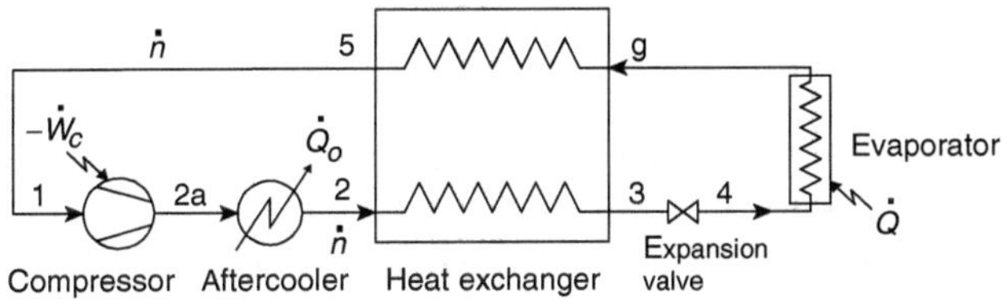

Fig. 4.3. Mixed refrigerant Linde–Hampson refrigerator without phase separators. Condensation does not occur in the aftercooler in the GRS mode, whereas partial condensation occurs in the aftercooler in the LRS mode.

[1] Also known as a pull-down process in the refrigeration industry.

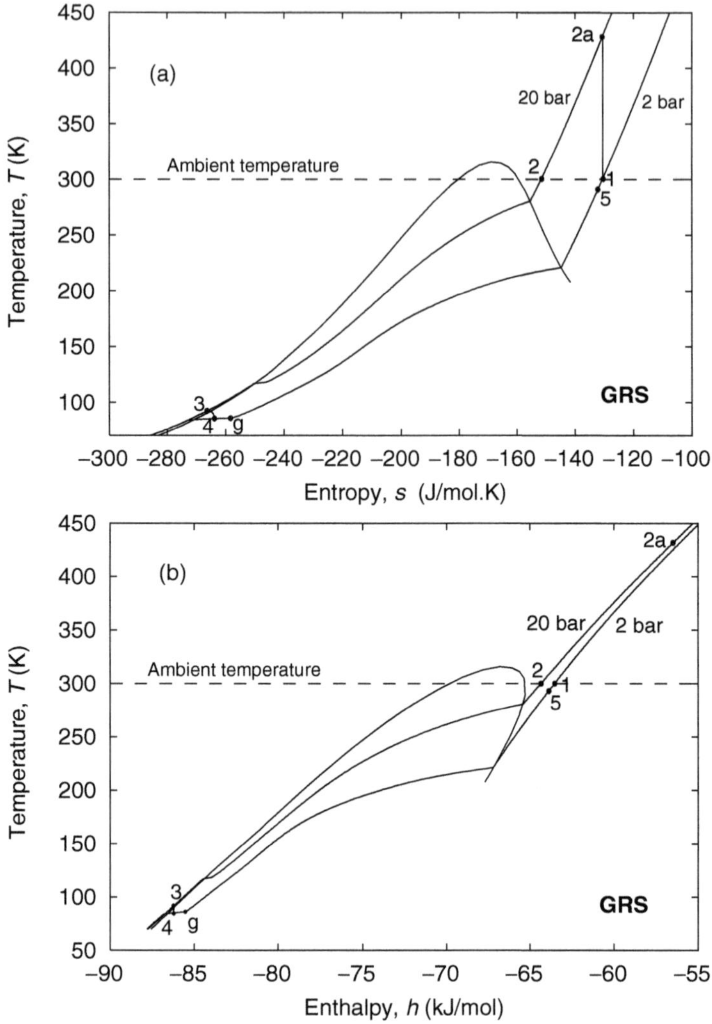

Fig. 4.4. Typical Linde–Hampson refrigerator shown in Fig. 4.3 operating with mixtures in GRS mode.

Figure 4.4 shows a typical mixed refrigerant Linde–Hampson refrigerator operating in GRS mode. The process is shown on *T–s* and *T-h* planes in Fig. 4.4 for a typical nitrogen-hydrocarbon mixture. A nearly constant temperature evaporation is possible because of the formation of a second immiscible liquid phase (VLLE) in the evaporator (see page 69).

The heat absorbed by the refrigerator shown in Fig. 4.3 can be expressed in terms of the enthalpy of the refrigerant from an energy balance across the heat exchanger, throttle valve, and evaporator as follows:

$$\dot{Q} = \dot{n}(h_5 - h_2). \tag{4.1}$$

The exergy efficiency of the refrigerator is based on the temperature of the refrigerant leaving the evaporator (T_g) and is given by the expression

$$\eta_{\text{ex}} = \frac{\dot{Q}}{-\dot{W}_c}\left(\frac{T_o}{T_g} - 1\right). \tag{4.2}$$

The exergy efficiency of a refrigerator can also be expressed in terms of the exergy efficiency of the cold box (all components other than the compressor and aftercooler) ($\eta_{\text{ex, cb}}$) and the exergy efficiency of the compressor section (compressor and aftercooler) ($\eta_{\text{ex, cs}}$) as follows:

$$\eta_{\text{ex}} = \eta_{\text{ex, cb}}\,\eta_{\text{ex, cs}}. \tag{4.3}$$

The exergy efficiency of the cold box ($\eta_{\text{ex, cb}}$) is given by the expression

$$\eta_{\text{ex, cb}} = \frac{\dot{Q}}{\dot{n}(\text{ex}_2 - \text{ex}_5)}\left(\frac{T_o}{T} - 1\right) = \frac{h_5 - h_2}{\text{ex}_2 - \text{ex}_5}\left(\frac{T_o}{T} - 1\right) \tag{4.4}$$

$$or,\ \eta_{\text{ex, cb}} = \frac{\left(\frac{T_o}{T} - 1\right)}{\left(T_o\frac{s_2 - s_5}{h_2 - h_5} - 1\right)}. \tag{4.5}$$

The exergy efficiency of the compressor section consisting of the compressor and aftercooler/condenser ($\eta_{\text{ex, cs}}$) is given by the expression

$$\eta_{\text{ex, cs}} = \frac{\dot{n}(\text{ex}_2 - \text{ex}_5)}{-\dot{W}_c}. \tag{4.6}$$

The exergy efficiency of a Linde–Hampson refrigerator is thus dependent on the relative variation of enthalpy and entropy of the refrigerant with pressure at room temperature. The mixture composition should be chosen to maximize the ratio $(h_5 - h_2)/(s_5 - s_2)$.

The temperature change in the evaporator will be very small, typically less than 1 K only when the mixture chosen exhibits a vapor-liquid-liquid equilibria (VLLE) at low temperatures. The second liquid phase occurs only with appropriate mixture compositions.

The optimum performance of different mixed refrigerant process refrigerators has been determined using the methods described in Chapter 5.

The performance of the refrigerators can be improved by varying the composition even slightly in some cases. The reason for this variation can be understood by analyzing the exergy loss in different components. The maximum exergy loss in all mixed refrigerant processes occurs in the heat exchanger. The temperature profiles of the hot and cold fluid streams in different heat exchangers, and the variation of the temperature approach with different refrigerant mixtures, are described in the following sections.

Most of the refrigerant compositions described are probably the optimum compositions for the particular process and operating conditions. The different cases described help in understanding the subtle differences that can lead to improved performance.

4.3 Mixed refrigerant Linde–Hampson refrigerator operating at 90 K in GRS mode

Consider a mixed refrigerant Linde–Hampson refrigerator operating at a temperature of 90 K in GRS mode. The design specifications of the system are shown in Table 4.1. The optimum mixtures that satisfy these requirements have been determined using the optimization procedure summarized in Table 5.3. The performance of the system with four different mixture compositions is summarized in Fig. 4.5 and the details are shown in Table 4.2.

The sequential quadratic programming (SQP) method was used to maximize the exergy efficiency. The result in Case 1 (Table 4.2) was obtained by limiting the nitrogen concentration to greater than 30 mol% [20]. No such limits were imposed on nitrogen in the other cases [92]. However, limits were imposed on the maximum concentration of propane. The main difference between Case 2 and Case 3 is the amount of propane used. The minimum temperature approach in the heat exchanger is 5 K in the first three

Table 4.1. Design specifications for a mixed refrigerant Linde–Hampson refrigerator operating in GRS mode

Exergy efficiency of compressor and aftercooler, $\eta_{ex,cs}$	100%
Volumetric efficiency of the compressor , η_v	100%
Maximum operating pressure, p_2	20 bar
Minimum temperature approach in the heat exchanger, ΔT_{min}	5 K
Pressure drop in the heat exchanger, Δp	0

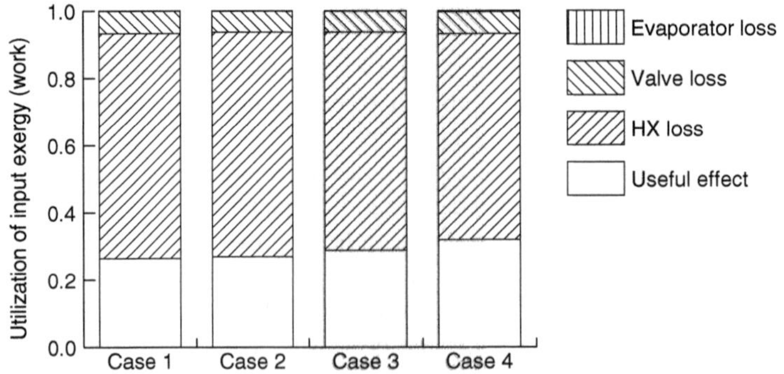

Fig. 4.5. Utilization of input exergy (work) in different cases of a mixed refrigerant Linde–Hampson refrigerator operating in GRS mode (Table 4.2).

Table 4.2. Performance of a GRS Linde–Hampson refrigerator with different mixtures

	Case 1	Case 2	Case 3	Case 4
Mixture				
Nitrogen (mol%)	30.0	28.7	28.4	29.0
Methane (mol%)	16.3	17.1	19.4	15.6
Ethane (mol%)	23.7	24.2	14.2	17.4
Propane (mol%)	30.0	30.0	38.0	38.0
Performance parameters				
Exergy efficiency, $\eta_{\text{ex,cb}}$ (%)	26.8	27.3	29.0	32.3
Specific refrigeration effect, Δh_{ref} (J/mol)	504.0	514.0	557.0	618.0
Volumetric cooling capacity, Q_v (J/l)	63.9	64.7	68.3	75.3
Operating pressures				
Compressor discharge pressure, p_2 (bar)	20.00	20.00	20.00	20.00
Compressor suction pressure, p_1 (bar)	3.06	3.04	2.96	2.94
Temperatures				
T_2 (K)	300.0	300.0	300.0	300.0
T_3 (K)	96.4	95.1	95.0	95.0
T_4 (K)	89.8	89.8	89.7	89.3
T_g (K)	90.0	90.0	90.0	90.0
T_5 (K)	295.0	295.0	295.0	295.7
ΔT_{min} (K)	5.0	5.0	5.0	3.0
Bubble, Dew point temp.				
$T_{\text{dew}}(p_2)$ (K)	280.1	280.4	286.3	287.0
$T_{\text{dew}}(p_1)$ (K)	230.1	230.0	233.7	233.8
$T_{\text{bub}}(p_2)$ (K)	118.7	118.8	119.2	118.5
$T_{\text{bub}}(p_1)$ (K)	89.7	89.7	89.6	89.2

cases, and 3 K in the fourth case (Case 4). A mixture of nitrogen, methane, propane, and ethane is used in all four cases. The exergy efficiency of the cold box varies from 26.8% to 32.3% for the four cases studied. The exergy loss in each of the components was determined using the expressions presented in Table 1.1. The fraction of exergy input used for meeting the useful effect (providing required cooling) or the exergy efficiency was determined using Eq. (4.5). The exergy efficiency of the cold box is also given by the expression

$$\eta_{\text{ex,cb}} = 1 - \frac{\sum \text{exergy loss}}{\text{ex}_2 - \text{ex}_5}. \tag{4.7}$$

It can be seen from Fig. 4.5 that the exergy loss in the heat exchanger is the highest in Case 1, resulting in the lowest exergy efficiency for that case. The mixture composition of Cases 1 and 2 differs only by a small margin. The nitrogen concentration is

30% in the first case and less than 30% in all other cases [92]. However, the exergy efficiency of the cold box ($\eta_{ex,cb}$), the volumetric cooling capacity[2] (Q_v), and the specific refrigeration effect (Δh_{min}) are all higher for Case 2 compared to Case 1.

The dew point temperatures of the high- and low-pressure refrigerants are about the same in Cases 1 and 2. A higher concentration of propane is used in the third case. This results in an increase of the dew point temperature of the high-pressure refrigerant to 286 K. The Joule–Thomson coefficient of the refrigerant at the warm end of the heat exchanger also increases with an increase in the propane content. Consequently, the specific refrigeration effect, the exergy efficiency, and the volumetric cooling capacity are all higher in Case 3 compared to Cases 1 and 2. In general, the higher the dew point of the high-pressure refrigerant, the higher will be the exergy efficiency and specific refrigeration effect. The performance of a Linde–Hampson refrigerator is the highest when the dew point of the high-pressure refrigerant is higher than the ambient temperature. Such refrigerators are known as LRS refrigerators and are described in the following sections.

A minimum temperature approach of 3 K is used in the fourth case. The exergy loss in the heat exchanger is therefore smaller in Case 4 compared to that in Case 3, leading to a higher exergy efficiency. The specific refrigeration effect also increases from 557 J/mol in Case 3 to 618 J/mol in Case 4.

It can be observed from Fig. 4.5 that the exergy loss in the expansion valve is nearly the same in all the cases. It can also be observed from Table 4.2 that the temperature change of the refrigerant in the evaporator is less than 1 K in all cases. The exergy loss in the evaporator is therefore negligible in all the cases.

Figures 4.6 and 4.7 show the temperature profiles and the temperature approach in the heat exchanger, respectively. The minimum temperature approach occurs at the warm end, as well as in between the ends, closer to the cold end of the heat exchanger in all cases except in Case 4, where the temperature approach at the warm end of the heat exchanger is higher than that at the pinch point. It can be seen from Fig. 4.7 that the temperature approach is less than 10 K for only a small part of the heat exchanger, leading to a large exergy loss in the heat exchanger. The large temperature approach between the streams in the heat exchanger close to the dew point temperature of the low-pressure refrigerant limits the maximum attainable exergy efficiency to about 40% in the case of ideal Linde–Hampson refrigerators operating in GRS mode with mixtures and a zero temperature approach between streams in the heat exchanger.

The exergy efficiency of the compressor section (compressor and aftercooler) typically varies between 35 to 40% in small cryocoolers. The overall exergy efficiency of well-built mixed refrigerant Linde–Hampson cryocoolers is typically about 4–6% due to pressure drop, heat leak from ambient, etc. The refrigeration ($\dot{Q}$) that can be obtained is given by the expression

$$\dot{Q} = \dot{n}(h_5 - h_2) = \eta_v \, \dot{V}_c \, \rho_5(h_5 - h_2) = \eta_v \, \dot{V}_c \, Q_v, \tag{4.8}$$

where $\dot{V}_c$ refers to the displacement rate of the compressor, η_v is the volumetric efficiency of the compressor, and Q_v is the volumetric cooling capacity. The vol-

[2] $Q_v = \rho_5(h_5 - h_2)$.

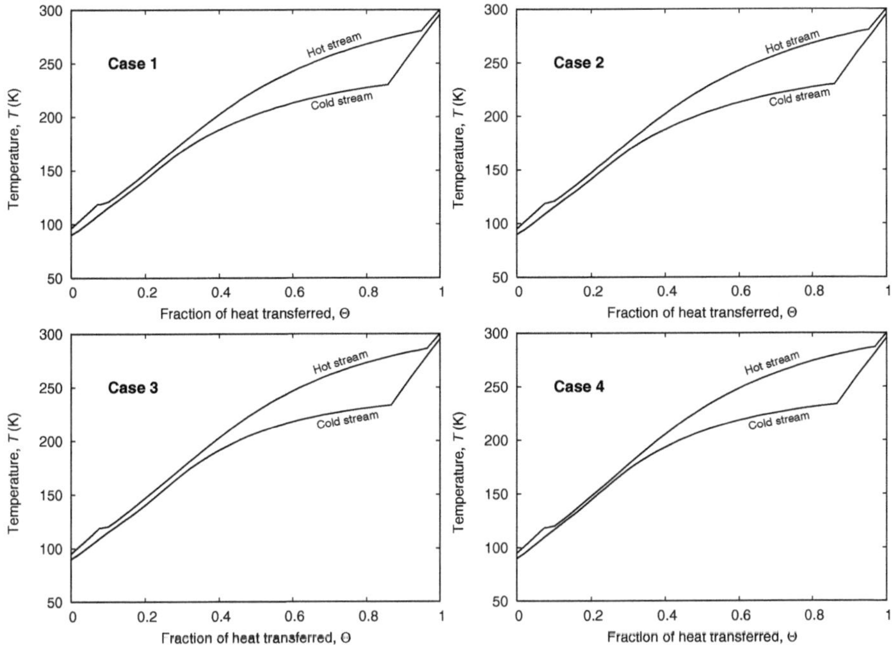

Fig. 4.6. Temperature profiles of the streams in the heat exchanger of a Linde–Hampson refrigerator operating in GRS mode with different mixtures of Table 4.2.

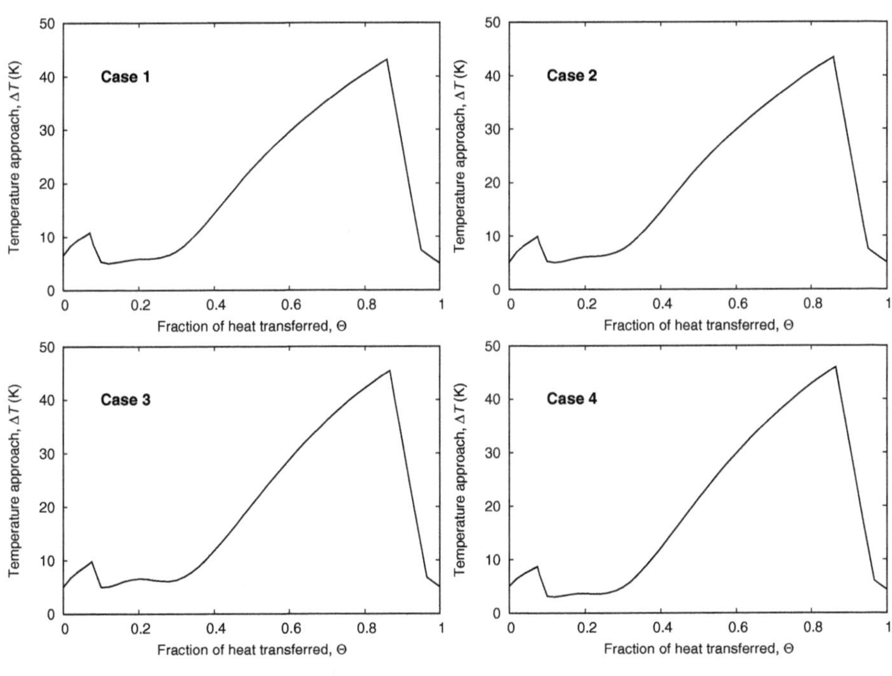

Fig. 4.7. Temperature approach between the streams in the heat exchanger of a Linde–Hampson refrigerator operating in GRS mode with different mixtures of Table 4.2.

umetric efficiency of an ideal compressor is dependent on the compression ratio (p_2/p_1) and the adiabatic index ($\gamma = c_p/c_v$) of the mixture as follows:

$$\eta_{v,\text{ideal}} = 1 + C - C\left(\frac{p_2}{p_1}\right)^{1/\gamma}, \tag{4.9}$$

where C refers to the volumetric clearance ratio. The volumetric efficiency of a real compressor is also dependent on other factors such as the size of ports and the pressure drop in the valves. The volumetric efficiency of a real compressor can be expressed as follows [21, 54, 76]:

$$\eta_v = \alpha - \beta\left(\frac{p_2}{p_1}\right)^{1/\gamma}, \tag{4.10}$$

where α and β are constants that depend on the construction of the compressor and mixture composition. Constant α has been found[3] to be nearly independent of γ in some of the small refrigeration compressors we tested [54, 76], while constant β has been found to be a function of the adiabatic index of the refrigerant mixture.

The true performance of a mixed refrigerant Linde–Hampson refrigerator also depends on other factors such as the pressure drop in the heat exchangers.

4.3.1 Effect of pressure drop in the heat exchanger

The pressure drop in the heat exchanger was assumed to be zero in the examples presented in Table 4.2. Pressure drop in the heat exchanger will essentially result in a reduction of both exergy efficiency and volumetric cooling capacity. This reduction can be minimized by making changes to the mixture composition and operating pressures. Consider Case 2 in Table 4.2 as the base case. The minimum temperature approach between the streams is fixed at 5 K. Table 4.3 gives the performance of a Linde–Hampson refrigerator operating in GRS mode at a temperature of 90 K for three different cases:

Case 1: with zero pressure drop in both channels (the same as Case 2 in Table 4.2),
Case 2: with a pressure drop of 0.6 bar in both the high- and low-pressure channels, using the same mixture as in Case 1 (above),
Case 3: with a pressure drop of 0.6 bar in both the channels, but a different mixture.

A pressure drop of 0.6 bar in both channels of the heat exchanger results in a decrease of exergy efficiency from 27.3% in Case 1 to 23.3% in Case 2 when the same mixture is used. Similarly, the volumetric cooling capacity decreases from 64.7 J/l to 51.6 J/l due to pressure drop. The mixture composition has been optimized taking into account the pressure drop in Case 3. With a higher propane content, the exergy efficiency improves to 26.1% and the volumetric cooling capacity increases to 54.8 J/l. A comparison of Cases 1 and 3 shows that the effect of pressure drop on volumetric cooling capacity is much larger (about 15%) than that on the exergy efficiency (about 6%).

[3] $\alpha < 1$.

Table 4.3. Effect of pressure drop in the heat exchanger on the performance of a mixed refrigerant Linde–Hampson refrigerator operating at 90 K in GRS mode (Fig. 4.3)

	Case 1	Case 2	Case 3
Nitrogen (mol%)	28.7	28.7	28.8
Methane (mol%)	17.1	17.1	19.0
Ethane (mol%)	24.2	24.2	17.4
Propane (mol%)	30.0	30.0	34.8
Exergy efficiency, $\eta_{ex,cb}$ (%)	27.3	23.3	26.1
Specific refrigeration effect, Δh_{ref} (J/mol)	514.0	516.0	558.0
Volumetric cooling capacity, Q_v (J/l)	64.7	51.6	54.8
p_2 (bar)	20.00	20.00	20.00
p_3 (bar)	20.00	19.40	19.40
$p_4 = p_g$ (bar)	3.03	3.03	2.97
$p_5 = p_1$ (bar)	3.03	2.43	2.37
T_2 (K)	300.0	300.0	300.0
T_3 (K)	95.1	95.1	95.0
T_4 (K)	89.8	89.8	89.8
T_g (K)	90.0	90.0	90.0
T_5 (K)	295.0	294.5	295.0

The exergy efficiency ($\eta_{ex,cs}$) of a small refrigeration compressor and aftercooler typically varies from 35 to 40%. The overall exergy loss due to pressure drop is therefore typically on the order of 1.5% (0.04*0.4). The choice of an appropriate compressor is therefore more vital than adoption of a higher-pressure drop in the case of Linde–Hampson refrigerators operating with refrigerant mixtures.

4.3.2 Effect of compressor discharge pressure

The performance of a Linde–Hampson refrigerator operating with mixtures is strongly dependent on the operating pressures. Consider the Linde–Hampson refrigerator shown in Fig. 4.3 and operating at 90 K in GRS mode (see Table 4.4). Figure 4.8 shows the variation of the exergy efficiency and the volumetric cooling capacity (Q_v) with compressor discharge pressure.[4] The compressor suction pressure is close to 3 bar in all cases. It can be seen that both the volumetric cooling capacity and the exergy efficiency increase monotonically with an increase in the compressor discharge pressure.

Consider the Linde–Hampson refrigerator operating with a single-stage compressor shown in Fig. 4.3. The mass flow rate through the compressor is given by the expression

$$\dot{n} = \rho_5 \dot{V}_c \eta_v, \tag{4.11}$$

[4] $Q_v = \rho_5\, \Delta h_{min}$.

Table 4.4. Design specifications for the Linde–Hampson refrigerator operating at different operating pressures and an evaporator temperature of 90 K

Exergy efficiency of compressor and aftercooler, $\eta_{ex,cs}$	100%
Volumetric efficiency of the compressor, η_v	100%
Maximum operating pressure, p_2	40 bar
Maximum operating pressure, p_1	5 bar
Minimum temperature approach in the heat exchanger, ΔT_{min}	5 K
Pressure drop in the heat exchanger, Δp	0

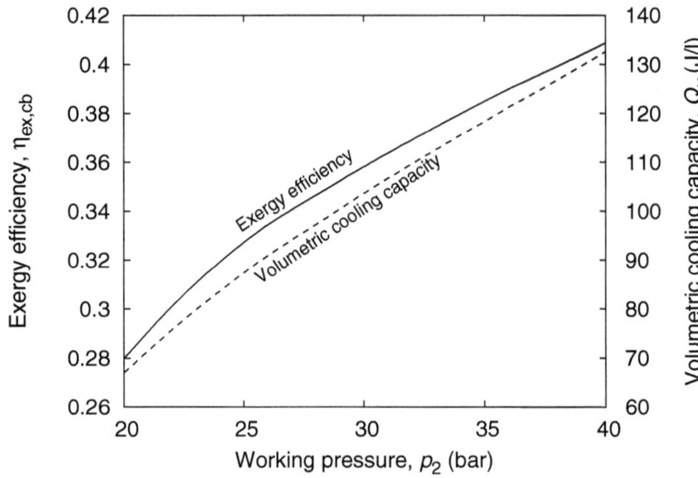

Fig. 4.8. Effect of operating pressure on the performance of a Linde–Hampson refrigerator operating with different mixtures in GRS mode at an evaporator temperature of 90 K.

where ρ_5 is the density of the refrigerant at compressor inlet, $\dot{V}_c$ is the volume displacement rate of the compressor, and η_v is the volumetric efficiency of the compressor.

The refrigeration capacity of the system taking into account the volumetric efficiency of the compressor is given by the expression

$$\dot{Q} = \dot{n}\ \Delta h_{min} = \rho_5 \dot{V}_c \eta_v \Delta h_{min} = Q_v \dot{V}_c \eta_v. \tag{4.12}$$

Since the volumetric efficiency decreases with an increase in the compression ratio [see Eq. 4.10], the cooling capacity does not increase considerably with an increase in the discharge pressure beyond 20 bar (see Fig. 4; Ref. [19]). Further, the compressor discharge temperatures increase with an increase in the compression ratio. Temperatures beyond 130–150 °C can lead to deterioration of the compressor lubricating oils. Single-stage Linde–Hampson systems are therefore best limited to an operating pressure of about 20 bar.

It is evident from the above discussion that two-stage compression may be required for operation beyond 20 bar in many cases so that the pressure ratio is in the range of 3 to 5 in each compression stage. It will be beneficial to adopt high operating pressures and two-stage compression for large systems. On the other hand, it will be economical to use operating pressures less than 20 bar and single-stage compression in small refrigerators.

4.4 Mixed refrigerant Linde–Hampson refrigerator operating at 100 K in LRS mode

Consider a mixed refrigerant Linde–Hampson refrigerator operating at a temperature of 100 K in LRS mode. The design specifications of the system are shown in Table 4.5. The performance of the system with four different mixture compositions is summarized in Fig. 4.9 and the details are shown in Table 4.6. The minimum temperature approach in the heat exchanger is 0.1 K in the first two cases (Cases 1 and 2) and 8 K for the last two cases (Cases 3 and 4). Ethylene is used in the first three cases, while ethane is used in the last case. No pentane is used in the third case, and no propane is used except in the first case.

The highest exergy efficiency (75.7%) is obtained in the first case (Case 1). However, there's a possibility of freezing the pentane at 100 K. In the second case (Case 2), the amount of pentane used is reduced.

The prediction of freezing point of mixtures is not straightforward, and very few methods exist in the literature [73]. The freezing point of the mixture components was estimated using a flash routine in which the fugacity of the solid phase is estimated using the liquid-phase fugacity and the difference in the specific heat at constant

Table 4.5. Design specifications for the Linde–Hampson refrigerator operating at 100 K in LRS mode

Exergy efficiency of compressor and aftercooler, $\eta_{ex,cs}$	100%
Volumetric efficiency of the compressor, η_v	100%
Maximum operating pressure, p_2	20 bar
Maximum operating pressure, p_1	5 bar

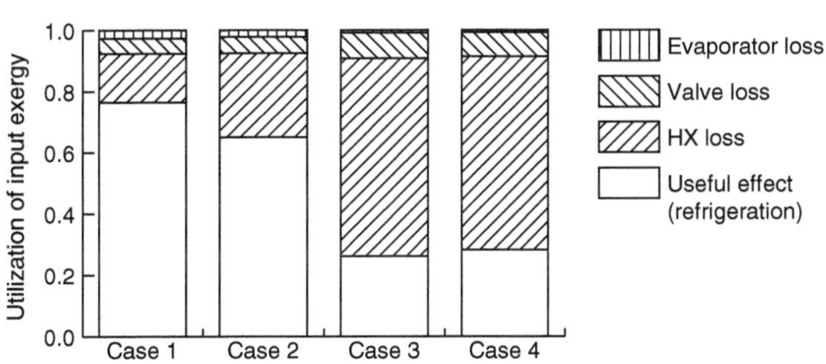

Fig. 4.9. Utilization of input exergy (work) in the cold box of an LRS Linde–Hampson refrigerator with different cases (Table 4.6).

Table 4.6. Performance of a Linde–Hampson refrigerator operating at 100 K with different mixtures in LRS mode (Fig. 4.3)

	Case 1	Case 2	Case 3	Case 4
Mixture				
Nitrogen (mol%)	32.6	28.4	30.8	33.1
Methane (mol%)	33.7	33.8	34.6	34.6
Ethane (mol%)	–	–	–	13.5
Ethylene (mol%)	16.4	12.8	9.9	–
Propane (mol%)	5.0	–	–	–
*i*Butane (mol%)	4.8	22.2	24.7	16.8
*n*Pentane (mol%)	7.4	2.9	–	2.1
Exergy efficiency, $\eta_{ex,cb}$ (%)	75.7	65.0	26.2	28.3
Specific refrigeration effect, Δh_{ref} (J/mol)	1121.0	900.0	541.0	552.0
Volumetric cooling capacity, Q_v (J/l)	201.0	162.0	80.9	90.5
Pressures				
p_1 (bar)	4.37	4.36	3.51	3.88
p_2 (bar)	14.87	13.99	20.00	20.00
p_3 (bar)	14.87	13.99	19.40	19.40
p_4 (bar)	4.37	4.36	4.11	4.48
p_g (bar)	4.37	4.36	4.11	4.48
p_5 (bar)	4.37	4.36	3.51	3.88
Temperatures				
T_2 (K)	300.0	300.0	300.0	300.0
T_3 (K)	100.2	100.7	108.1	108.1
T_4 (K)	97.0	97.1	97.0	97.8
T_g (K)	100.0	100.0	100.0	100.0
T_5 (K)	298.8	298.1	288.3	289.7
$\Delta T_{evap} = T_g - T_4$ (K)	3.0	2.9	3.0	2.2
$\Delta T_{J-T} = T_3 - T_4$ (K)	3.2	3.6	11.1	10.3
$\Delta T_{we} = T_2 - T_5$ (K)	1.2	1.9	11.7	10.3
$\Delta T_{we} = T_3 - T_g$ (K)	0.2	0.7	8.1	8.1
ΔT_{min} (K)	0.1	0.1	8.0	8.0
Bubble, dew point temperatures				
T_{dew} at pressure p_2 (K)	310.9	305.1	303.1	305.5
T_{dew} at pressure p_1 (K)	280.9	274.6	256.5	264.7
T_{bub} at pressure p_2 (K)	117.2	115.9	122.9	122.6
T_{bub} at pressure p_1 (K)	96.7	96.8	93.9	95.0

pressure of the solid and liquid phases at the triple point and the heat of fusion as follows [41, 70, 95]:

$$\ln\left(\frac{f_{\text{liq}}}{f_{\text{solid}}}\right) = \frac{\Delta h_{\text{fusion}}}{RT_{\text{tp}}}\left(\frac{T_{\text{tp}}}{T}-1\right) - \frac{\Delta c_{p,\,\text{fusion}}}{R}\left(\frac{T_{\text{tp}}}{T}-1\right) + \frac{\Delta C_{p,\,\text{fusion}}}{R}\ln\left(\frac{T_{\text{tp}}}{T}\right) + \frac{\Delta v_{\text{fusion}}}{RT_{\text{tp}}}\left(p-p_{\text{tp}}\right), \tag{4.13}$$

where Δh_{fusion} is the enthalpy of fusion, $\Delta c_{p,\,\text{fusion}}$ is the change in specific heat upon fusion, and Δv_{fusion} is the change in volume upon fusion, all taken for the solute at its triple point (tp).

Most commercial process simulators can estimate the freezing point using a similar approach. For example, one can use the "TFREEZE" routine of the Aspen Plus simulation program [10, 11] to determine the freezing point of one of the components in the vapor, liquid-1, or liquid-2 phases.

Any compressor lubricating oil carried over to low-temperature regions can result in its freezing and a consequent reduction in the performance of the system over a period of time. It has been observed, however, in the case of lubricating oil refrigerant mixtures that the slush ice with finely dispersed solid oil phase can continue to circulate through the capillary tube [96]. Predicting the solidification of mixtures accurately is beyond the scope of this work. The data given in Table 4.6 as well as other tables should therefore be used with caution. The examples in Table 4.6 are intended only to provide an insight into the working of Linde–Hampson refrigerators operating in LRS mode from a thermodynamic point of view. Most of the problems associated with the freezing of compressor lubricating oil and high boilers can be overcome by using other processes described in the following sections.

The decrease of pentane results in a lower dew point for the low-pressure stream, as shown in Fig. 4.10 and Table 4.6. The exergy efficiency of the cold box decreases to 65% in Case 2. Similarly, both specific refrigeration effect and volumetric cooling capacity are lower in Case 2 than in Case 1. The reasons for the decrease in the exergy efficiency can be understood from a distribution of the exergy input to the cold box (Fig. 4.9). The exergy loss in the heat exchanger is higher in Case 2 compared to Case 1, as shown in Fig. 4.9. Consequently, the exergy efficiency of an ideal Linde–Hampson refrigerator operating in the LRS mode with the mixture in Case 2 is nearly 10% lower than that operating with the mixture in Case 1. It can be observed from Fig. 4.9 that the exergy loss in the evaporator is also not negligible compared to that in the expansion valve, as in the case of GRS systems (Fig. 4.5), because of a large temperature change of 3.2 K and 3.6 K, respectively, in Cases 1 and 2 compared to only 0.2 to 0.7 K in the case of GRS systems (Table 4.2).

Case 3 corresponds to a system with 0.6 bar pressure drop in the hot and cold stream channels, and with a temperature approach of 8 K in the heat exchanger. The temperature drop in the expansion valve is also higher at 11.1 K. The exergy efficiency of the system in Case 3 is only 26.2%, with most of the exergy loss in the heat exchanger (Fig. 4.9). Replacement of ethylene by ethane and a readjustment of the composition of other components essentially result in an increase in exergy efficiency

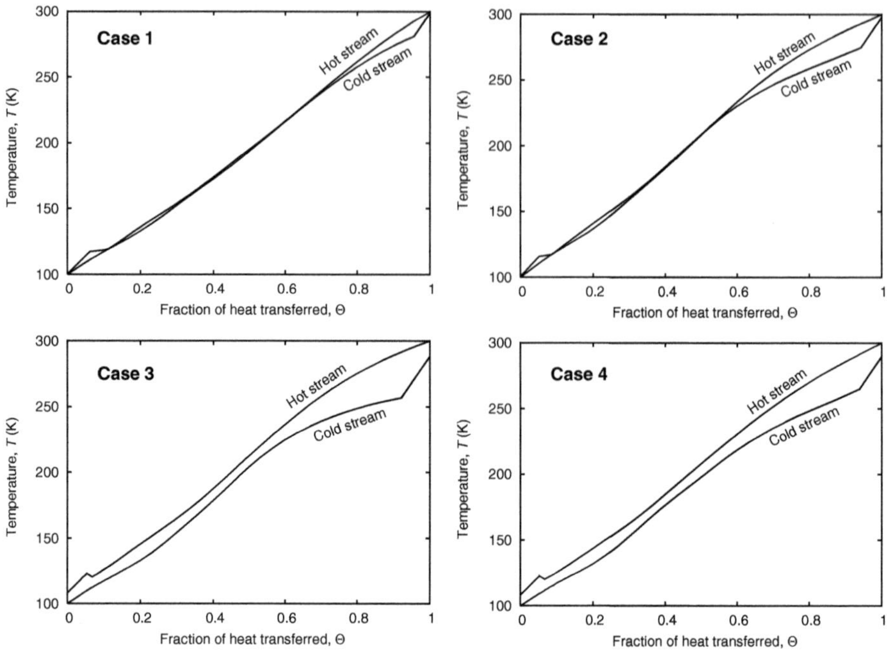

Fig. 4.10. Temperature profiles of the streams in the heat exchanger of a Linde–Hampson refrigerator operating in LRS mode with different mixtures of Table 4.6.

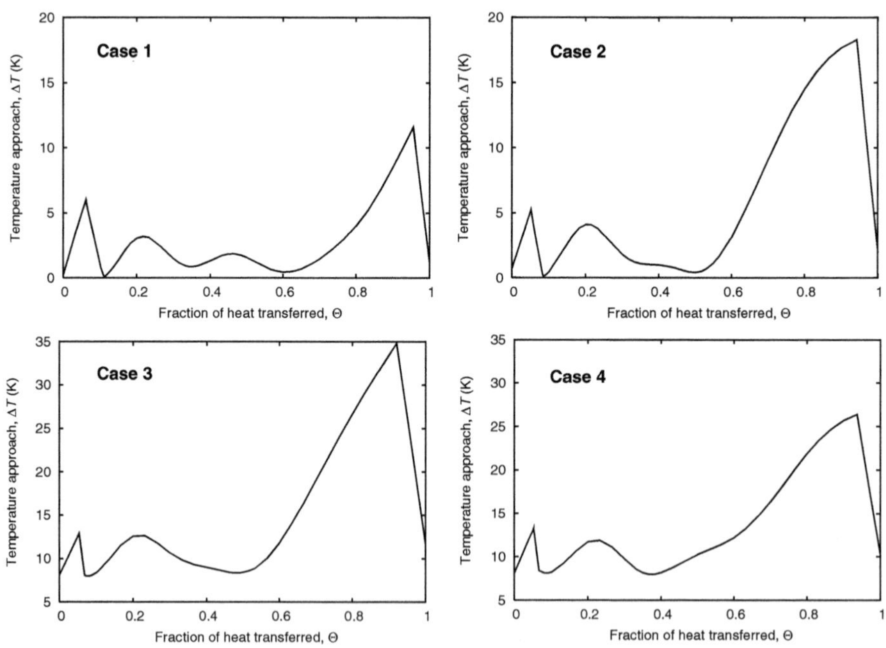

Fig. 4.11. Temperature approach between the streams in the heat exchanger of a Linde–Hampson refrigerator operating in LRS mode with different mixtures of Table 4.6.

to 28.3% in Case 4, largely due to a smaller temperature approach between the streams (Fig. 4.11) and a consequent reduction in the exergy loss in the heat exchanger as seen in Fig. 4.9. The volumetric cooling capacity is higher than in Case 3 by nearly 12%.

A comparison of the temperature profiles and temperature approach in a mixed refrigerant Linde–Hampson refrigerator operating in GRS mode (Figs. 4.6 and 4.7) with that in LRS mode (Figs. 4.10 and 4.11) shows that the temperature approach between the streams is small, over typically 30 to 40% of the length in GRS systems compared to LRS systems (typically 50 to 60%), resulting in smaller exergy loss in the heat exchangers of LRS systems.

It can also be observed from Fig. 4.6 that the temperature change across the evaporator is on the order of 3 K compared to about 0.2 K in GRS systems (Table 4.2). This leads to a higher exergy loss in the evaporator, as shown in Fig. 4.9. This loss, however, is very small (about 0.75%) in practical systems (Cases 3 and 4).

It can be observed from Figs. 4.10 and 4.11 that the minimum temperature approach between the streams in the heat exchanger occurs at several locations. The maximum temperature approach between the streams (Fig. 4.7) is much smaller in LRS systems (Fig. 4.11) compared to GRS systems. Though the mixed refrigerant Linde–Hampson refrigerators operating in LRS mode offer a high exergy efficiency, several problems need to be overcome, particularly in small systems:

- Oil separators need to be placed between the compressor and the condenser in the case of LRS systems, whereas they need to be placed after the aftercooler in GRS systems. Because of higher operating temperatures, more efficient oil separators are required for LRS systems.
- The composition of the refrigerant in circulation changes due to the accumulation of the liquid phase in the pipelines between the partial condenser and the cryostat in the case of LRS systems.
- Sometimes mixed refrigerant systems need to be transported from the manufacturer's site to the customer's site by air. The amount of hydrocarbons (refrigerant) that can be carried on an aircraft is regulated in some countries. Since LRS systems require a higher refrigerant charge, they are not preferred in small systems due to transport regulations. A higher refrigerant charge needs to be used in an LRS system compared to a GRS system for the same operating conditions due to condensation of refrigerant in the LRS condenser. Consequently, the operating pressure at start-up (pull-down) is also higher in the case of LRS systems.

The problem of oil carryover in mixed refrigerant Linde–Hampson systems operating in LRS mode can be overcome in two different ways:

- Use a precooling process to remove oil at temperatures below ambient.
- Use phase separators to remove the oil carried over.

Both of these methods improve the reliability of the mixed refrigerant systems and are discussed in the next sections. Both methods also help in reducing the possibility of freezing of high boilers at low temperature. High boilers that freeze above the operating temperatures are not used in precooling processes, and high boilers that can freeze at operating temperatures are removed in the phase separators in the phase separator processes.

4.5 Effect of the addition of neon or helium

The examples presented in Tables 4.2 and 4.6 contain only nitrogen and hydrocarbon fluids. Sometimes it is beneficial to add neon or helium to these fluids. Since both neon and helium are noncondensable at the usual operating temperatures of mixed refrigerant fluids (70–130 K), they reduce the partial pressure of nitrogen in the liquid phase and result in lower evaporating temperatures. Evaporating temperatures lower than 77 K can therefore be obtained with nitrogen-hydrocarbon mixtures at an evaporating pressure of 1 bar. The benefits of using helium or neon were elegantly presented by Grezin and Zacharov [49].

Let the design evaporating pressure be p (say 3 bar). Let T_{N_2-HC} be the evaporating temperature when a nitrogen-hydrocarbon refrigerant is used. The addition of helium or neon to a nitrogen-hydrocarbon mixture will result in a lower evaporating temperature, say $T_{N_2-HC-He}$. $T_{N_2-HC-He}$ can be lower than T_{N_2-HC} by 1 to 10 K depending on the amount of helium or neon added. A higher evaporating pressure will need to be used if the same refrigeration temperature (T_{N_2-HC}), is required in the presence of helium or neon.

The volumetric cooling capacity[5] (Q_v) is given by the expression:

$$Q_v = \rho_5 \Delta h_{\text{min}} \eta_v. \tag{4.14}$$

The density of the refrigerant ρ_5 will be higher in the case of nitrogen-hydrocarbon-helium mixtures because of the higher pressure, but the specific refrigeration effect Δh_{min} will be lower since the Joule–Thomson coefficient of helium or neon is negative at room temperature. The decrease in Δh_{min} is normally lower than the increase in evaporating pressure when neon or helium is used. Thus, a higher amount of cooling can be obtained for a given compressor when helium or neon is used. A higher evaporating pressure also results in the decrease of reexpansion losses or in the increase of volumetric efficiency (η_v) of the compressor [49]. The compressor work will, however, increase with the addition of helium or neon, due to the higher adiabatic index (γ) of helium and neon (1.67) compared to about 1.25 to 1.3 for nitrogen-hydrocarbon mixtures.

4.5.1 Mixed refrigerant Linde–Hampson refrigerator operating at 85 K in GRS mode with N_2-He-HC mixtures

Consider a Linde–Hampson refrigerator operating in GRS mode with neon-nitrogen-hydrocarbon mixtures. The design specifications of the system are shown in Table 4.7. The neon content was increased from zero to 12.5 mol%. The performance of the

[5] Q_v is the refrigeration obtained per unit displacement volume of the compressor.

Table 4.7. Design specifications for the Linde–Hampson refrigerator operating in GRS mode with neon-nitrogen-hydrocarbon mixtures

Exergy efficiency of compressor and aftercooler, $\eta_{ex,cs}$	100%
Volumetric efficiency of the compressor, η_v	100%
Maximum operating pressure, p_2	20 bar
Maximum operating pressure, p_1	4 bar

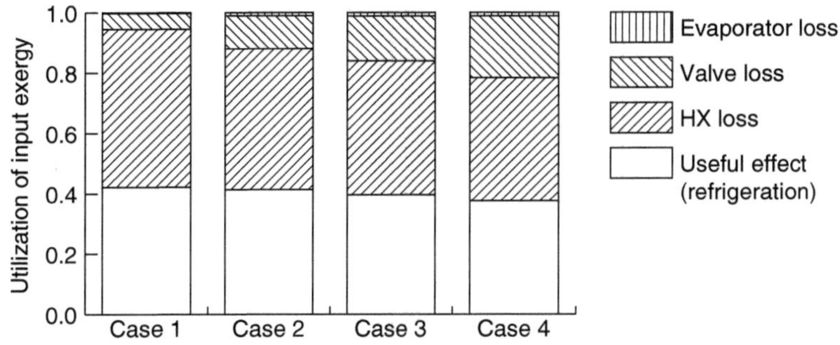

Fig. 4.12. Utilization of input exergy in the cold box of a Linde–Hampson refrigerator operating in GRS mode with mixtures in Table 4.8.

system with four different mixture compositions is summarized in Fig. 4.12, and the details are shown in Table 4.8. A heat exchanger temperature approach of 0.1 K has been used in all cases.

The compressor suction pressure and optimum mixture composition have been determined for each of these cases. It can be seen from Table 4.8 that the evaporating pressure increases with an increase in neon content. Similarly, the vapor fraction at the inlet of the expansion valve increases from zero in Case 1 to 11.5% in Case 4 with an increase in the neon mole fraction. The increase in vapor fraction essentially results in an increase in the exergy loss in the expansion device with an increase in neon quantity and consequently a decrease in overall exergy efficiency. The exergy loss in the heat exchanger decreases with an increase in neon content due to a smaller temperature approach between the streams, as shown in Figs. 4.13 and 4.14.

Though the specific refrigeration effect (J/mol) decreases with an increase in the neon quantity, the increase in evaporating pressure is much more than the drop in specific refrigeration effect, resulting in an increase in the volumetric cooling capacity (Q_v)[6] of the refrigerant (see Table 4.8).

The volumetric cooling capacity increases with an increase in the neon quantity (evaporating pressure), while the exergy efficiency decreases with an increase in the neon quantity, as shown in Fig. 4.15. An increase in the evaporating pressure also results in an increase in the density of the refrigerant at compressor suction and consequently a higher mass flow rate through the compressor. A higher mass flow rate results in a higher refrigeration capacity as well as a higher heat load on the

[6] $Q_v = \rho_5 \Delta h_{min}$.

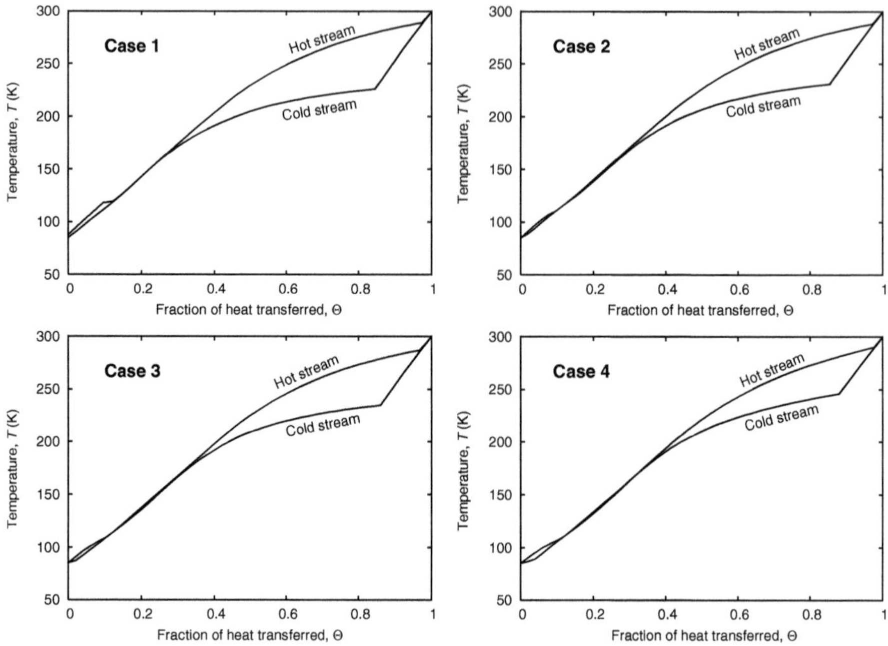

Fig. 4.13. Temperature profiles of the streams in the heat exchanger of a Linde–Hampson refrigerator operating in GRS mode with different mixtures of Table 4.8.

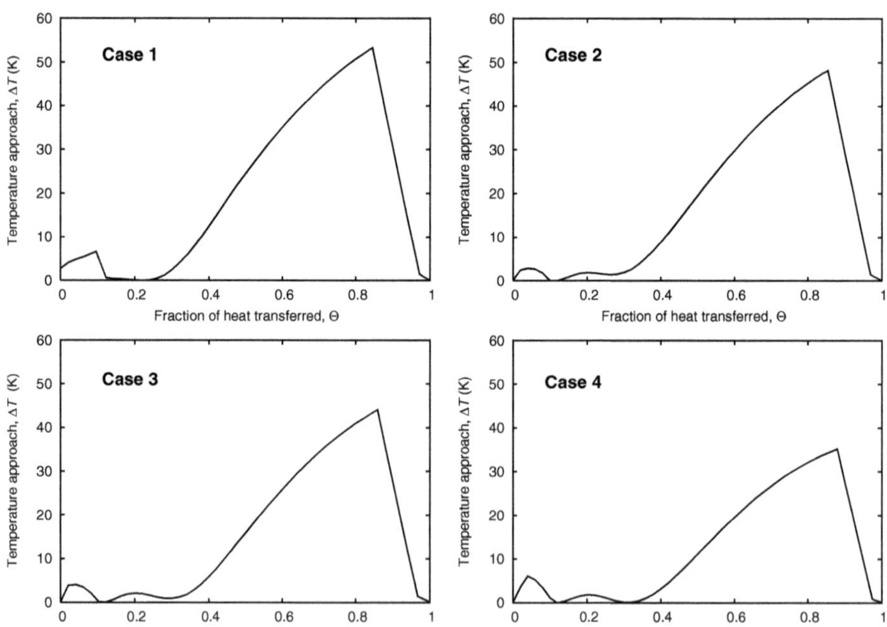

Fig. 4.14. Temperature approach between the streams in the heat exchanger of a Linde–Hampson refrigerator operating in GRS mode with different mixtures of Table 4.8.

Table 4.8. Effect of neon on the performance of a mixed refrigerant Linde–Hampson refrigerator operating in GRS mode with neon-nitrogen-hydrocarbon mixtures

	Case 1	Case 2	Case 3	Case 4
Mixture				
Nitrogen (mol%)	29.7	26.2	27.2	29.2
Methane (mol%)	13.5	17.0	17.1	15.5
Ethane (mol%)	16.8	12.2	7.9	5.6
Ethylene (mol%)	–	–	–	–
Propane (mol%)	40.0	40.0	40.0	37.3
Neon (mol%)	–	4.6	7.8	12.5
Exergy efficiency, $\eta_{ex,cb}$ (%)	41.9	41.3	39.7	37.3
Specific refrigeration effect, Δh_{ref} (J/mol)	909.0	806.0	707.0	564.0
Volumetric cooling capacity, Q_v (J/l)	73.0	82.0	86.0	92.0
Compressor discharge pressure, p_2 (bar)	20.00	20.00	20.00	20.00
Compressor suction pressure, p_1 (bar)	1.98	2.50	3.00	4.00
Temperatures				
T_2 (K)	300.0	300.0	300.0	300.0
T_3 (K)	87.8	85.2	85.1	85.1
T_4 (K)	84.8	81.5	81.1	81.1
T_g (K)	85.0	85.0	85.0	85.0
T_5 (K)	299.9	299.9	299.9	299.9
$\Delta T_{evap} = T_g - T_4$ (K)	0.2	3.5	3.9	3.9
$\Delta T_{J-T} = T_3 - T_4$ (K)	3.0	3.7	4.0	4.0
$\Delta T_{we} = T_2 - T_5$ (K)	0.1	0.1	0.1	0.1
$\Delta T_{we} = T_3 - T_g$ (K)	2.8	0.2	0.1	0.1
ΔT_{min} (K)	0.1	0.1	0.1	0.1
Dew point temperatures				
T_{dew} at pressure p_2 (K)	288.7	287.8	286.9	283.8
T_{dew} at pressure p_1 (K)	226.0	230.9	234.7	239.5

heat exchanger. For example, the heat load of the heat exchanger in Case 4 is nearly 1.8 times that of Case 1 (without neon). The compressor power input also increases with an increase in the neon mole fraction, due to an increase in the suction pressure, the volumetric efficiency (lower pressure ratio), and the adiabatic index (γ). The maximum amount of helium or neon to be added is largely decided by the heat load of the heat exchanger and the maximum power that can be drawn by the compressor. The method to be adopted for the optimization of a mixture composition that includes a noncondensable gas such as neon or helium is presented in Section 5.3.

The addition of neon or helium to nitrogen hydrocarbon mixtures inhibits the formation of the second (immiscible) liquid phase in the evaporator in many mixtures

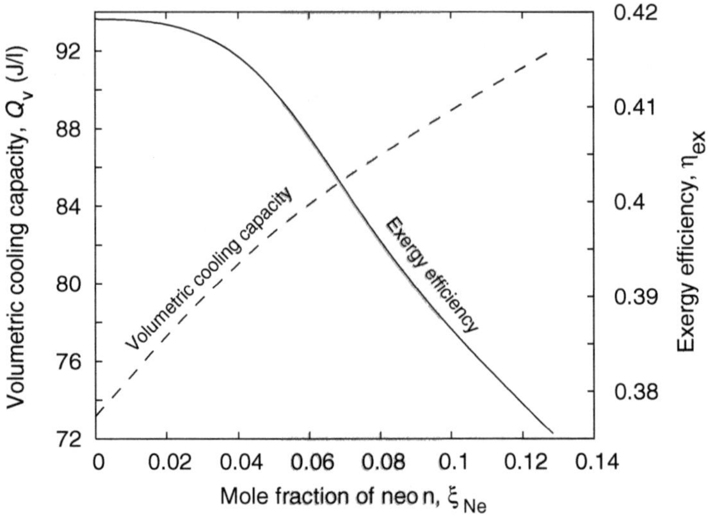

Fig. 4.15. Variation of volumetric cooling capacity and exergy efficiency with an increase in neon content of different cases discussed in Table 4.8.

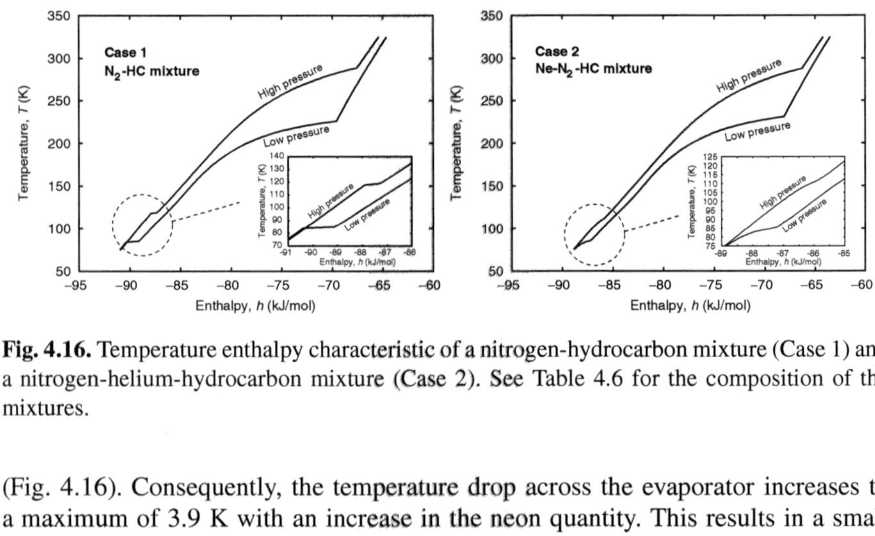

Fig. 4.16. Temperature enthalpy characteristic of a nitrogen-hydrocarbon mixture (Case 1) and a nitrogen-helium-hydrocarbon mixture (Case 2). See Table 4.6 for the composition of the mixtures.

(Fig. 4.16). Consequently, the temperature drop across the evaporator increases to a maximum of 3.9 K with an increase in the neon quantity. This results in a small increase in exergy loss in the evaporator with an increase in the neon content (Case 1 to Case 4), as shown in Fig. 4.12.

Table 4.9 shows the performance of a compressor with a displacement volume of 28 l/min. The volumetric efficiency of the compressor has been assumed to be 1.0 and the minimum temperature approach in the heat exchanger to be 0.2 K. The pressure drop in the heat exchanger has been assumed to be zero. The first two cases are that of a refrigerator operating at 82 K, while the next two are that of a refrigerator operating at 85 K. The amount of neon used is different in the first two cases (82 K), while no neon is used in Case 4 (85 K). Assuming that the exergy efficiency of the compressor section is 40%, which is typical of a small compressor, the compressor work varies from 515 to 587 watts, depending on the amount of neon used. The power consumed in a real compressor will, however, be smaller because of the smaller mass flow rate corresponding to $\eta_v < 1$.

The exergy efficiency and refrigeration capacity values presented in Table 4.9 match closely with those presented by Boiarski et al. [19] for similar operating conditions, except for Case 4, which gives a slightly lower amount of refrigeration, but at a much higher exergy efficiency and a smaller heat exchanger heat load.

Consider a practical system operating with a neon-nitrogen-hydrocarbon mixture at 85 K. Table 4.10 gives the mixture composition that can be used when the minimum temperature approach in the heat exchanger is increased to 3 K and the temperature drop across the throttle is increased to 10.1 K. With this increase, the exergy efficiency decreases from 40% to 27% and the volumetric cooling capacity decreases from 77 J/l to 56 J/l (Case 3, Table 4.9).

The main advantage, however, in the use of helium or neon along with nitrogen-hydrocarbon or fluorocarbon mixtures is the possibility of maintaining a nearly constant evaporating pressure when a fixed throttle is used. Consider a Linde–Hampson GRS refrigerator with a capillary tube expansion device. Figure 4.17 shows the variation of compressor suction pressure measured in a Linde–Hampson refrigerator operating with nitrogen-hydrocarbon and nitrogen-neon-hydrocarbon refrigerants [76]. The evaporation pressure remains relatively stable when a neon- or helium-based mixture is used instead of decreasing steadily with an increase in heat load as in the case of nitrogen-hydrocarbon mixtures. The compressor exit pressure increases with an increase in heat load because of the evaporation of some of the liquid on the high-pressure side in both cases. A higher evaporating pressure (compressor suction pressure) ensures a higher amount of refrigeration when a noncondensable component such as helium or neon is used. The alternative is to use expensive automatic throttle devices. It therefore makes economic sense to use neon or helium in refrigerators operating with nitrogen-hydrocarbon mixtures to regulate the performance with heat load. Similarly, the addition of nitrogen or argon is beneficial in refrigerators operating with fluorocarbon mixtures such as R14-R23-R134a, etc.

From the above discussion, it can be concluded that the use of helium or neon in nitrogen-hydrocarbon mixtures depends on the following conditions:

- the size of the heat exchanger used (heat load that can be met),
- the variation of volumetric efficiency of the compressor with pressure ratio,
- the maximum power that can be drawn by the compressor,
- the heat load-evaporating temperature characteristic that is desired in the particular application.

The above conditions are also valid for the use of nitrogen or argon with mixtures of perfluorocarbon-fluorocarbon such as R14-R23-R22 or perflurocarbon-fluorcarbon-hydrocarbon mixtures such as R14-R23-R290.

The addition of a noncondensable fluid is preferred from a heat-load regulation point of view when a fixed orifice/throttle is used, as already discussed, and should be preferred where ever possible.

Table 4.9. Performance of a GRS Linde–Hampson refrigerator with different nitrogen-neon-hydrocarbon mixtures and a compressor with a displacement rate of 28 l/min. $\Delta T_{\mathrm{min}} = 0.2$, $\eta_v = 1.0$, $\Delta p = 0$ and $\eta_{\mathrm{ex,comp}} = 0.4$.

	Case 1 82 K	Case 2 82 K	Case 3 85 K	Case 4 85 K
Mixture				
Nitrogen (mol%)	26.3	25.4	25.1	29.5
Methane (mol%)	15.3	15.3	16.9	12.5
Ethane (mol%)	13.8	18.6	19.6	23.0
Propane (mol%)	35.0	35.0	35.0	35.0
Neon (mol%)	9.6	5.7	3.5	–
Exergy efficiency, $\eta_{\mathrm{ex,cb}}$ (%)	36.2	38.1	40.0	40.0
Specific refrigeration effect, Δh_{ref} (J/mol)	681.0	787.0	806.0	874.0
Volumetric cooling capacity, Q_v (J/l)	68.5	64.0	77.0	70.0
Compressor discharge pressure, p_2 (bar)	20.00	20.00	20.00	20.00
Compressor suction pressure, p_1 (bar)	2.48	2.00	2.36	1.99
Temperatures				
T_2 (K)	300.0	300.0	300.0	300.0
T_3 (K)	82.3	82.4	85.3	87.7
T_4 (K)	78.1	78.4	81.7	84.7
T_g (K)	82.0	82.0	85.0	85.0
T_5 (K)	299.8	299.8	299.8	299.8
ΔT_{evap} (K)	3.9	3.6	3.3	0.3
$\Delta T_{\mathrm{J-T}}$ (K)	4.2	4.0	3.6	3.0
ΔT_{we} (K)	0.2	0.2	0.2	0.2
ΔT_{we} (K)	0.3	0.4	0.3	2.7
Dew point temperatures				
T_{dew} at pressure p_2 (K)	283.3	284.4	284.6	285.3
T_{dew} at pressure p_1 (K)	227.9	223.7	227.3	223.8
Performance with 28 l/min compressor and $\eta_v = 1$				
Refrigeration, $\dot{Q}$ (W)	32.0	29.8	36.0	32.7
Heat exchanger load, $\dot{Q}_{\mathrm{hx}}$ (W)	1010.0	850.0	1010.0	860.0
Work of compression, $-\dot{W}_c$ with $\eta_{\mathrm{ex,cs}} = 0.4$ (W)	587.0	520.0	568.0	515.0

Table 4.10. Performance of a GRS Linde–Hampson refrigerator with a neon-nitrogen-hydrocarbon mixture operating at 85 K ($\Delta T_{\text{min}} = 3.0$, $\Delta T_{\text{J-T}} = 10.1$)

Mixture	
Nitrogen (mol%)	29.6
Methane (mol%)	15.4
Ethane (mol%)	11.5
Propane (mol%)	35.0
Neon (mol%)	8.6
Exergy efficiency, $\eta_{\text{ex, cb}}$ (%)	27.0
Specific refrigeration effect, Δh_{ref} (J/mol)	517.0
Volumetric cooling capacity, Q_v (J/l)	56.3
Compressor discharge pressure, p_2 (bar)	20.00
Compressor suction pressure, p_1 (bar)	2.66

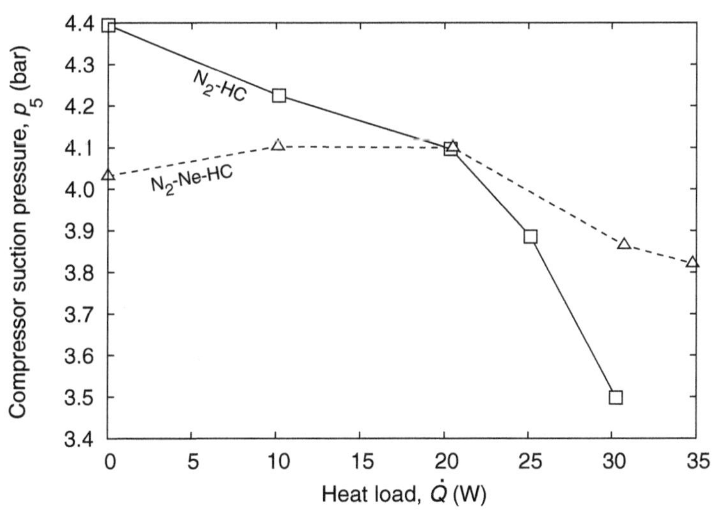

Fig. 4.17. Variation of compressor suction pressure with heat load in a Linde–Hampson GRS refrigerator operating with a nitrogen-hydrocarbon mixture and a nitrogen-neon-hydrocarbon mixture [76]. The uncertainty in pressure measurement is 0.125 bar.

4.6 Effect of precooling

The exergy efficiency (η_{ex}) and the specific refrigeration effect (Δh_{min}) of a mixed refrigerant process reduce sharply with an increase in the ambient temperature, as shown in Fig. 4.18, because of the decrease in the Joule–Thomson coefficient with an increase in temperature at the warm end of the heat exchanger. One of the easiest ways of making a process less dependent on the ambient temperature variations is to use a two-stage system as shown in Fig. 4.19.

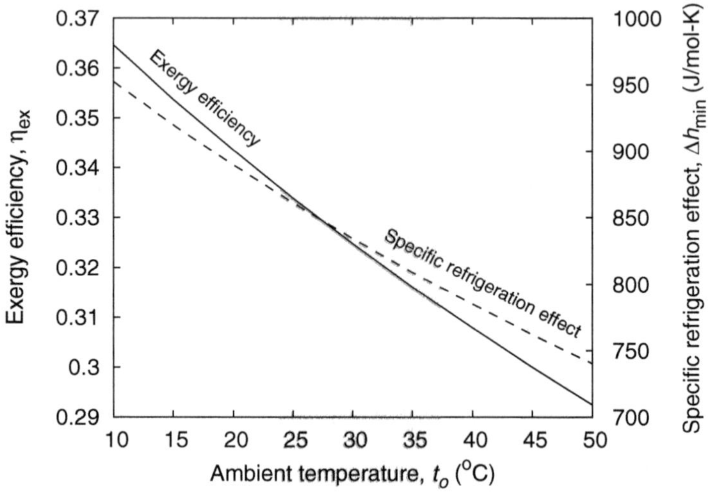

Fig. 4.18. Variation of ideal GRS Linde cycle exergy efficiency and specific refrigeration effect as a function of ambient temperature for mixture M1 (Table 3.3).

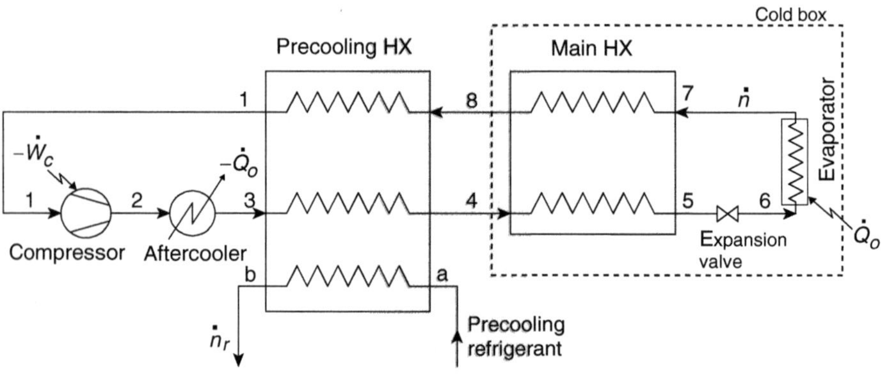

Fig. 4.19. Precooled Linde–Hampson refrigerator.

The refrigerant used in the precooler can be either a pure fluid, an azeotropic, or a zeotropic mixture. High exergy efficiencies similar to a Linde–Hampson refrigerator operating in LRS mode can be obtained with precooled processes. There are several advantages to using an independent precooling process:

- The effects of variation of ambient temperature can be minimized.
- The reliability of oil separation is improved by separating the oil at temperatures below the ambient temperature.
- A larger amount of refrigeration can be obtained with the same compressor.
- Very small amounts of high boiling point components such as pentanes and butanes need not be used when precooling is used. In most cases, butanes and

pentanes need not be used at all. The possibility of freezing the refrigerant is therefore much less likely in precooled processes.

4.6.1 Precooled mixed refrigerant process refrigerator operating at 100 K

Consider a precooled mixed refrigerant process shown in Fig. 4.19 and operating at a temperature of 100 K. The design specifications of the system are shown in Table 4.11. The performance of the system with four different mixture compositions is summarized in Fig. 4.20 and the details are shown in Table 4.12. The minimum temperature approach in the heat exchanger is 0.2 K in the first two cases, and 8 K in the third and fourth cases.

The exergy efficiency of the cold box consisting of the main (two-stream) heat exchanger, expansion valve, and evaporator is defined as follows:

$$\eta_{ex,\,cb} = \frac{\text{minimum power required}}{\text{exergy input}} = \frac{\dot{Q}}{\dot{n}(ex_4 - ex_8)}\left(\frac{T_a}{T_7} - 1\right). \tag{4.15}$$

Mixtures of nitrogen, methane, ethylene, and propane are used in all cases. The highest exergy efficiency (72.8%) is obtained in the first case (Case 1), with a minimum temperature approach of 0.2 K between the streams in the two-stream heat exchanger. The pressure drop in the heat exchangers is also assumed to be zero in this case. In the second case, a pressure drop of 0.6 bar is assumed, keeping the temperature approach the same as before (0.2 K). With these assumptions, the exergy efficiency of the cold

Table 4.11. Design specifications for a precooled Linde–Hampson refrigerator operating with mixtures

Exergy efficiency of compressor and aftercooler, $\eta_{ex,\,cs}$	100%
Volumetric efficiency of the compressor, η_v	100%
Maximum operating pressure, p_2	20 bar
Maximum operating pressure, p_1	5 bar
Precooling temperature, T_4	253 K

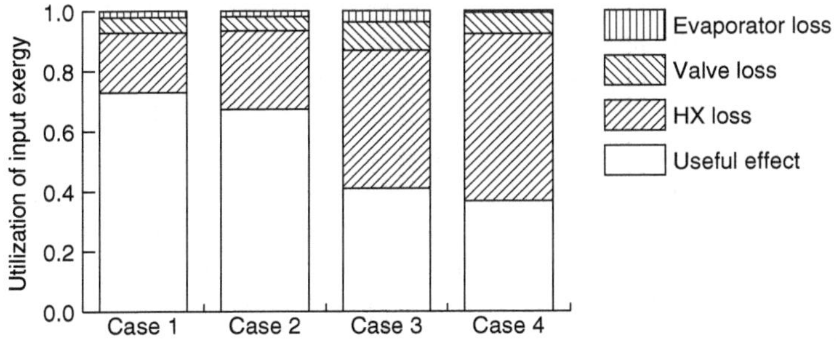

Fig. 4.20. Utilization of input exergy in different cases of the precooled mixed refrigerant process (Table 4.12).

Table 4.12. Performance of a precooled mixed refrigerant Linde refrigerator shown in Fig. 4.19 and operating with different mixtures and design conditions shown in Table 4.11

	Case 1	Case 2	Case 3	Case 4
Mixture				
Nitrogen (mol%)	36.8	38.1	36.8	38.1
Methane (mol%)	27.9	27.7	31.1	32.5
Ethylene (mol%)	17.0	16.3	13.5	12.4
Propane (mol%)	18.3	17.9	18.6	17.0
Exergy efficiency, $\eta_{ex,cb}$ (%)	72.8	67.2	40.9	36.7
Specific refrigeration effect, Δh_{ref} (J/mol)	1203.6	1269.5	1129.3	681.1
Volumetric cooling capacity, Q_v (J/l)	247.0	230.0	98.0	122.0
Pressures				
p_4 (bar)	18.7	20.0	19.9	20.0
p_5 (bar)	18.7	19.4	19.3	19.4
$p_6 = p_7$ (bar)	5.0	5.0	2.7	5.0
p_8 (bar)	5.0	4.4	2.1	4.4
Temperatures				
T_4 (K)	253.0	253.0	253.0	253.0
T_5 (K)	100.3	100.4	108.1	108.4
T_6 (K)	97.7	97.7	90.9	98.5
T_7 (K)	100.0	100.0	100.0	100.0
T_8 (K)	250.5	248.6	242.1	241.7
ΔT_{min} (K)	0.2	0.2	8.0	8.0

box reduces to 67.2%. The volumetric cooling capacity reduces from 247 J/l in Case 1 to 230 J/l in Case 2.

Case 3 is similar to Case 2, but with a temperature approach of 8.0 K in the heat exchanger. Because of the larger exergy loss in the heat exchanger, the exergy efficiency further reduces to 40.9%. The volumetric cooling capacity reduces from 247 J/l in Case 1 to 98 J/l in Case 3 because of the pressure drop and a larger temperature approach in the heat exchanger. Case 4 is similar to Case 3 except that the mixture composition and pressure are optimized for the highest volumetric cooling capacity, and not the highest exergy efficiency as in Case 3. It must, however, be understood that an increase in cooling capacity also results in a higher heat exchanger load, because of the smaller specific refrigeration capacity of the mixture in Case 4 (681 J/mol) compared to that in Case 3 (1129 J/mol). The volumetric efficiency of the compressor will be much higher in Case 4 compared to Case 3 because of the smaller compression ratio, leading to a much higher cooling capacity in Case 4 for a given compressor.

The exergy loss in the evaporator is quite small in the fourth case since the temperature difference across the evaporator is limited to 1.5 K, compared to 9.1 K in Case 3.

The exergy loss in the expansion valve is also smaller in Case 4 ($\Delta T_{\text{valve}} = 9.9$K) compared to that in Case 3 ($\Delta T_{\text{valve}} = 17.2$ K).

It is interesting to note that the composition of the mixture in Case 3 is close to that in Case 4. The different operating pressures adopted in the two cases lead to different exergy losses in the heat exchanger, expansion valve, and evaporator. Some authors [6, 32, 48, 57] have attempted optimization of mixture composition for specified operating pressures, say 20 bar and 3 bar. A higher amount of refrigeration, however, can be obtained with optimum mixtures at operating pressures of 20/5 bar. The highest evaporating pressure that can be adopted is normally determined by the maximum power that can be drawn by the compressor motor. Ideally, both the compressor suction and discharge pressures must be adopted as the variables of the optimization problem. Case 4 is a good example of such an approach.

The temperature profiles in the heat exchanger in the precooled case (Figs. 4.21 and 4.22) are similar to those in the LRS Linde–Hampson refrigerator (Figs. 4.10 and 4.11). In both of these examples, the minimum temperature approach (pinch point) in the heat exchanger occurs in between the two ends of the heat exchanger, and at more than one point. In all cases, the pinch points that occur between the two ends of the heat exchanger are closer to the cold end than the warm end of the heat exchanger.

Table 4.13 gives the details of the precooling system, and Fig. 4.23 shows the temperature profiles of the hot and cold streams of the precooling (three-stream) heat exchanger of Fig. 4.19 for Case 4 of Table 4.12. Propane is used as the refrigerant in the precooler. It can be observed that the temperature profiles of the streams are nearly parallel over a large length of the heat exchanger, resulting in a high exergy efficiency of the precooler.

The overall exergy efficiency of the cold box and the precooling heat exchanger shown in Fig. 4.19 is given by the expression

$$\eta_{\text{ex, cb+pre}} = \frac{\dot{Q}(T_a/T_7 - 1)}{\dot{n}(\text{ex}_3 - \text{ex}_1)/\eta^{\text{mr}}_{\text{ex, cs}} + \dot{n}_r(\text{ex}_a - \text{ex}_b)/\eta^{\text{C3}}_{\text{ex}}}, \tag{4.16}$$

where $\dot{Q}$ is the heat absorbed in the evaporator, $\eta^{\text{mr}}_{\text{ex, cs}}$ is the exergy efficiency of the mixed refrigerant (main) compressor and aftercooler, and $\eta^{\text{C3}}_{\text{ex}}$ is the exergy efficiency of the propane precooler. In well-built small systems, $\eta^{\text{mr}}_{\text{ex, cs}}$ is on the order of 40% and $\eta^{\text{C3}}_{\text{ex}}$ is on the order of 20 to 25%. If an exergy efficiency of the propane precooling system ($\eta^{\text{C3}}_{\text{ex}}$) is assumed to be 50% and that of the main (low-temperature) refrigerant compressor and condenser ($\eta^{\text{mr}}_{\text{ex, cs}}$) to be 100%, then the overall exergy efficiency of the entire precooling system is 34.7% compared to the exergy efficiency of 36.7% of the cold box consisting of the three-stream heat exchanger, expansion valve, and evaporator alone. The power consumed by the propane precooling system will be only a small fraction (typically 7–10%) of that consumed by the mixed refrigerant compressor.

The main advantage of precooling is the elimination of high boilers such as butanes and pentanes, which may freeze at low temperatures. The exergy efficiency of the cold box of the precooled system is higher than that of an LRS system (see Case 4,

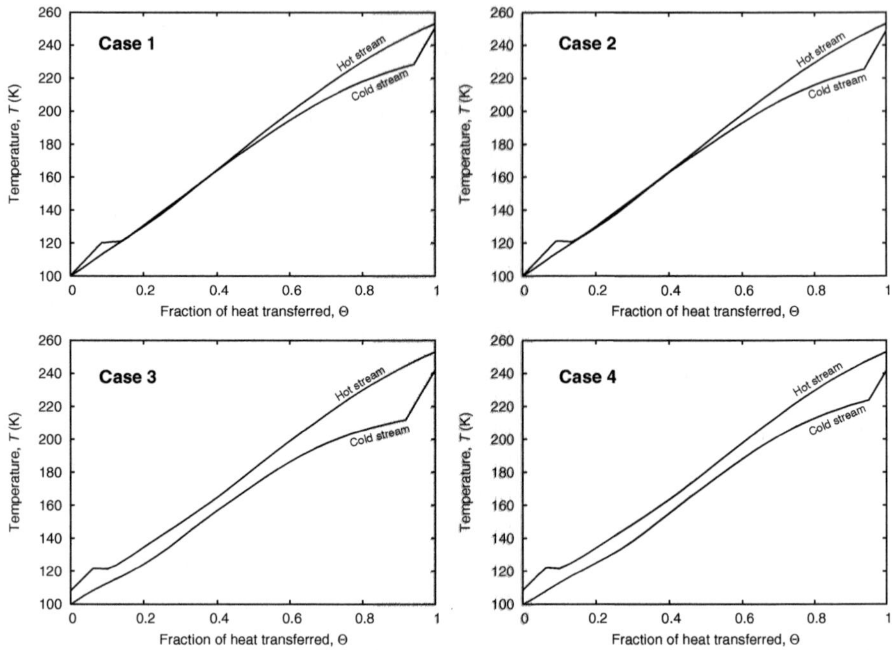

Fig. 4.21. Temperature profiles of the streams in the heat exchanger of a Linde–Hampson refrigerator operating in LRS mode with different mixtures of Table 4.12.

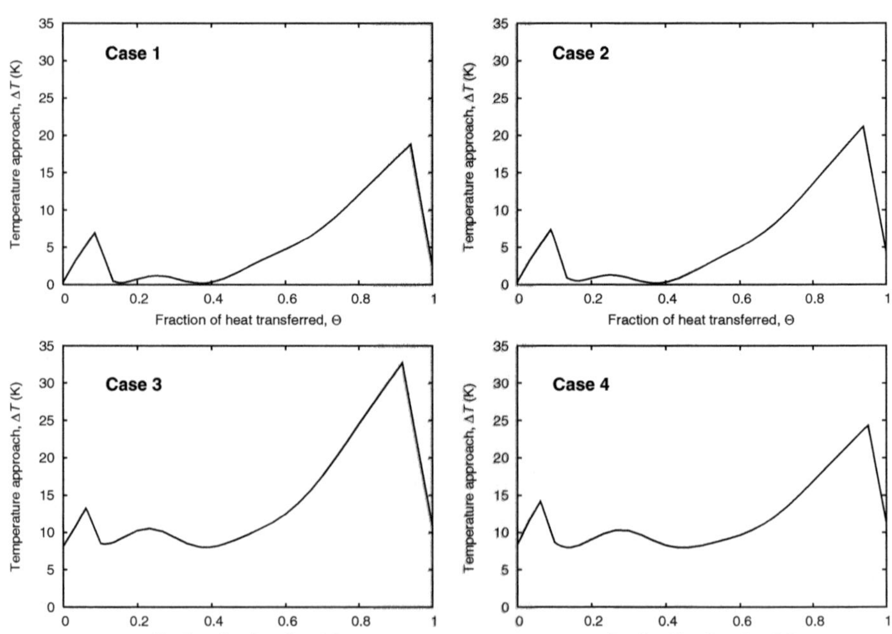

Fig. 4.22. Temperature approach between the streams in the low-temperature (two-stream) heat exchanger of the precooled Linde–Hampson refrigerator with different mixtures from Table 4.12.

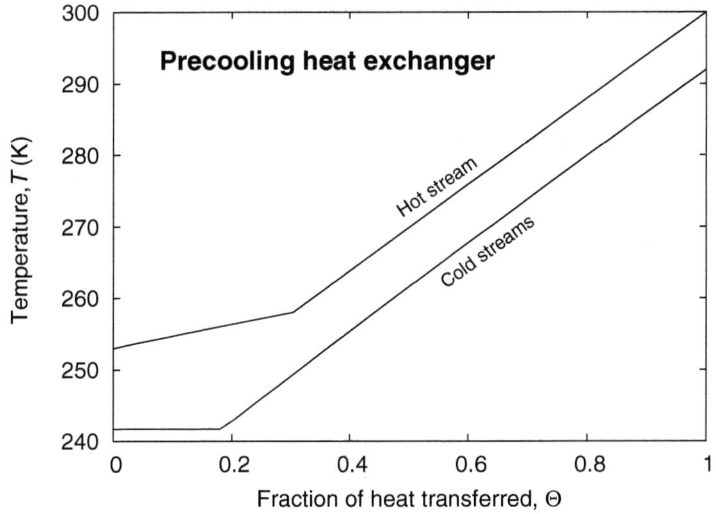

Fig. 4.23. Temperature profiles of hot and cold fluid streams in the precooling (three-stream) heat exchanger of a precooled Linde–Hampson refrigerator with the mixture shown in Case 4 of Table 4.12 and propane as the precooling refrigerant.

Table 4.13. Temperature and pressure of different streams in the three-stream heat exchanger on the performance of a precooled mixed refrigerant Linde–Hampson refrigerator (Fig. 4.19) and Table 4.12

	Case 4
Pressures	
$p_1 = p_8 = p_9$ (bar)	4.4
$p_2 = p_3 = p_4$ (bar)	20.0
$p_a = p_b$ (bar)	1.6
Temperatures	
T_3 (K)	300.0
T_4 (K)	253.0
T_8 (K)	241.7
T_9 (K)	292.0
T_b (K)	292.0
$\Delta T_{\min}$ (K)	8.0
Flow rates	
$\dot{n}$ (mixture) (mol/s)	1.00
$\dot{n}_r$ (propane) (mol/s)	0.04

Tables 4.6 and 4.12). When a larger quantity of butanes and pentanes can be tolerated, the exergy efficiency of the LRS system can be close to that of the precooled system. However, the volumetric cooling capacity as well as the specific refrigeration effect of the precooled system will always be higher than those of a Linde–Hampson refrigerator operating in the LRS mode.

One of the main advantages in using a precooled system is that oil separation can be done at low temperatures, thus improving the system's reliability. One such system patented by Alexeev and Quack [5] is shown in Fig. 4.24. The use of a precooler also results in a shorter startup (pull-down) time.

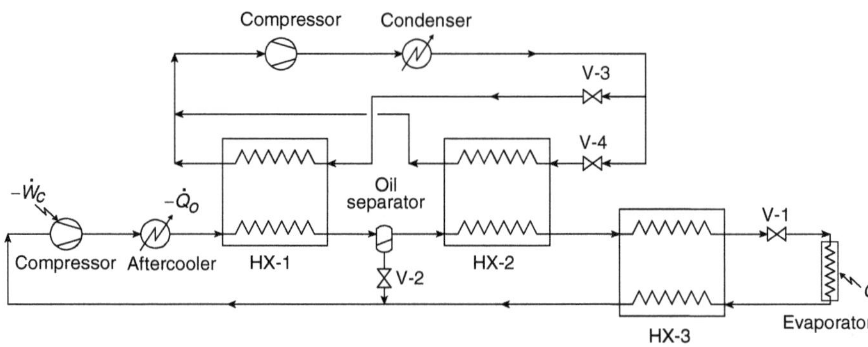

Fig. 4.24. Precooled mixed refrigerant cycle refrigerator of Alexeev and Quack [5] (Adapted from German Patent no. 19922364.)

4.7 Mixed refrigerant process refrigerator with a phase separator

Figure 4.25 shows a mixed refrigerant cascade refrigerator with a phase separator. The process is quite similar to the Claude process shown in Fig. 1.47, except that the flow divider and turbine of a Claude process are replaced by a phase separator and a throttling device in a mixed refrigerant cascade (MRC) refrigerator process with a phase separator. The process is also commonly known as the Kleemenko refrigerator process or auto-refrigerant cascade (ARC) process. The phase separator process can be operated in both GRS and LRS modes.

The exergy efficiency of the phase separator process operating in GRS mode, however, will be slightly lower than that of a Linde–Hampson process without any phase separator, because of the exergy loss associated with the mixing of streams 12 and 10 of different compositions.

Most of the temperature drop of the high-pressure stream occurs in the coldest heat exchanger (HX-3). The phase separation temperature in many cases is on the order of 250 to 270 K. The high boilers such as pentane and a significant part of butanes are separated in the phase separator and expanded to temperatures on the order of 230 to 250 K. This results in a much lower flow rate of the refrigerant to heat

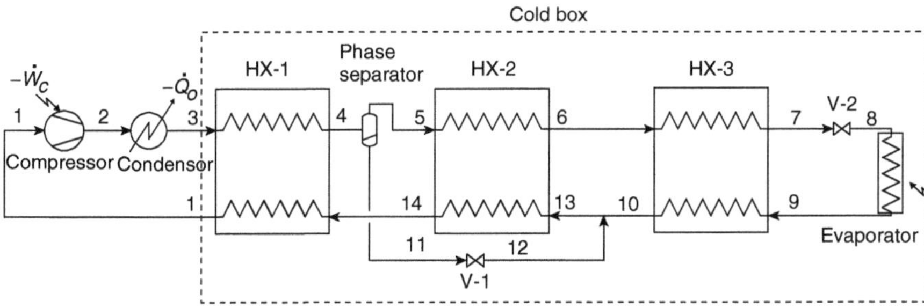

Fig. 4.25. Mixed refrigerant process refrigerator with a phase separator, also known as auto-refrigerant cascade (ARC) or Kleemenko refrigerator.

exchanger HX-3, resulting in smaller equipment as well as decreasing the chance of the high boilers freezing at low temperatures.

The liquid condensate removed from the phase separator can also be subcooled before condensation, as in U.S. Patent 3,203,194 [40], to improve the exergy efficiency slightly.

The greatest advantage in using a phase separator, however, is in the separation of oil along with the high boilers. The improvement in the reliability of the system (oil separation), albeit an increased process complexity, makes the single-phase separator process shown in Fig. 4.25 attractive for large refrigeration applications.

In some cases, it may also be possible to eliminate the warmest heat exchanger (HX-1). The process would then be similar to a Heylandt process. This will, however, result in oil separation at ambient temperature, and not at much lower temperatures, as in the MRC process with three heat exchangers and a phase separator shown in Fig. 4.25. The process with only heat exchangers HX-2 and HX-3 (Fig. 4.25) will therefore be less reliable than a process with phase separation at low temperatures.

A higher heat exchanger area is required in the case of a phase separator process compared to a Linde–Hampson refrigerator operating in the LRS mode due to the change in liquid fraction of the high-pressure fluid from 20–30% for stream 4 to 0% for stream 5 due to the phase separator.

4.7.1 Mixed refrigerant process with a phase separator operating at 100 K

Consider a mixed refrigerant process refrigerator (Fig. 4.25) operating at a temperature of 100 K. The design specifications of the system are shown in Table 4.14. The minimum temperature approach in the heat exchanger has been assumed to be 8 K in all heat exchangers. The performance of the system has been evaluated for a mixture shown in Table 4.15. The composition of the mixture circulating through the compressor is the same as that of stream 4. The details of the pressure, temperature, vapor fraction, and mole flow rate of each stream are shown in Table 4.16.

It can be seen from Table 4.15 that the vapor stream (stream 5) leaving the phase separator and entering the final heat exchanger (HX-3) has a very small quantity of pentane. This ensures that the mixture does not freeze at low temperatures. It can also

Table 4.14. Design specifications for a mixed refrigerant cascade refrigerator with a phase separator shown in Fig. 4.25

Exergy efficiency of compressor and aftercooler, $\eta_{ex,cs}$	100%
Volumetric efficiency of the compressor, η_v	100%
Maximum operating pressure, p_2	20 bar
Maximum operating pressure, p_1	5 bar
Minimum temperature approach in the heat exchanger, ΔT_{min}	8 K
Pressure drop in the heat exchangers, Δp	0
Phase separation efficiency, η_{ps}	100%

Table 4.15. Composition of the refrigerant of the MRC refrigerator with the phase separator shown in Fig. 4.25 at the entry and exit of the phase separator

	Stream		
(mol%)	4	5	11
Nitrogen	35.95	35.58	0.37
Methane	26.77	25.53	1.24
Ethane	30.62	20.10	10.51
*n*Pentane	6.66	0.08	6.58

Table 4.16. Temperature, pressure, vapor fraction, and flow rate of the different streams of the MRC refrigerator with the phase separator in Fig. 4.25

	Stream						
	1	3	4	5	6	7	8
Temperature, K	300.0	300.0	230.0	230.0	195.4	108.1	96.6
Pressure, bar	4.36	20.00	20.00	20.00	20.00	20.00	4.36
Vapor fraction	1.00	0.97	0.81	1.00	0.80	0.00	0.15
Flow rate, mol/s	1.00	1.00	1.00	0.81	0.81	0.81	0.81
	9	10	11	12	13	14	15
Temperature, K	100.0	178.4	230.0	215.2	187.4	213.5	292.0
Pressure, bar	4.36	4.36	20.00	4.36	4.36	4.36	4.36
Vapor fraction	0.29	0.90	0.00	0.16	0.77	0.90	1.00
Flow rate, mol/s	0.81	0.81	0.19	0.19	1.00	1.00	1.00

be observed that a large fraction of ethane is also removed in the liquid phase in the phase separator. Any traces of lubricating oil can also be effectively removed along with the liquid phase in the phase separator.

The exergy input to the cold box of the system ($ex_3 - ex_1$) is used in meeting the refrigeration load (useful effect) and overcoming the losses in the system. Figure 4.26 summarizes the exergy expenditure across different components. As in

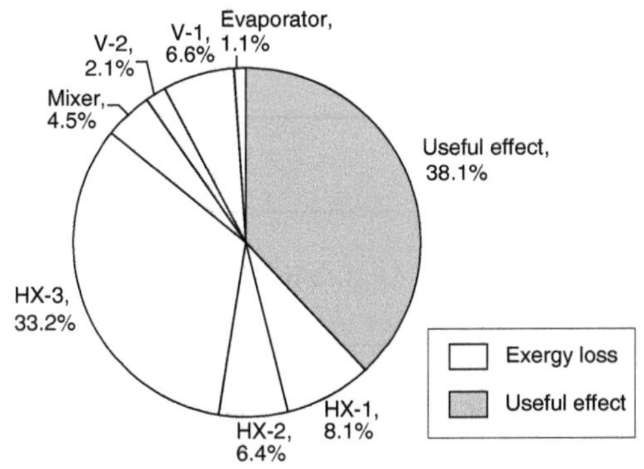

Fig. 4.26. Utilization of the input exergy in the phase separator process shown in Fig. 4.25 and operating with the mixture shown in Table 4.15.

other systems, a significant part of the exergy loss occurs in the heat exchangers. In the present system, there's an additional exergy loss in the mixer due to the large temperature and composition difference between streams getting mixed (streams 10 and 12). About 40% of the exergy loss in the mixer is due to a composition difference and the remaining due to a temperature difference between streams 10 and 12. The break in the temperature of the cold stream shown in Fig. 4.27 is also due to this temperature difference. The temperature profiles of the hot and cold fluid streams are nearly parallel along the length of all the heat exchangers, as shown in Fig. 4.27.

Table 4.17 compares the performance of an LRS refrigerator without phase separator (Fig. 4.3) and that with a phase separator (Fig. 4.25) with a minimum temperature

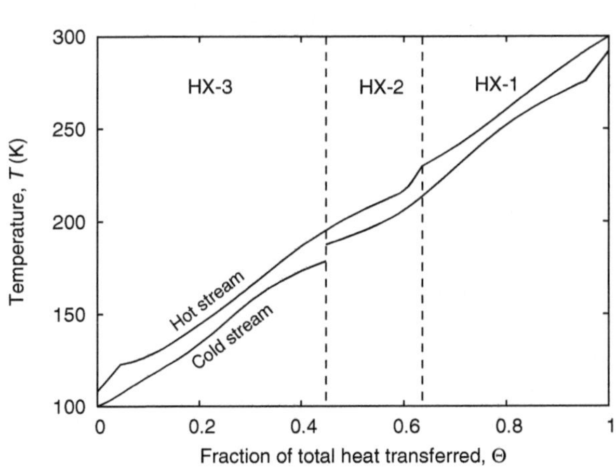

Fig. 4.27. Variation of stream temperature profiles in the different heat exchangers of the phase separator process shown in Fig. 4.25 and Table 4.15.

Table 4.17. Performance of an LRS refrigerator without phase separator (Fig. 4.3) and that with a phase separator (Fig. 4.25) with a minimum temperature approach of 8 K and zero pressure drop in all heat exchangers

	Fig. 4.3 Linde–Hampson LRS	Fig. 4.25 Phase separator
Exergy efficiency, $\eta_{ex,cb}$ (%)	31.2	38.0
Specific refrigeration effect, Δh_{ref} (J/mol)	575.0	693.0
Volumetric cooling capacity, Q_v (J/l)	103.0	123.0

Table 4.18. Composition of refrigerant passing through the compressor in the two cases discussed in Table 4.17

	Mol %	
Component	Linde–Hampson LRS Fig. 4.3	Phase separator Fig. 4.25
Nitrogen	32.6	36.0
Methane	33.3	26.8
Ethane	14.8	30.6
*i*Butane	17.4	0.0
*n*Pentane	1.9	6.7

approach of 8 K and zero pressure drop in all heat exchangers. The composition of the refrigerant circulating in the two systems is shown in Table 4.18. Only a small amount of pentane can be used in a Linde–Hampson LRS system due to the possibility of freezing at low temperature. On the other hand, a larger amount of pentane can be used in the phase separator system since most of the pentane is returned to the compressor at relatively high temperature (230 K) from the phase separator (see Table 4.15). The use of a larger amount of pentane (higher μ_{J-T} at ambient temperature) results in a higher exergy efficiency and higher refrigeration effect in the case of the phase separator processes (Table 4.17).

Phase separator systems are therefore superior to systems without phase separators both from oil separation and efficiency points of view. Phase separators are used in most mixed refrigerant natural gas liquefaction processes (see Chapter 6).

4.7.2 Effect of separation efficiency

The performance of the phase separator process shown in Fig. 4.25 depends strongly on the efficiency of the phase separators employed or the amount of liquid that is entrained by the vapor stream leaving the phase separator. The phase separator efficiency is defined as follows:

$$\eta_{ps} = \frac{\dot{n}_{11}}{\dot{n}_4(1 - x_4)}, \tag{4.17}$$

where $\dot{n}$ refers to the mole flow rate and x refers to the vapor fraction at the inlet.

The performance of the phase separator process shown in Fig. 4.26 operating with the mixture shown in Table 4.15 and different phase separator efficiencies is summarized in Tables 4.19 and 4.20. Both the exergy efficiency and the refrigeration produced decrease with a decrease in the separation efficiency. As the separation efficiency decreases, the amount of nPentane concentration needs to be decreased to avoid freezing at low temperature. nPentane may need to be completely replaced by nButane below a certain efficiency, as shown in Table 4.20.

The amount of nButane used was chosen such that the freezing point of nButane is at least 5 K lower than the operating temperature (100 K). As in previous LRS examples, the freezing point was evaluated using the method described on page 103. In spite of the uncertainty in the freezing point, the examples do demonstrate the drop in the exergy efficiency of the cold box with a drop in separation efficiency.

Table 4.19. Effect of liquid separation efficiency of the phase separator on the overall performance of a phase separator refrigeration system shown in Fig. 4.25

	Case 1	Case 2	Case 3
Liquid separation efficiency, η_{ps} (%)	100	90	75
Exergy efficiency, $\eta_{ex,cb}$ (%)	38	36	33
Specific refrigeration effect, Δh_{ref} (J/mol)	693	630	543
Volumetric cooling capacity, Q_v (J/l)	123	117	111

Table 4.20. Optimum mixture composition at the inlet and exit of the phase separator of the system shown in Fig. 4.25 for different phase separator efficiencies

	Mol%								
	$\eta_{ps} = 100\%$			$\eta_{ps} = 90\%$			$\eta_{ps} = 75\%$		
Stream	4	5	6	4	5	6	4	5	6
Nitrogen	35.95	35.58	0.37	34.77	34.47	0.29	35.38	35.10	0.28
Methane	26.77	25.53	1.24	28.59	27.71	0.88	28.74	27.89	0.85
Ethane	30.62	20.10	10.51	20.11	16.46	3.65	19.28	15.81	3.47
nButane	–	–	–	16.01	4.15	11.85	16.31	5.71	10.60
nPentane	6.66	0.08	6.58	0.53	0.08	0.45	0.30	0.08	0.21

4.8 Mixed refrigerant process refrigerators with multiple phase separators

Mixed refrigerant process refrigerators with a phase separator can be extended to multiple phase separators, as shown in Fig. 4.28. MRC refrigerators with two phase separators are somewhat similar to the Collins liquefaction process shown in Fig. 1.48, except that the stream divider and turbine combinations in the Collins process are replaced by a phase-separator-throttle combination in MRC refrigerators with phase separators.

The phase separator efficiency can range from 50% for low-efficiency systems to about 99% for high-efficiency systems. Sometimes a complete removal of lubricating oil or high-boiling components is not achievable when only one phase separator with low separation efficiency is used. Multiple phase separators are quite attractive to use in such cases.

Missimer [59] states in his patent that in MRC refrigerators with multiple phase separators, the lowest temperature is sometimes reached in one of the heat exchangers rather than in the evaporator. This happens when the liquid fraction at the inlet of the throttle is insufficient. Missimer notes that this situation can be corrected by further addition of more of the lowest boiling point refrigerant. However, this upsets the required temperature as well as the balance required for startup. Missimer's invention essentially was to replace the phase separators with bypass circuits, as shown in Fig. 4.29. According to the patent, not only does this eliminate the problem mentioned above, but it also reduces the cooldown time substantially.

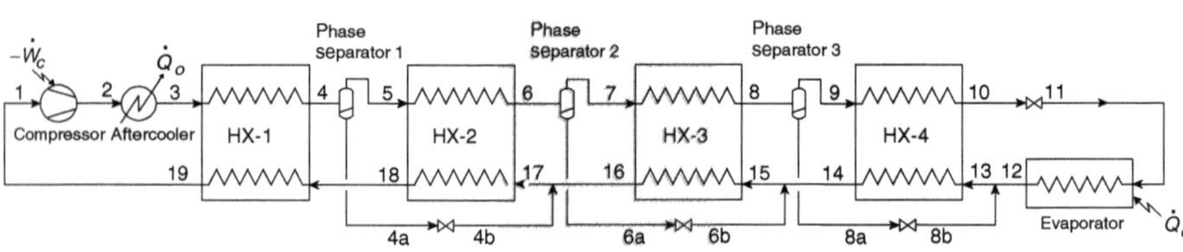

Fig. 4.28. Mixed refrigerant cycle refrigerator with multiple phase separators.

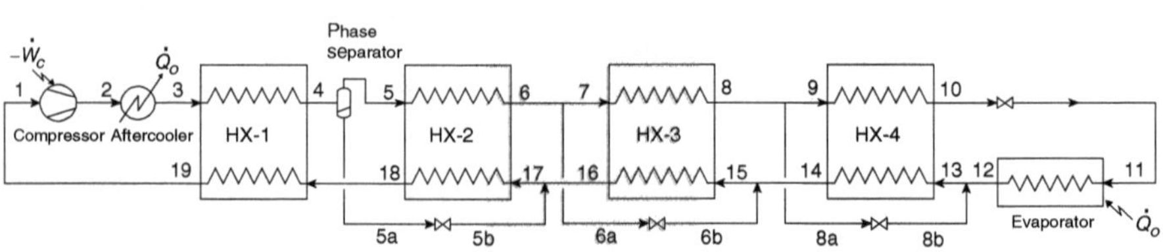

Fig. 4.29. Mixed refrigerant cycle refrigerator with bypass circuits [59].

4.9 Summary

Mixed refrigerant processes can be largely divided into two groups: (1) gas refrigerant supply (GRS) systems in which no condensation of the refrigerant occurs in the aftercooler/condesner and (2) liquid refrigerant supply (LRS) systems in which the refrigerant condenses partially in the condenser. The exergy efficiency of LRS systems is higher than that of corresponding GRS systems due to the higher enthalpy change in the condenser/aftercooler, for the same compressor power input. Several practical problems, however, need to be solved to make the LRS systems operate efficiently (see page 105). LRS systems are therefore preferable only for large refrigeration loads.

Prevention of compressor lubricant oil from getting carried over to the low-temperature region of the system is the most critical aspect of the design of a Linde–Hampson refrigeration system operating with mixtures in LRS mode. Many of the oil separation problems can be overcome by using precooled and phase separator systems.

Single-stage Linde–Hampson refrigeration systems operating in GRS mode with refrigerant mixtures are superior to phase separator systems operating in GRS mode from an exergy efficiency point of view since additional exergy losses are associated with mixing of the liquid and vapor phases separated in the phase separator. Phase separator systems are superior from an oil separation point of view since any lubricating oil carried over from the compressor is likely to be separated in the phase separator. Phase separator process refrigerators operating in LRS mode are superior to Linde–Hampson refrigerators from both exergy efficiency and lubricant oil separation points of view since a higher concentration of high boilers can be used in the mixture in the case of the former.

The choice of a process depends not only on the exergy efficiency but also on the method used to separate compressor lubricating oil from the mixture and the refrigeration load.

5

Optimum mixture composition

5.1 Choice of mixture constituents

The exergy efficiency of any mixed refrigerant process depends on the mixture's constituents and their concentration. The exergy efficiency of MRC refrigerators will be high when a second liquid phase occurs in the evaporator. Liquid-liquid immiscibility is observed at low temperatures in multicomponent mixtures of nitrogen–hydrocarbon, fluorocarbon–hydrocarbon, fluorocarbon–hydrochlorofluorocarbon and fluorocarbon–hydrofluorocarbon refrigerants [34, 55, 87, 89]. This immiscibility can be exploited to obtain a near-constant temperature evaporation with mixtures, as shown in Figs. 3.4 and 3.6 with both binary and multicomponent mixtures. The liquid-liquid immiscibility also allows us to reach temperatures close to the boiling point of the low boiler in the mixture (see inset in Fig. 3.4).

Consider a refrigerator operating with nitrogen-hydrocarbon mixtures. The power drawn by the compressor is related to the flow rate or the refrigerant pressure at compressor suction. Most off-the-shelf domestic refrigeration and air conditioning compressors are designed to operate below 5–6 bar to limit the amount of current drawn by the compressor motor. Applying the 5-bar limit, one can realize that the maximum evaporating temperature is 94 K in the case of nitrogen-hydrocarbon mixtures (Fig. 5.1). Argon-hydrocarbon mixtures can be used for higher refrigerating temperatures. Because of the lower vapor pressure of argon, the volumetric cooling capacity would always be lower in the case of argon-hydrocarbon mixtures than nitrogen-hydrocarbon mixtures at refrigeration temperatures lower than 94 K, as is the case with Linde–Hampson refrigerators operating with pure argon and nitrogen. In most cases, the upper limit for nitrogen-hydrocarbon mixtures to obtain a near-constant temperature evaporation is about 104 K when the pressure drop in the heat exchangers and the difference between the boiling point of nitrogen and the bubble point of nitrogen-hydrocarbon mixtures are also taken into account. Argon-hydrocarbon mixtures need to be used above this temperature. Nitrogen-hydrocarbon mixtures can be used at much higher temperatures if a small temperature change is tolerated in the evaporator.

G. Venkatarathnam, *Cryogenic Mixed Refrigerant Processes*,
DOI: 10.1007/978-0-387-78514-1_5,

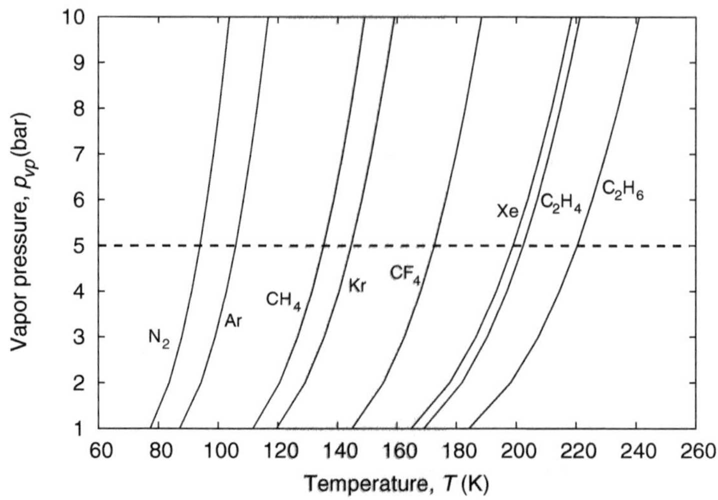

Fig. 5.1. Vapor pressure of different fluids.

Temperatures lower than the boiling point of the low boiler (for example, nitrogen) can also be obtained by adding a noncondensable fluid such as helium, hydrogen, or neon. These noncondensable fluids, however, inhibit the formation of liquid-liquid immiscibility at low temperature (see Fig. 4.16). Propane freezes below a temperature of 69 K in most nitrogen-helium-methane-ethane-propane mixtures. A similar limit is observed in mixtures where neon is used instead of helium. High boilers such as isobutane and pentane freeze at much higher temperatures. The lowest operating temperature is thus decided by the concentration of the high boilers as well as by the compressor lubricating oil in the refrigerant.

Mixtures of tetrafluoromethane (refrigerant R14), trifluoromethane (refrigerant R23), and butane or propane show liquid-liquid immiscibility at low temperatures and can be used for temperatures above 145 K [55]. Liquid-liquid immiscibility is also observed in mixtures of refrigerants R14-R23-R22 [55]. Refrigerant R22, however, is a hydrochlorofluorocarbon and cannot be used in new refrigerators in many countries. Refrigerant R22 can be replaced by one or more hydrofluorocarbon refrigerants such as R218, R125, R143a, R134a, R227ea, RC318, etc. Temperatures below 145 K can be obtained by adding a noncondensable fluid such as argon, nitrogen, neon, or helium to fluorocarbon mixtures of R14-R23-R22, etc.

The method for determining the most basic components of a nitrogen-hydrocarbon mixture was first given by Alfeev et al. in their patent [7]. These principles can be extended to other fluid mixtures also. The guidelines for choosing the components of a mixture are as follows:

- Choose a first fluid whose boiling point temperature at 1.5 bar is less than the desired refrigerating temperature. For example, nitrogen can be used for temperatures between 80 and 105 K, tetrafluoromethane (Refrigerant R14) between 150 K and 180 K). A mixture of nitrogen and argon can be used between 105 and 140 K, and argon between 120 and 150 K, etc.

- Choose a second fluid whose boiling point is about 30 to 60 K above that of the basic fluid and that does not exhibit liquid-liquid immiscibility at low temperature with the primary fluid. For example, one can choose methane with argon and nitrogen, trifluoromethane (R23) with tetrafluoromethane (R14), etc.
- Choose a third fluid that exhibits a liquid-liquid immiscibility at low temperature with the first fluid and whose boiling point is at least 30 K above that of the second fluid, for example ethane, ethylene, etc., which exhibit liquid-liquid immiscibility at low temperatures with nitrogen. Ethylene also exhibits a liquid-liquid immiscibility at low temperatures with argon [87]. Propane, butanes, and chlorodifluoromethane (R22) exhibit a liquid-liquid immiscibility with R14 (tetrafluoromethane) [55].
- Choose a fourth fluid that exhibits a liquid-liquid immiscibility at low temperatures with the first fluid, for example, propane or butanes with nitrogen.
- Choose an optional fifth fluid that exhibits a liquid-liquid immiscibility at low temperatures with the first fluid, for example, pentanes with nitrogen.

A minimum quantity of third and fourth fluids is required for the multicomponent mixture to exhibit a liquid-liquid immiscibility at low temperatures. It is also possible to use fluids that do not exhibit liquid-liquid immiscibility at low temperatures with the first fluid as the components of the mixture. For example, a mixture of argon, methane, ethane, and propane does not exhibit liquid-liquid immiscibility at low temperatures. However, the bubble point of the mixture will be substantially higher than that of the first fluid. The refrigeration temperature of such a mixture can be decreased by adding a noncondensable fluid, for example, neon, or helium in the case of argon-hydrocarbon mixtures, or nitrogen or argon in the case of fluorocarbon mixtures. It is also advantageous to use noncondensable fluids with nitrogen-hydrocarbon mixtures (see Section 4.5).

Some authors [57] have advocated the use of more than five components. The choice of the optimum mixture components and concentration, however, can only be determined using an optimization model and a process simulation program.

5.2 Optimization of mixture composition for refrigeration processes

5.2.1 Optimization methods proposed in the literature

A number of methods have been described in the open literature and in patents to choose the mixture composition used in refrigerators. Little [57] describes a method for liquid refrigerant supply (LRS) systems in which the refrigerant mixture condenses partially in the condenser or aftercooler. Little suggests that the constant-pressure lines corresponding to compressor suction and discharge pressures must be parallel on a *T-h* diagram in the two-phase region, as shown in Fig. 5.2, to obtain high exergy efficiency.

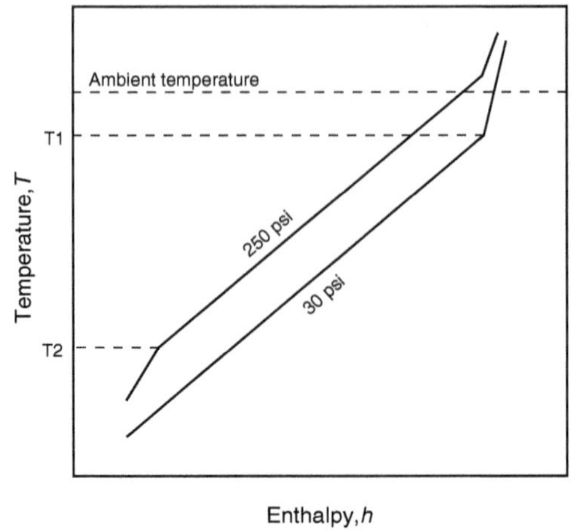

Fig. 5.2. Desired variation of enthalpy with temperature for optimized mixtures proposed by Little in his U.S. patent [57].

Little suggests that the above condition can be obtained by maximizing the following function:

$$\text{maximize} \quad f(\xi_i) = \frac{\Delta h_{\min}}{|c| + \text{variance}\ (\Delta h)}, \tag{5.1}$$

where c is a constant, and the variance Δh is the variance of the enthalpy difference between the two curves in the two-phase region between temperatures T1 and T2. ξ_i is the mole fraction of the different components of the refrigerant mixture. Little suggests the use of NIST property programs such as REFPROP, DDMIX, SUPERTRAPP, etc. to determine the enthalpy of mixtures.

The exergy efficiency of the system depends on the enthalpy difference as well as the entropy difference between the low- and high-pressure fluids, as shown in Eq. (4.5). The method of Little described above, however, considers only the enthalpy of the mixture for optimizing the mixture composition. The operating pressures of the system (compressor suction and discharge pressures) are assumed in the above method. We will show in the following pages that it is beneficial to determine the operating pressures along with the mixture composition simultaneously. Little suggested the use of a mixture of propane, nButane, the refrigerants R14, R23, R123, R134a, and the atmospheric gases nitrogen and argon. The binary interaction parameters used in the mixing rules of different equations of state are not available in the open literature for the refrigerant combinations Little used. The efficacy of this method can only be determined when the binary interaction parameters become available in the open literature. Little's method is applicable only for refrigeration systems operating in LRS mode with partial condensation in the condenser/aftercooler.

Alexeev and Quack [6] reported mixtures for a precooled refrigerator operating with a precooling temperature of 233 to 243 K and a refrigeration temperature of 90 to 110 K. The specific refrigeration effect ($\Delta h_{\min}$) is maximized in their method. As in the case of Little [57], Alexeev and Quack [6] also assume the operating pressures. A number of examples of optimum mixture compositions have been presented for a precooling temperature of 243 K. The highest specific refrigeration effect ($\Delta h_{\min}$) in the examples presented in Ref. [6] is 1514 J/mol, with compressor discharge and suction pressures of 16 and 1.6 bar, respectively. The minimum temperature approach in the heat exchanger has not been stated in Ref. [6].

The performance of a Linde–Hampson system has been reestimated with a mixture that has the highest specific refrigeration effect in Ref. [6], but with a minimum temperature approach of 0.1 K. Table 5.1 shows the comparison of the best mixture in the patent of Alexeev and Quack [6] with mixtures optimized by considering both the suction and discharge pressures as design parameters along with mixture composition. The volumetric efficiency of the compressor has been assumed to be 100% for all cases.

It can be seen that the volumetric cooling capacity in the first case (Case 1) is about half of that in the other two cases (Case 2 and Case 3). The same compressor discharge pressure is used in Case 1 and Case 2, and the same compressor suction pressure in Cases 2 and 3. The compression ratio is 10 in Case 1, compared to 3.2 in Cases 2 and 4 in Case 3. The volumetric efficiency of a compressor operating with the mixture in Case 1 will be considerably smaller than that in Cases 2 and 3 because of the high compression ratio. The amount of refrigeration obtained with mixtures in Case 2 and Case 3 will, therefore, be many times that in Case 1 when the volumetric efficiency of the compressor is also considered. Mixtures in Cases 2 and 3 are therefore superior for a given compressor size, in spite of having a lower specific refrigeration effect

Table 5.1. Comparison of performance of a precooled Linde–Hampson system operating at 100 K with different mixtures and a precooling temperature of 243 K

	Alexeev and Quack [6]	Method in Section 5.2.2	
	Case 1	Case 2	Case 3
Mixture composition			
Nitrogen (mol%)	35.9	36.2	39.3
Methane (mol%)	31.9	30.1	27.6
Ethane (mol%)	18.2	0.0	0.0
Ethylene (mol%)	0.0	17.3	18.1
Propane (mol%)	14.0	16.4	15.0
Compressor suction pressure, p_1 (bar)	1.6	5.0	5.0
Compressor discharge pressure, p_2 (bar)	16.0	16.0	20.0
Specific refrigeration effect, $\Delta h_{\min}$ (J/mol)	1825	1113	1308
Volumetric cooling capacity, Q_v (J/l)	152	299	353

* The specific refrigeration effect was estimated in Case 1 assuming a temperature approach of 0.1 K between the streams in the heat exchanger.

compared to that in Case 1. The above example clearly demonstrates the importance of varying the compressor pressures during the optimization of mixture composition.

Small compressors are available only in specific sizes commercially. The following objective function is recommended to maximize the cooling capacity of a refrigerator operating with a given compressor size:

$$\text{maximize } f(\xi_i, p_2, p_1) = \frac{\dot{Q}}{\dot{V}_c} = \rho_1 \Delta h_{\text{min}} \eta_v, \tag{5.2}$$

where $\dot{V}_c$ represents the displacement rate of the compressor, ρ_1 the refrigerant density at compressor suction, Δh_{min} the specific refrigeration effect, and η_v the volumetric efficiency of the compressor.

When the compressor volumetric efficiency characteristics of the compressor are not known a priori, the above function can be used with a volumetric efficiency of 100%.

Dobak et al. [32] presented a method for choosing the mixture composition for refrigerators to be used as catheters or cryosurgical devices. In their patent, they use the following function to determine the mixture composition:

$$\text{maximize } f(\xi_i) = \frac{\dot{Q}_{\text{min}}}{\dot{Q}_{\text{hx, max}}} = \frac{\Delta h_{\text{min}}}{\Delta h_{\text{hx, max}}}, \tag{5.3}$$

where $\Delta h_{\text{hx, max}}$ is the maximum heat load in the heat exchanger and Δh_{min} is the specific refrigeration effect. ξ_i is the mole fraction of the components in the mixture. The operating pressures in the system are assumed as 1 and 21 bar, respectively, and the temperature approach between the streams is assumed to be zero.

Gong et al. [48] presented a method for the selection of mixture composition for a Linde–Hampson refrigerator. The following function is maximized in their method for a chosen compressor suction and discharge pressure:

$$\text{maximize } f(\xi_i) = \frac{\Delta h_{\text{min}}}{\text{specific compressor work}} \left(\frac{T_o}{T} - 1 \right). \tag{5.4}$$

When the compressor efficiencies are considered, the above equation reduces to the following:

$$\text{maximize } g(\xi_i) = \frac{\Delta h_{\text{min}}}{\text{specific compressor work}} \left(\frac{T_o}{T} - 1 \right) \eta_v \eta_m \eta_{fc}. \tag{5.5}$$

where η_v, η_m, and η_{fc} refer to the volumetric efficiency, motor efficiency, and frictional efficiency of the compressor, respectively. The specific compressor work is the work supplied per unit mass flow rate assuming adiabatic compression. The minimum temperature approach between the streams, the operating pressures of the system, as well as the ambient temperature are design constants. The function $f(\xi_i)$ in Eq. (5.4) is nothing but the exergy efficiency of the refrigerator.

Boiarskii et al. [18] presented a method for the optimization of mixture composition used in Linde–Hampson refrigerators. They defined two different objective functions for optimizing the mixture, based on the cooling capacity of the refrigerator ($\dot{Q}$), the desired cooling ($\dot{Q}_{\text{specified}}$), the power drawn by the compressor ($-\dot{W}_c$), and the maximum power that can be drawn by the compressor motor ($-\dot{W}_{c,\max}$) as follows:[1]

$$\text{maximize } \dot{Q} \quad \text{if } -\dot{W}_c < -\dot{W}_{c,\max}, \tag{5.6}$$

$$\text{minimize } -\dot{W}_c \quad \text{if } \dot{Q} < \dot{Q}_{\text{specified}}. \tag{5.7}$$

Unlike other authors, Boiarskii et al. [18] determined the optimum operating pressures along with the optimum mixture composition simultaneously.

5.2.2 Maximization of exergy efficiency

Since cryogenic refrigeration processes are very energy-intensive, it is ideal to minimize the compressor power input or maximize the exergy efficiency of the process. The two proposals of Boiarskii et al. [Eqs. (5.6) and (5.7)] for different conditions can actually be combined into a single condition that maximizes the exergy efficiency (η_{ex}) of the refrigerator while keeping the maximum power drawn by the compressor below the required limit and satisfying the minimum cooling capacity required, as follows:

$$\text{maximize } \eta_{\text{ex}} = \frac{\dot{Q}}{-\dot{W}_c}\left(\frac{T_o}{T} - 1\right) \tag{5.8}$$

subject to

$$-\dot{W}_c \leq -\dot{W}_{c,\max}, \tag{5.9}$$

$$\dot{Q} \geq \dot{Q}_{\text{specified}}. \tag{5.10}$$

The composition of the mixture and the operating pressures are varied simultaneously to determine the maximum exergy efficiency (η_{ex}). The minimum temperature approach in the heat exchanger ($\Delta T_{\min}$) is specified as a constraint.

The method for choosing the optimum mixture can be expressed mathematically as follows:

Objective function:

maximize exergy efficiency (η_{ex}).

---

[1] $-\dot{W}_c > 0$.

Constraints:

- Power drawn by compressor should be less than the maximum power limit of the compressor (motor):

$$-\dot{W}_c < -\dot{W}_{c,\,\max}.$$

- The temperature approach in the heat exchanger should be less than that specified:

$$\Delta T_{\min} \geq \Delta T_{\min,\,\text{specified}}.$$

- The dew point temperature of the high-pressure refrigerant should be less than the ambient temperature in the case of a GRS system and more than the ambient temperature in the case of an LRS system (see section 4.1):

$$T_{\text{dew, hp}} < T_o \quad \text{(GRS)},$$
$$T_{\text{dew, hp}} > T_o \quad \text{(LRS)}.$$

- The pressure drop in the high-pressure refrigerant channel of the heat exchanger should be less than that specified:

$$\Delta p_{\text{hx, hp}} < \Delta p_{\text{hx, hp, specified}}.$$

- The pressure drop in the low-pressure refrigerant channel of the heat exchanger should be less than that specified:

$$\Delta p_{\text{hx, lp}} < \Delta p_{\text{hx, lp, specified}}.$$

5.2.3 Design variables

- mole fraction of different components of the mixture such as nitrogen, methane, ethane, propane, etc. (ξ_i, i = 1 to n)
- refrigerant pressure at compressor exit (p_2)
- refrigerant pressure at compressor inlet (p_1)

5.2.4 Example: Linde–Hampson refrigerator operating in GRS mode at 92 K with a mixture of nitrogen, methane, ethane, and propane

Consider a Linde–Hampson refrigerator operating in GRS mode (Fig. 5.3). The refrigerator needs to provide refrigeration at a near-constant temperature of 92 K. The components of the mixture have been chosen as nitrogen, methane, ethane, and propane. The minimum temperature approach in the heat exchanger has been chosen as 5 K. The efficiency of the compressor motor varies with load, with partial load efficiencies being much smaller than full-load efficiency in fractional horsepower motors used in small refrigerator compressors. In order to simplify this analysis, the exergy efficiency of the compressor has been assumed to be constant and 100%. Similarly, the volumetric efficiency of the system varies with the pressure ratio. In order to simplify the

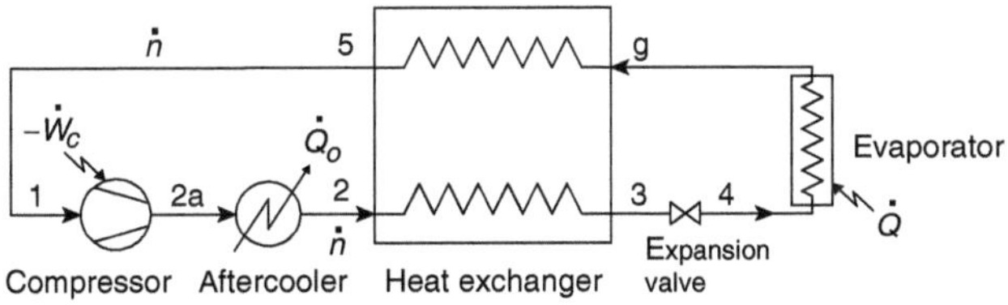

Fig. 5.3. Mixed refrigerant Linde–Hampson refrigerator operating in GRS mode.

analysis, the volumetric efficiency of the compressor has also been assumed to be 100%. The assumptions are summarized in Table 5.2.

With the compressor exergy efficiency assumed to be 100%, only the heat exchanger, expansion valve, and evaporator need to be simulated to determine the exergy efficiency for a given mixture composition (ξ_i) and operating pressures (p_1 and p_2). As mentioned in Section 2.4, the temperature of the high-pressure stream leaving the heat exchanger (T_3) needs to be varied to meet the minimum temperature approach condition in the heat exchanger. Temperature T_3 therefore needs to be included as a design variable, along with the mole fraction of nitrogen, methane, ethane, and propane, and pressures p_1 and p_2. The objective function, constraints, and design variables are summarized in Table 5.3. The simulation procedure described in Section 2.4 was used to simulate the heat exchanger.

Table 5.4 shows the lower and upper limits of the design variables and the initial estimated values used for different variables. The final optimum solution obtained at the end of the convergence is also shown. The solution was obtained using a common nonlinear optimization method known as the Sequential Quadratic Programming (SQP) method. The enthalpy and entropy (exergy) at different points were obtained using the Peng–Robinson equation of state. Figure 5.4 shows the variation of design variables and the objective function (exergy efficiency) during the optimization process using the ASPEN Plus process simulator. The optimum values are reached in fewer than 25 iterations.

The temperature, pressure, and vapor fraction at different points of the process (Fig. 5.3) are also shown in Table 5.5. The temperature change in the evaporator is limited to 0.9 K, due to the liquid-liquid immiscibility observed in the chosen mixture at low temperature.

Table 5.2. Design specifications for a mixed refrigerant Linde–Hampson refrigerator operating at 92 K in GRS mode

Exergy efficiency of compressor and aftercooler, $\eta_{\mathrm{ex,cs}}$	100%
Volumetric efficiency of the compressor, η_v	100%
Maximum operating pressure, p_2	20 bar
Minimum temperature approach in the heat exchanger, $\Delta T_{\min}$	5 K
Pressure drop in the heat exchanger, Δp	0

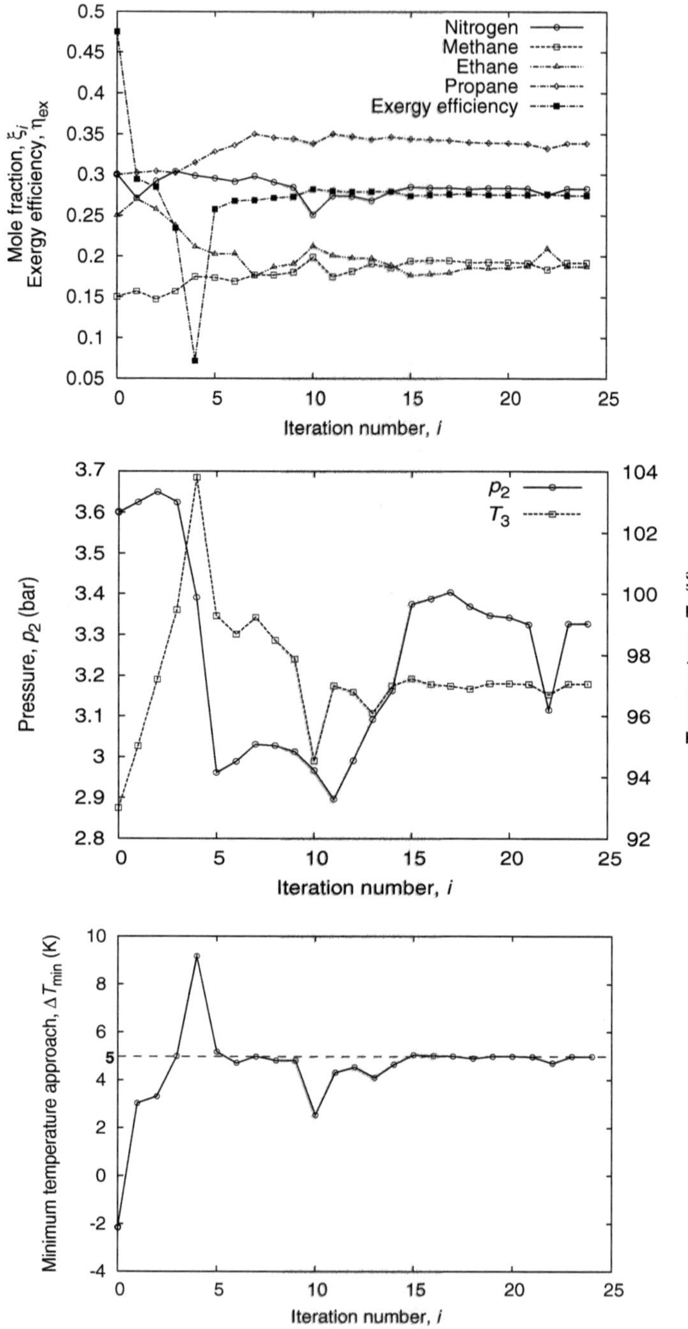

Fig. 5.4. Variation of design variables and the objective function (exergy efficiency) during the optimization of the Linde–Hampson refrigeration system operating in GRS mode at 92 K.

Table 5.3. Method for the optimization of mixture composition and operating pressures of a Linde–Hampson refrigerator operating in GRS mode and specified in Table 5.2

Objective function: Maximize exergy efficiency (η_{ex})

Subject to constraints:

- Minimum temperature approach between the streams, $\Delta T_{min} \geq \Delta T_{min,\,specified} = 5$ K
- No condensation should occur in the aftercooler, or $T_{dew,\,2} < T_o$ (300 K)

Design variables:

- Mole fraction of nitrogen (ξ_{N_2})
- Mole fraction of methane (ξ_{CH_4})
- Mole fraction of ethane ($\xi_{C_2H_6}$)
- Mole fraction of propane ($\xi_{C_3H_8}$)
- Compressor discharge pressure (p_2)
- Compressor suction pressure (p_1)
- Temperature of high-pressure refrigerant leaving the heat exchanger (T_3)

Constants:

- Pressure drop in the high-pressure refrigerant channel of the heat exchanger ($\Delta p_{hx,\,hp}$) is zero.
- Pressure drop in the low-pressure refrigerant channel of the heat exchanger ($\Delta p_{hx,\,lp}$) is zero.

Table 5.4. Lower limit, upper limit, initial estimate, and final solution obtained for different design variables of the Linde–Hampson refrigerator operating in GRS mode at 92 K (Fig. 5.3)

Design variable	Lower limit	Upper limit	Initial estimate	Optimal solution
Nitrogen mole fraction	0.0	0.40	0.30	0.280
Methane mole fraction	0.0	0.25	0.15	0.192
Ethane mole fraction	0.0	0.50	0.25	0.187
Propane mole fraction	0.3	0.35	0.30	0.338
Pressure, p_2	16.0	20.0	20.0	20.0
Pressure, p_1	1.5	4.0	3.6	3.327
Temperature, T_3	82.0	110.0	93.0	97.1

Table 5.5. Temperature and pressure of different streams of Fig. 5.3 while operating with optimum mixture composition and operating pressures (Table 5.4)

	Stream				
	2	3	4	g	5
Temperature, K	300.0	97.1	91.1	92.0	295.0
Pressure, bar	20.00	20.00	3.327	3.327	3.327
Vapor fraction	1.000	0.000	0.092	0.184	1.000

Table 5.6. Performance of a Linde–Hampson refrigerator operating at 92 K with optimum mixture composition (Table 5.4)

Performance parameter	Value
Exergy efficiency, $\eta_{\mathrm{ex,cb}}$ (%)	27.8
Specific refrigeration effect, $\Delta h_{\min}$ (J/mol)	513.0
Volumetric cooling capacity, Q_v (J/l)	71.0

Table 5.6 shows the performance parameters such as exergy efficiency, specific refrigeration effect, and volumetric cooling capacity, assuming that the volumetric efficiency and exergy efficiency of the compressor is 100%. The exergy efficiency of most compressors and aftercoolers (hot section) is normally about 35 to 40% in small refrigeration systems, and the volumetric efficiency is on the order of 60–75%. The next example demonstrates the use of the volumetric efficiency characteristics of a compressor in the optimization of the mixture composition and operating pressures of the compressor.

5.3 Example: Linde–Hampson refrigerator operating in GRS mode at 80 K

Consider a Linde–Hampson refrigerator operating in GRS mode with a compressor whose volume displacement rate is 28 l/min or 1 cu.ft/min (CFM). A heat load of 6 W needs to be met at 80 K. The following assumptions are made for the sake of simplicity.

Assumptions

- The exergy efficiency of the compressor and aftercooler is 35% and is independent of the flow rate through the compressor.
- The volumetric efficiency of the compressor is related to the compression ratio (p_2/p_1) as follows [16]:

$$\eta_v = 0.89 - 0.04\left(\frac{p_2}{p_1}\right). \tag{5.11}$$

- The high-pressure refrigerant is cooled to the ambient temperature (300 K) in the aftercooler.
- The pressure drop in the heat exchanger is negligible.

Consider two cases: (1) when the heat load to be met ($\dot{Q}_{\mathrm{specified}}$) is 3 W and (2) when the heat load to be met is 6 W. The design procedure followed is similar to that summarized in Table 5.3, except for an additional constraint:

$$\dot{Q} \geq \dot{Q}_{\mathrm{specified}}. \tag{5.12}$$

Additionally, neon is also used along with nitrogen, methane, ethane, and propane. The results obtained with the optimization algorithm for both the 3 W and 6 W cases are shown in Table 5.7. The evaporating pressure is only 1.162 bar when no neon is used. The neon concentration is zero when the heat load to be met is 3 W and 5.2 mol% when the heat load to be met is 6 W. The optimizer increases the neon concentration until the desired heat load can be met. The addition of neon results in a higher evaporating pressure (1.54 bar). The volumetric efficiency nearly doubles when neon is added, due to a 25% decrease in the compression ratio (p_2/p_1), making it possible to meet a heat load of 6 W.

It is evident from the above example that the optimization method described in Table 5.3 automatically adjusts the concentration of neon to meet the required heat loads and also achieves as high an exergy efficiency as possible.

The above example also shows that the concentration of some of the components will be zero in an optimum mixture when a good optimization algorithm is used. Ideally, one can start with a large number of possible mixture components as design variables. With a good algorithm, the concentration of some of these will become zero. However, the computation time will increase considerably with an increase in the number of design variables.

The addition of higher boilers such as pentane and butanes is beneficial in the case of LRS systems in which partial condensation of the refrigerant occurs in the aftercooler. Some of these high boilers can freeze at low temperature. Additional constraints on the freezing point of the high boilers can be used to limit the concentration of the high boilers in the mixture.

Table 5.7. Performance of a Linde–Hampson refrigerator operating at 80 K with a 28 l/min compressor with and without neon

	Without neon	With neon
Exergy efficiency, η_{ex} (%)	27.9	24.5
Refrigeration capacity, Q (W)	3.1	6.0
Specific refrigeration effect, Δh_{min} (J/mol)	691.0	547.0
Volumetric cooling capacity, Q_v (J/l)	33.0	35.0
Volumetric efficiency of compressor, η_v (%)	20.0	37.0
Mixture composition		
Nitrogen (mol%)	27.5	27.4
Methane (mol%)	13.1	15.6
Ethane (mol%)	24.5	16.7
Propane (mol%)	35.0	35.0
Neon (mol%)	–	5.2
Pressure, p_2 (bar)	20.00	20.00
Pressure, p_1 (bar)	1.162	1.541

5.4 Comparison of performance of a Linde–Hampson refrigerator operating in GRS mode at 92 K with mixtures obtained using the method of Dobak et al. [32] and the present method

Consider the optimization method proposed by Dobak et al. in their patent [32]. The results obtained when the objective function $f(\xi_i)$ in Eq. (5.3) is used to optimize the mixture composition and the operating pressures simultaneously are shown in Table 5.8. The characteristics of the compressor made in the previous example are also applicable to this example (see Section 5.3). The displacement volume of the compressor has been assumed to be 28 l/min in this example also. The ratio of the cooling capacity of the refrigerator to the heat exchanger heat load ($\dot{Q}/\dot{Q}_{hx}$) is maximized in Eq. (5.3).

It can be seen from Table 5.8 that the exergy efficiency (η_{ex}) and volumetric cooling capacity (Q_v) are higher with the present method. The ratio $\dot{Q}/\dot{Q}_{hx}$ is larger with the present method compared to that obtained with the Dobak et al. method in which $\dot{Q}/\dot{Q}_{hx}$ is maximized. This is essentially due to a smaller volumetric efficiency of the compressor when the Dobak et al. method is used. The refrigeration obtained is also much smaller when the Dobak et al. method is used even when the same compressor is used due to a smaller density (pressure) of the refrigerant at the compressor inlet. If the compressor suction pressure is assumed to be constant (1 bar) as originally assumed by Dobak et al., the refrigeration obtained would be even smaller. The method proposed

Table 5.8. Performance of a Linde–Hampson refrigerator operating at 92 K in GRS mode with mixtures optimized using algorithm proposed in Eqs. (5.8)–(5.10) and that proposed by Dobak et al. [Eq. (5.3)] [32]

	Present method Eq. (5.8)	Dobak et al. [32] Eq. (5.3)
Exergy efficiency, η_{ex} (%)	27.8	25.5
Refrigeration capacity, $\dot{Q}$ (W)	18.2	6.8
Specific refrigeration effect, Δh_{min} (J/mol)	513.0	666.0
Volumetric cooling capacity, Q_v (J/l)	71.0	41.0
Volumetric efficiency of compressor, η_v (%)	55.0	36.0
$\dot{Q}/\dot{Q}_{hx}$	0.013	0.0076
Mixture composition		
Nitrogen (mol%)	0.275	0.258
Methane (mol%)	0.131	0.147
Ethane (mol%)	0.245	0.268
Propane (mol%)	0.350	0.327
Neon (mol%)	–	–
Pressure, p_2 (bar)	20.00	20.00
Pressure, p_1 (bar)	3.33	1.50

in Eqs. (5.8)–(5.10) provides more exacting results than obtained from the method by Dobak et al. even for cryosurgical applications.

5.5 Optimization of mixture composition and operating pressures of liquefaction processes

Consider a simplified precooled process for the liquefaction of natural gas (Fig. 5.5). Natural gas is precooled (desuperheated) in the precooling heat exchanger (HX-1) to a temperature of 245 K and is condensed and subcooled in the main heat exchanger (HX-2). If we assume that the adiabatic efficiency of the compressor is independent of the mixture composition and compression ratio, the operating pressures and the composition of the main refrigerant can be optimized by considering the cold box alone. The optimization method to be followed is similar to that described in Table 5.3, except that the main refrigerant is partially condensed before it enters the

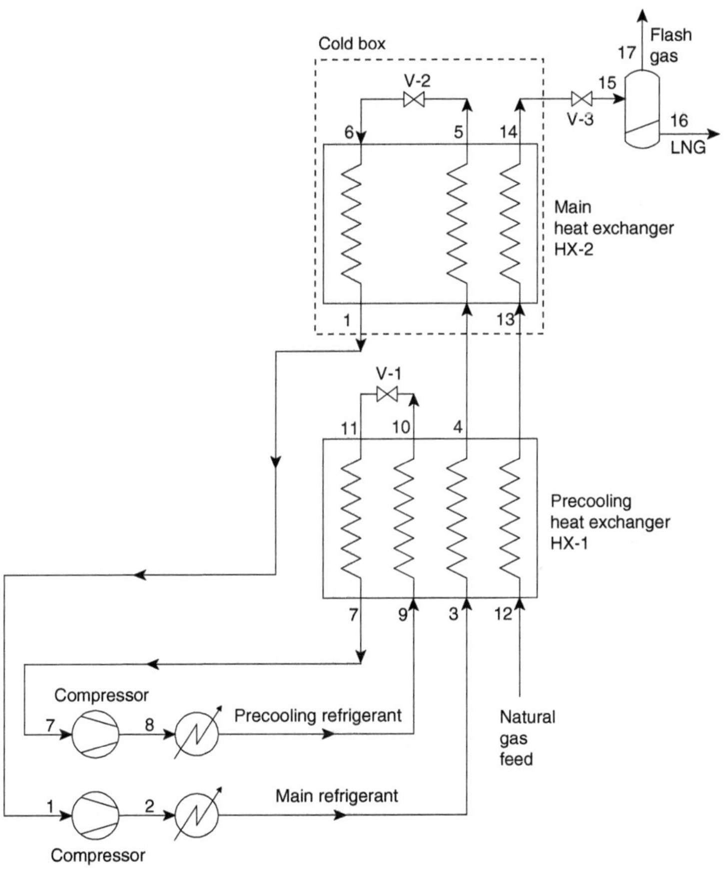

Fig. 5.5. Simplified precooled natural gas liquefaction process.

liquefaction heat exchanger (HX-2 in Fig. 5.5) in all natural gas liquefaction processes. The refrigerant can be partially condensed using a room temperature air/water condenser (aftercooler), or using a separate precooling refrigerant in a precooling heat exchanger. Only the refrigerant composition (mole fraction) needs to be varied during the optimization of a refrigeration process. On the other hand, both the refrigerant composition and the refrigerant flow rate need to be varied simultaneously in the case of liquefaction systems. Alternately, the component flow rate of different mixture constituents can be varied simultaneously.

Since liquefaction processes are very energy-intensive processes, it is ideal to minimize the compressor power input or maximize the exergy efficiency. The optimization method for optimizing a typical liquefaction process is similar to that described for the refrigeration process (Table 5.3) and is summarized in Table 5.9. The exergy efficiency of the cold box of the precooled process can be estimated using the following expression:

$$\eta_{ex,cb} = \frac{\text{minimum power for liquefaction}}{\text{exergy expenditure}} = \frac{\dot{n}_{13}(h_{14} - h_{13})}{\dot{n}_1(h_4 - h_1)}. \tag{5.13}$$

Table 5.10 shows the results of the initial and final values of the design variables of the optimization study. The upper and lower limits of the design variables are

Table 5.9. Method for the optimization of composition of the main refrigerant of a natural gas liquefaction process shown in Fig. 5.5

Objective function: Maximize exergy efficiency ($\eta_{ex,cb}$)

Subject to constraints:
- Minimum temperature approach between the streams, $\Delta T_{min} \geq \Delta T_{min,\,specified} = 3$ K.
- Partial condensation should occur in the precooling heat exchanger, or $T_{dew,4} > T_4$ (245 K).

Design variables:
- Mole flow rate of nitrogen (ξ_{N_2})
- Mole flow rate of methane (ξ_{CH_4})
- Mole flow rate of ethane ($\xi_{C_2H_6}$)
- Mole flow rate of propane ($\xi_{C_3H_8}$)
- Compressor discharge pressure (p_4)
- Compressor suction pressure (p_1)
- Temperature of high-pressure refrigerant leaving the heat exchanger (T_5)

Constants:
- Pressure drop in the high-pressure refrigerant channel of the heat exchanger ($\Delta p_{hx,\,hp}$) is zero.
- Pressure drop in the low-pressure refrigerant channel of the heat exchanger ($\Delta p_{hx,\,lp}$) is zero.
- Both high-pressure refrigerant and natural gas enter the main heat exchanger (HX-2) at 245 K.

also shown in Table 5.10. The composition of the optimum refrigerant mixture and the natural gas feed are presented in Table 5.11. The temperature, pressure, vapor fraction, and mole flow rate of different streams of the cold box of the precooled process (Fig. 5.5) are shown in Table 5.12.

The exergy efficiency of the cold box with the optimum mixture is 77% at the end of the optimization study. The high exergy efficiency is due to a small temperature difference between the streams along the length of the heat exchanger (Fig. 5.7).

Figure 5.6 shows the variation of design variables and the objective function (exergy efficiency) during the optimization process. A large number of iterations are needed to optimize a liquefaction process compared to refrigeration processes. The minimum temperature approach is close to 100 K at the start of the iterations, but it increases to close to the desired approach of 3 K in a few iterations. The above optimization method was solved using the commercial process simulator Aspen Plus. A nearly identical result was obtained with the CRYOSIM process simulator [74] along with the CFSQP optimization program [56].[2]

Table 5.10. Lower limit, upper limit, initial estimate, and final solution obtained for different design variables of the main refrigerant of a precooled natural gas liquefier shown in Fig. 5.5

Design variable	Units	Lower limit	Upper limit	Initial estimate	Optimal solution
Nitrogen flow rate	mol/s	0.0	0.3	0.1	0.204
Methane flow rate	mol/s	0.0	0.8	0.4	0.548
Ethane flow rate	mol/s	0.0	0.5	0.4	0.439
Propane flow rate	mol/s	0.0	0.4	0.1	0.345
Pressure, p_2	bar	20.0	60.0	50.0	43.78
Pressure, p_1	bar	3.0	5.0	3.0	3.449
Temperature, T_5	K	113.0	118.0	113.0	113.9

Table 5.11. Composition of the natural gas feed and the main refrigerant of a precooled natural gas liquefier shown in Fig. 5.5

Component	Natural gas feed[a] (mole fraction)	Main refrigerant (mole fraction)
Nitrogen	4.0	13.3
Methane	87.5	35.7
Ethane	5.5	28.6
Propane	2.1	22.5
*n*Butane	0.5	–
*i*Butane	0.3	–
*i*Pentane	0.1	–

[a] Feed composition as in Ref. [65].

[2] Courtesy Prof. A. L. Tits, AEM Design and University of Maryland.

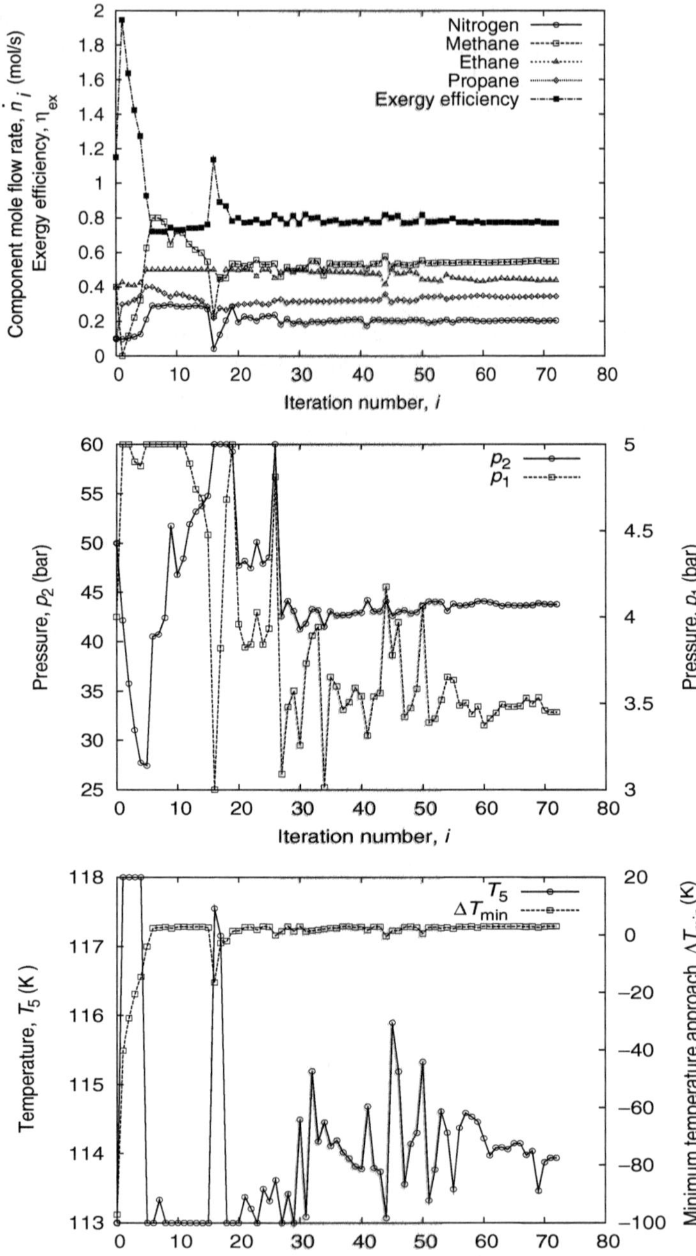

Fig. 5.6. Variation of design variables and the objective function (exergy efficiency) during the optimization of the precooled natural gas liquefaction process.

Table 5.12. Temperature, pressure, and flow rate of different streams of the main refrigerant for the precooled natural gas liquefier shown in Fig. 5.5

	Stream					
	13	14	4	5	6	1
Temperature, K	245.0	113.0	245.0	113.9	109.6	240.9
Pressure, bar	65.0	65.0	43.78	43.78	3.449	3.449
Vapor fraction	1.0	0.0	0.431	0.000	0.094	1.000
Flow rate, mol/s	1.0	1.0	1.536	1.536	1.536	1.536

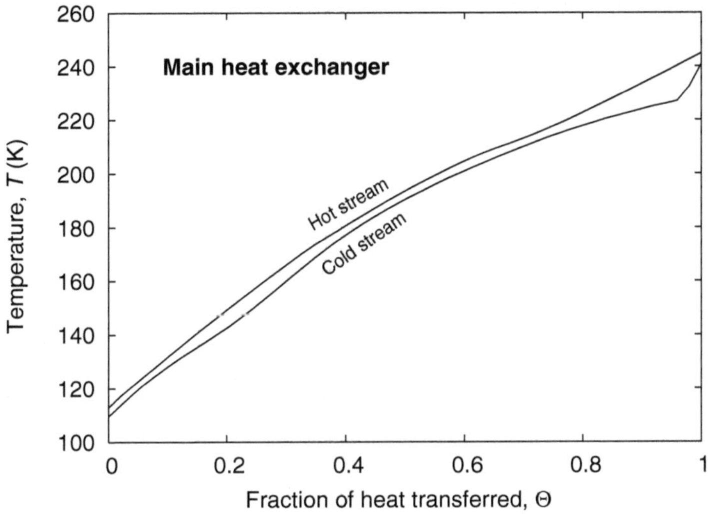

Fig. 5.7. Temperature profiles of the hot and cold fluid streams of the main heat exchanger (HX-2) of the precooled natural gas liquefier operating with optimum mixtures (Table 5.11).

The example clearly demonstrates that optimum refrigerant mixture composition, refrigerant flow rate, temperature of the high-pressure refrigerant at the entry of the expansion valve, and operating pressures of the refrigerant can be obtained despite assuming design variables far from the optimum solution when the optimization problem is formulated correctly, and when a robust optimization method such as the sequential quadratic programming (SQP) as well as a robust process simulator are used.

Specialized compressors can be used in large liquefiers. However, off- the-shelf compressors of a given displacement volume need to be used with small liquefiers and gas coolers. It will be beneficial to liquefy or cool as much of the process gas as possible using the given compressor. A useful objective function in such cases is the maximization of the liquid yield for a given compressor displacement volume rate ($\dot{V}_c$):

$$\text{objective function:}\quad \text{maximize}\ \frac{\dot{n}_{\text{gas}}}{\dot{V}_c}. \tag{5.14}$$

Consider the Linde gas cooler (Fig. 3.15). The above objective function can be expressed for a Linde gas cooler using Eq. (3.12) as follows:

$$\text{objective function:}\quad \text{maximize}\ \frac{\dot{n}}{\dot{V}_c}\left(\frac{h_5 - h_2}{h_a - h_b}\right) = \frac{\rho_5(h_5 - h_2)}{h_a - h_b}, \tag{5.15}$$

where $\rho_5(h_5 - h_2)$ is the volumetric cooling capacity of the refrigerator. Since the conditions of gas being cooled are specified at the inlet and outlet of the liquefier, and are independent of the mixture composition, the above objective function can be simplified as follows:

$$\text{objective function:}\quad \text{maximize}\ Q_v = \rho_5(h_5 - h_2). \tag{5.16}$$

However, the limit on the power that the compressor can draw should be used as a constraint whenever the volumetric cooling capacity is used as the objective function since a large compressor suction pressure (density) can result in a large mass flow rate through the compressor, and consequently a large compressor input power.

The methods described above for choosing the mixture composition, operating pressures, and temperatures of gas coolers and liquefiers can be used for both single-stage and multistage systems with different refrigerants in different stages. The mixture composition of each of the refrigerants and the operating pressures of each of the processes should be included as design variables in multistage processes. When the compressor adiabatic efficiency is fixed, it is advantageous to optimize the exergy efficiency of the cold box alone.

6

Natural gas liquefaction processes

The liquefaction of natural gas using a mixed refrigerant process was first proposed by Kleemenko in 1959 [50]. Mixed refrigerant processes were subsequently adopted for the commercial liquefaction of natural gas nearly 40 years ago. Over 95% of the base-load LNG plants operate on mixed refrigerant processes, with the remaining few operating on conventional cascade processes. The enthalpy of natural gas varies nonlinearly with temperature (at constant pressure), with points of inflection on the enthalpy temperature curve. A simple Linde process cooler shown in Fig. 3.15 therefore does not ensure a close temperature approach between the streams over the entire length of the heat exchanger, resulting in low exergy efficiency. Complex mixed refrigerant natural gas liquefaction processes have therefore been designed to accomplish the close temperature approach between streams in the superheated vapor, subcooled liquid, and two-phase regions, most of which are described in international patents.

Turbine-based processes such as the reverse Brayton process that operate with nitrogen as the working fluid are also used in the liquefaction of natural gas on a small scale. The efficiency of many of the turbine-based processes can be improved by using mixtures of nitrogen and methane as the refrigerant instead of pure nitrogen. Some of the fundamental turbine-based processes are also discussed in this chapter for the sake of completeness.

A number of patents exist on different processes that can be used for the liquefaction of natural gas. Some of them have also been used in practice. The choice of a particular process depends on a number of considerations such as cost and availability of equipment, size and transportability of heat exchangers, choice and efficiency of prime movers (gas turbines, steam turbines, electric drives), etc., apart from the work of compression or exergy efficiency. A number of articles exist in the literature on the comparison of different LNG processes on a different basis such as work of compression, heat exchanger area required, etc. [38, 65].

The aim of this monograph is to bring out the evolution of the LNG processes from the simple to the very sophisticated processes used currently in large base-load plants, and to help the reader understand the differences between the different processes. The order in which the different LNG processes are presented is, therefore, different from the chronological order in which they were invented and has been chosen to help the

G. Venkatarathnam, *Cryogenic Mixed Refrigerant Processes*,
DOI: 10.1007/978-0-387-78514-1_6, © Springer Science+Business Media, LLC 2008

reader understand the fundamental differences between the processes as well as the evolution of LNG processes in a logical fashion. Optimum refrigerant compositions and compressor operating pressures and temperatures have been determined for the processes discussed using the methods described in Chapter 5. The performance of different processes has been discussed from an exergy utilization point of view.

6.1 Classification of natural gas liquefaction processes

LNG processes can be broadly classified into three groups based on the liquefaction process used as described in Fig. 6.1:

1. cascade liquefaction processes,
2. mixed refrigerant processes,
3. turbine-based processes.

The first few natural gas liquefaction plants and a few current plants are based on the classical cascade processes operating with pure fluids such as methane, ethylene, and propane. Most existing base-load natural gas liquefaction plants operate on the

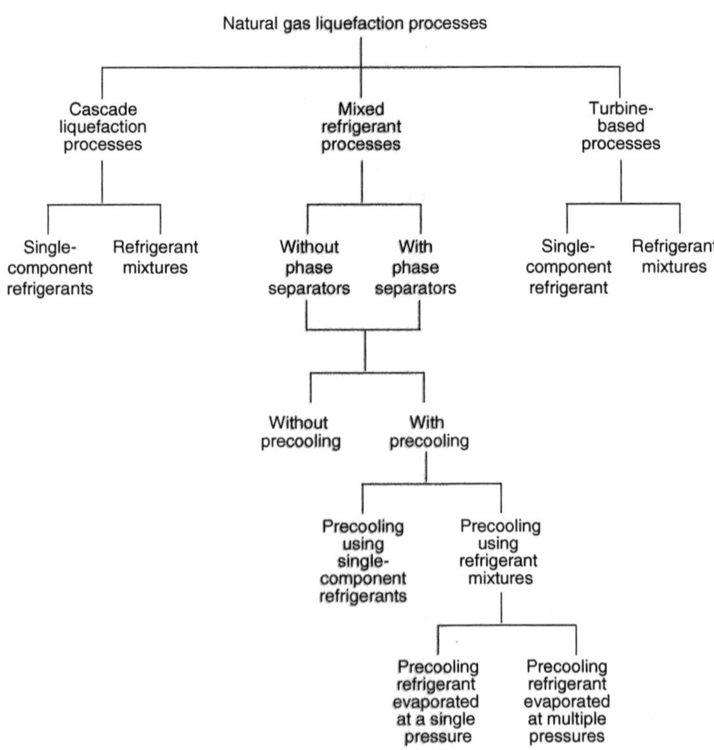

Fig. 6.1. Classification of natural gas liquefaction processes.

mixed refrigerant processes, with the propane precooled mixed refrigerant process being the most widely used. Cascade processes operating with mixtures have also been recently developed.

Mixed refrigerant processes can be further classified into those that use phase separators and those that do not. They can also be classified into processes that use precooling and those that do not. A variety of processes can be used for the precooling processes.

Turbine-based processes are preferred for use in peak shaving plants, because of their simplicity and the possibility of a quick startup. Some of the important turbine-based natural gas liquefaction processes are also discussed in this chapter.

Natural gas liquefaction plants can also be classified into (1) peak shaving and (2) base-load processes based on the type of usage. Any of the processes shown in Fig. 6.1 can be used for these applications.

Peak shaving plants store liquefied natural gas (LNG) in large, insulated containers. LNG is evaporated during the winter season, when natural gas consumption is very high. Peak shaving plants typically operate for 150 to 200 days a year, with storage capacity of 50,000 to 100,000 m^3 [71]. The large difference in the peak and off-peak price of natural gas makes the peak shaving plants economically viable. Most peak shaving plants are located far from natural gas sources. These plants receive the natural gas feed from a pipeline and serve a local community. Both turbine-based and mixed refrigerant liquefaction processes have been used in peak shaving plants. Turbine-based processes have lower efficiency, but can be started relatively quickly. Turbine-based plants have also been proposed for reliquefaction of boil-off gas in large LNG ships [44].

Base-load natural gas liquefaction plants, on the other hand, are typically located near the natural gas source and are operated continuously throughout the year, except for an occasional maintenance shutdown. Most modern base-load plants have a liquefaction capacity of 1 to 5 million tons per annum (MTPA). Even larger plants are planned in the near future. Most modern base-load plants use the propane precooled mixed refrigerant process. Other processes such as the cascade system operating with pure fluids as well as mixed refrigerants in each stage and dual mixed refrigerant processes (mixed refrigerant in the precooling as well as the liquefaction stages) are being used in plants currently under construction.

6.2 Classical cascade processes

Figure 6.2 shows a simplified classical cascade process for the liquefaction of natural gas [38]. The classical cascade process was used in the first commercial peak shaving natural gas liquefier in Cleveland, Ohio, in 1941. The cascade process was also used in some LNG base-load plants in the 1960s until it was essentially replaced by different mixed refrigerant processes. The classical cascade process has, however, been used recently in the first four trains of the Atlantic LNG plant in Trinidad and the first two trains of the Egyptian LNG and Bayu Udan LNG plants in Indonesia after almost a 30-year gap by ConocoPhillips.

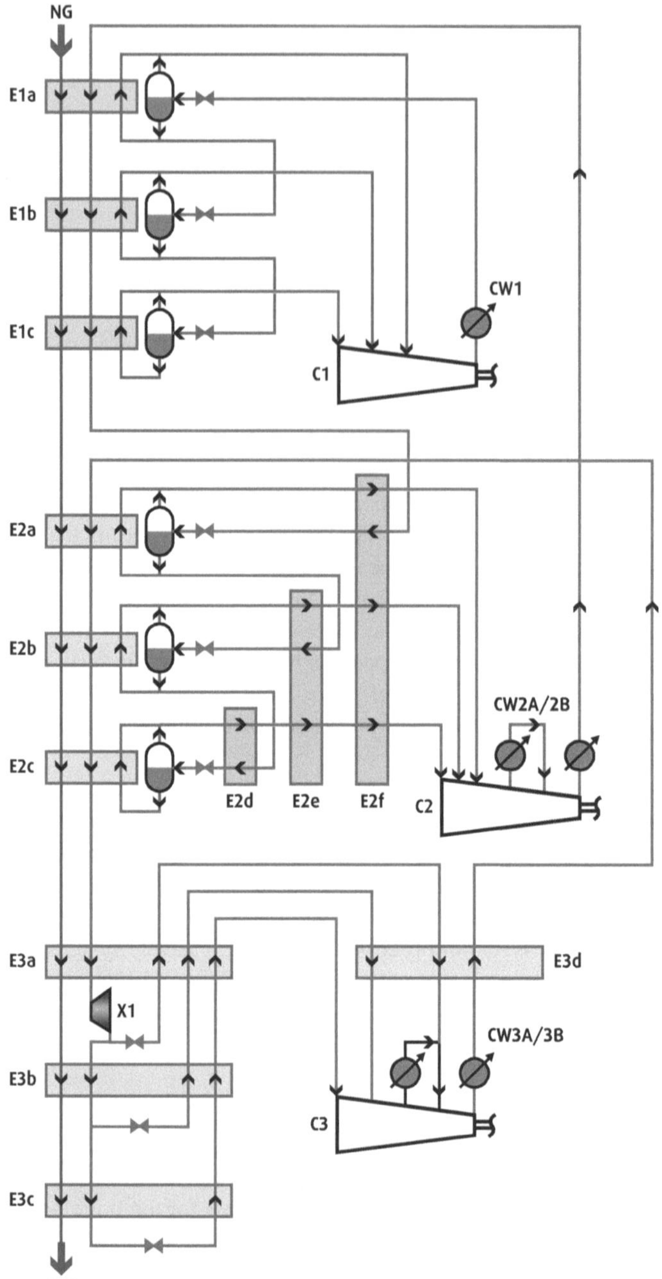

Fig. 6.2. Classical cascade process for the liquefaction of LNG. (Reproduced from Ref. [38] with kind permission of Linde AG, Germany.)

In the classical cascade process, the natural gas feed is cooled and liquefied using three different pure refrigerants: propane, ethylene, and methane. Each of these refrigerants is evaporated at three or four pressures as shown in Fig. 6.2, to provide refrigeration at nine or 10 temperature levels. A large number of heat exchangers are necessary, as shown in Fig. 6.2. An open-loop methane process is used in the ConocoPhillips optimized cascade process [75], with the feed flash gas generated after expansion of the subcooled natural gas feed serving as the refrigerant for the open-loop methane refrigeration stage.

The cascade process operating with pure fluids provides refrigeration at a constant temperature at different temperature levels, while mixed refrigerant processes provide continuous refrigeration between room temperature and the desired LNG subcooling temperature (110–130 K), resulting in a higher efficiency.

Consider the integration of a continuously differentiable function $f(x)$. An analytical integration method provides an exact solution, while a numerical method provides an approximate solution by adding the area under the $f(x)$ curve in an $f(x) - x$ diagram using trapezoid elements. The comparison between the classical cascade process and mixed refrigerant processes is analogous to the above comparison. Mixed refrigerant processes provide refrigeration over a range of temperatures continuously, analogous to analytical integration methods. A classical cascade process provides constant-temperature refrigeration at a few discrete temperature levels analogous to numerical integration methods. The temperature approach between the natural gas and the low-pressure refrigerant stream is therefore small and nearly uniform along the length of the heat exchangers in the case of mixed refrigerant processes, whereas the temperature approach between the hot and cold fluid streams is small only at the cold end of all heat exchangers with a classical cascade process. The exergy efficiency of mixed refrigerant processes is therefore higher than that of classical casade processes operating with single-component refrigerants.

Mixed refrigerant processes have been adopted on very large scale in base-load plants due to their higher exergy efficiency. The different mixed refrigerant processses that can be used for the liquefaction of natural gas are discussed in the following sections.

6.3 Assumptions

The following assumptions are made in all the examples provided in this chapter:

- The pressure drop in all heat exchangers and phase separators is zero.
- The ambient temperature is 300 K.
- The minimum temperature approach between the hot and cold streams is 3 K in all cold heat exchangers.
- The adiabatic efficiency of all compressors is 80% and that of all pumps is 90%.
- The heat inleak from ambient is negligible.

6.4 Single-stage mixed refrigerant LNG process without phase separators

Figure 6.3 shows the simplest of the mixed refrigerant natural gas liquefaction processes that do not have any phase separators [58]. This process is also sometimes called the single refrigerant process. Figure 6.4 shows a minor variation of the fundamental process with a phase separator in which propane and higher boilers are removed [88]. In this process (Fig. 6.4), propane and higher hydrocarbons are removed at about −30 °C. The process is more commonly known as the PRICO process and was commercialized by the Pritchard Company.

The composition of the refrigerant used in the process is a strong function of the feed composition, feed pressure, ambient temperature, and operating pressures used. The composition of the natural gas feed and the corresponding refrigerant mixture in the example presented in U.S. Patent 4,033,735 [88] is shown in Table 6.1. According to Swenson [88] the refrigerant does not freeze at the operating temperatures in spite of the large quantity of isopentane present. Table 6.2 shows the optimum composition range claimed in that patent [88].

The design specifications for the single-stage LNG process without a phase separator shown in Fig. 6.3 are shown in Table 6.3. The minimum temperature approach in the cold box heat exchanger has been assumed to be 3.0 K and that in the condensers as 5.0 K. A two-stage compression process has been assumed for the system, as shown in Fig. 6.5.

The performance of the system has been evaluated for a feed and refrigerant specified in Table 6.4. The feed is assumed to be available at a pressure of 40 bar and is subcooled to a temperature of 113 K. The composition of the refrigerant,

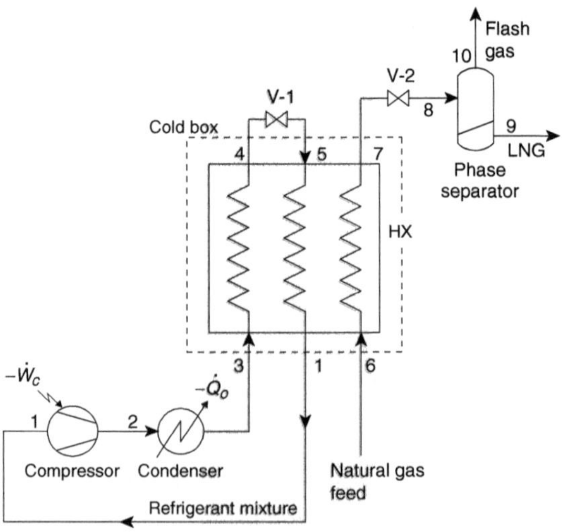

Fig. 6.3. Single-stage mixed refrigerant natural gas liquefaction process without a phase separator. (Adapted from U.S. Patent no. 3,914,949).

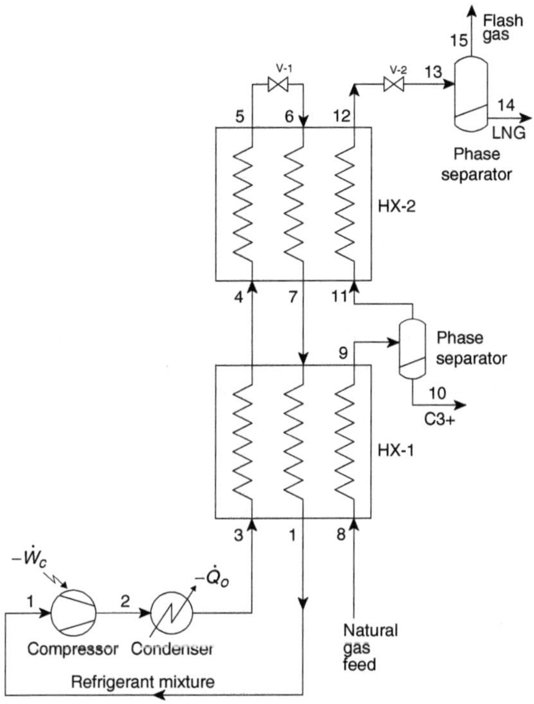

Fig. 6.4. Single-stage mixed refrigerant natural gas liquefaction process without a phase separator in the refrigeration process, but with separation of propane and higher hydrocarbons from the natural gas feed [88]. (Adapted from U.S. Patent no. 4,033,735.)

Table 6.1. Example of LNG feed and refrigerant compositions presented in U.S. Patent 4,033,735 [88]

Component	Natural gas feed (mol%)	Refrigerant (mol%)
Helium	0.20	trace
Nitrogen	5.80	10.6
Methane	83.20	35.6
Ethane	7.10	28.2
Propane	2.25	3.4
*i*Butane	0.40	8.0
*n*Butane	0.60	2.1
*i*Pentane	0.12	11.4
*n*Pentane	0.15	0.7
Hexane	0.10	trace
C_7 and above	0.08	trace

Table 6.2. Refrigerant composition claimed in U.S. Patent 4,033,735 [88]

Component	Mol%
N_2	0–12
C_1	20–36
C_2	20–40
C_3	2–12
C_4	6–24
C_5	2–14

Table 6.3. Design specifications for the single-stage natural gas liquefier shown in Fig. 6.3

Compressor discharge pressure, p_2	24 bar
Compressor suction pressure, p_1	3 bar
Feed (natural gas) operating pressure, p_6	40 bar
Minimum temperature approach in the cold box heat exchangers, $\Delta T_{\min}$	3 K
Minimum temperature approach in the condensers, aftercoolers, $\Delta T_{\min}$	5 K
Pressure drop in the heat exchanger, Δp	0 bar
LNG temperature before expansion, T_7	113 K
Adiabatic efficiency of compressor, $\eta_{ad,c}$	80%

Table 6.4. Feed and refrigerant composition for the process shown in Fig. 6.3

Component	NG feed[a] (mol%)	Refrigerant (mol%)
Nitrogen	4.0	11.6
Methane	87.5	28.4
Ethane	5.5	30.7
Propane	2.1	14.0
*n*Butane	0.5	5.7
*i*Butane	0.3	0.0
*i*Pentane	0.1	9.5

[a] Feed composition as in Ref. [65].

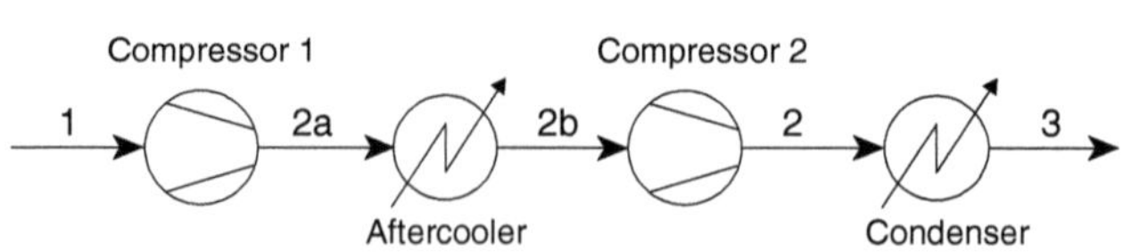

Fig. 6.5. Compressor arrangement for the single-stage process shown in Fig. 6.3.

the discharge and suction pressure of the compressor, and the temperature of the high-pressure refrigerant leaving the heat exchanger were considered as the design variables of the optimization study, as in Table 5.9.

The temperature, pressure, vapor fraction, and flow rate of each of the streams obtained in the optimization study are shown in Table 6.5. It can be observed that the composition of the refrigerant (Table 6.4) is very close to that claimed in Table 6.2, except for propane and *n*Butane, which are slightly outside the limits claimed.

If a saturated liquid natural gas at 40 bar is expanded to a pressure of 1 bar, 38% of the liquid would be converted to vapor during expansion. The amount of flash vapor generated during expansion can be decreased by subcooling the liquefied natural gas to a temperature of 100 to 125 K. The choice of subcooling temperature depends on the amount of nitrogen present in the feed. Figure 6.6 shows the relationship among the temperature of the natural gas at the entry of the expansion valve, the flow rate of flash vapor generated on expansion, and the concentration of nitrogen in the liquid. It can be seen from Fig. 6.6 that a decrease in nitrogen concentration below 1% can be achieved only when the natural gas feed in the present example is cooled to a temperature greater than 120 K. This will, however, result in a large amount of flash vapor. Nitrogen rejection units that use distillation columns are therefore used to reduce the nitrogen concentration in the liquid product. The concentration of nitrogen in the liquified natural gas should be reduced to less than 1% in the nitrogen rejection units to prevent rollover in LNG tanks.

In this monograph, nitrogen rejection units have not been included in any of the examples. Similarly, LPG (C_{3+}) recovery units have not been included. Also, throttle valves have been used instead of expansion turbines. This has been done deliberately to reduce the complexity of the processes and to concentrate on understanding the differences between the different liquefaction processes.

Table 6.5. Temperature, pressure, vapor fraction and flow rate of different streams of the process shown in Figs. 6.3 and 6.5, and operating with the feed and refrigerants shown in Table 6.4

Cold box

	1	3	4	5	6	7
Temperature, K	297	305	113	110	300	113
Pressure, bar	4.0	24.0	24.0	4.0	40.0	40.0
Vapor fraction	1.000	0.817	0.000	0.070	1.000	0.000
Flow rate, mol/s	1.00	1.00	1.00	1.00	0.24	0.24

Compressor

	1	2a	2b	2	3
Temperature, K	297.0	347.6	305.0	358.6	305.0
Pressure, bar	4.0	9.8	9.8	24.0	24.0
Vapor fraction	1.000	1.000	1.000	1.000	0.817
Flow rate, mol/s	1.00	1.00	1.00	1.00	1.00

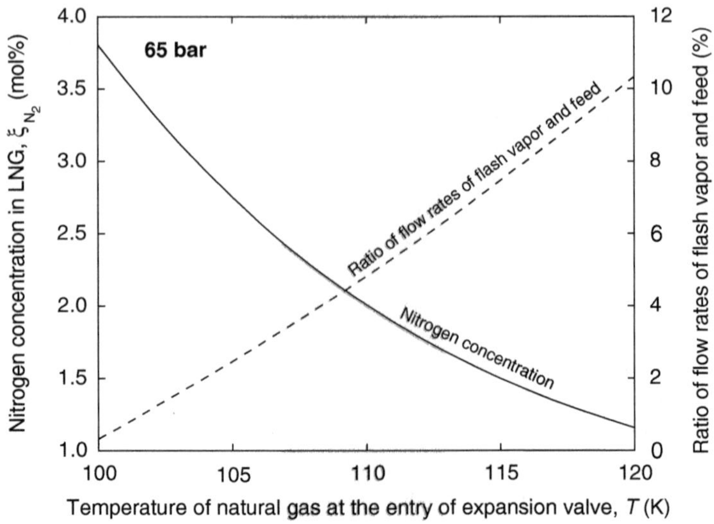

Fig. 6.6. Relationship among the temperature of the natural gas at the entry of the expansion valve (subcooling temperature), the nitrogen content in the liquefied natural gas product, and the ratio of flow rates of the flash gas and the feed natural gas for the natural gas feed shown in Table 6.4 and an operating pressure of 65 bar.

Figure 6.7 shows the enthalpy-temperature variation for the feed gas at a pressure of 40 and 65 bar. The enthalpy-temperature curves are largely linear in the superheated and subcooled regions but are non linear in the two-phase region, particularly at a pressure of 40 bar. The main objective of the different processes is to obtain a composite enthalpy-temperature variation of the hot streams (natural gas feed and high-pressure refrigerant) that is parallel to that of the cold streams (see Section 3.4). Since the slope $(\partial h/\partial T)_p$ is different in the superheated, subcooled and two-phase regions, ideally, three different mixtures need to be used for these three regions. The use of a single refrigerant to desuperheat (precool), condense (liquefy), and subcool significantly reduces the complexity of the process, albeit at the cost of efficiency.

Because of the small operating pressures and a pressure ratio of just 6, a single-stage compressor can be used in small scale liquefiers and a two-stage compressor in large liquefiers (Fig. 6.5). The pressure at the exit of the two stages is shown in Table 6.5. It can be seen from Table 6.5 that only 18.3% of the refrigerant condenses in the condenser (state point 3).

The refrigerant composition shown in Table 6.4 has been determined using the optimum procedure described in Chapter 5. The exergy efficiency of the cold box was chosen as the objective function to be maximized. The composition of the refrigerant, the flow rate of the natural gas feed, the operating pressures (p_1, p_2), and the temperature (T_4) are the design variables of the optimization problem.

The exergy efficiency of the cold box (Fig. 6.3) consisting of the heat exchanger, and expansion valve (V-1) is given by the expression

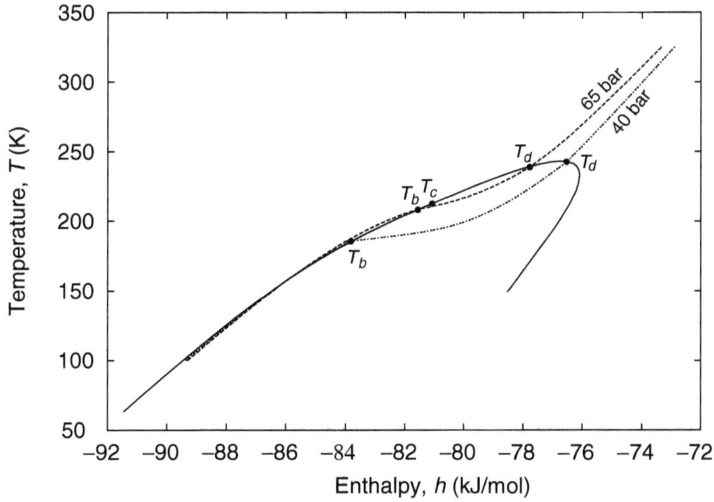

Fig. 6.7. Variation of enthalpy with temperature for the natural gas feed at pressures of 40 and 65 bar. T_d and T_b refer to the dew and bubble point temperature at the particular pressure respectively, and T_c refers to the critical temperature. See Table 6.4 for natural gas feed composition.

$$\eta_{\text{ex, cb}} = \frac{\text{minimum power for liquefaction}}{\text{exergy input}} = \frac{\dot{n}_8(\text{ex}_7 - \text{ex}_6)}{\dot{n}_3(\text{ex}_3 - \text{ex}_1)} \tag{6.1}$$

The utilization of exergy in the cold box, compressors, aftercoolers/condensers, and overall process is shown in Fig. 6.8. Only 53.5% of the exergy input to the cold box is used for the liquefaction of natural gas, as shown in Fig. 6.8. The rest of the input exergy is lost in the heat exchanger (38.2%) and expansion valve (8.3%). The reasons for the large exergy loss in the heat exchanger can be understood by analyzing the performance of the heat exchanger.

Figure 6.9 shows the variation of the composite temperature profile for the hot and cold streams in the heat exchanger. The temperature difference between the streams is shown in Fig. 6.10. It can be observed that the temperature approach between the streams varies between 3 and 18.2 K over the length of the heat exchanger. The minimum temperature approach between the streams occurs at the cold end as well as in the middle of the heat exchanger, as shown in Fig. 6.10. The average LMTD of the heat exchanger in this example is 7.4 K (Table 6.6). The relatively large temperature difference between the hot and cold streams results in a large exergy loss in the heat exchanger. The temperature difference between the streams is also large at the warm end, as shown in Table 6.6.

The exergy efficiency of the compressor and condenser (Fig. 6.5) is given by the expression

$$\eta_{\text{ex, cs}} = \frac{\text{minimum power for liquefaction}}{\text{compressor power input}} = \frac{\dot{n}_1(\text{ex}_3 - \text{ex}_1)}{-\dot{W}_c} \tag{6.2}$$

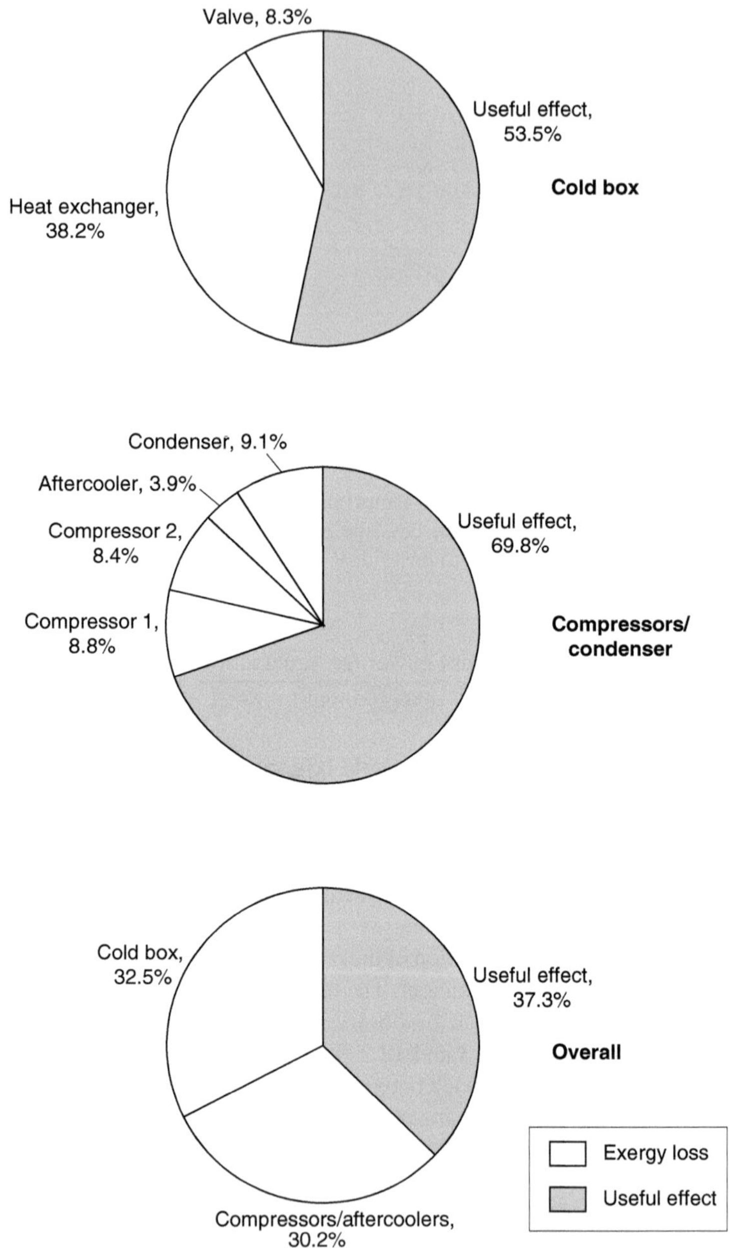

Fig. 6.8. Exergy utilization in the cold box and compressors of the single-stage mixed refrigerant natural gas liquefaction process shown in Fig. 6.3 with the feed and refrigerants shown in Table 6.4. The overall exergy efficiency is the product of the useful effect in the cold box and condensers ($\eta_{ex,o} = \eta_{ex,cb} * \eta_{ex,cs} = 53.5\% * 69.8\% = 37.3\%$).

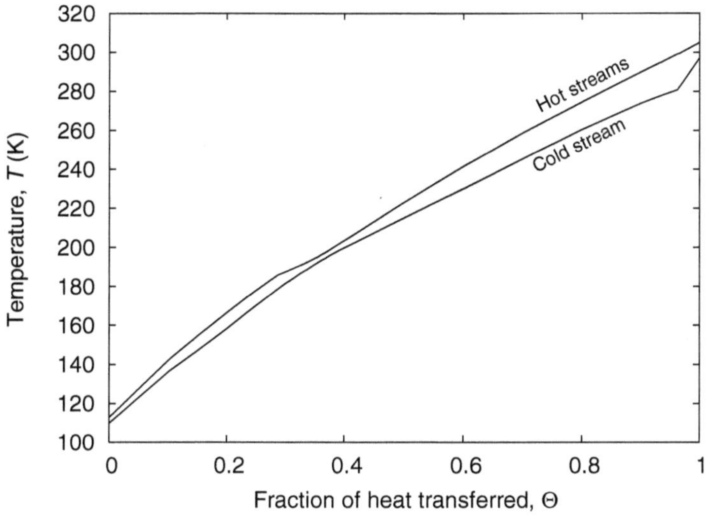

Fig. 6.9. Temperature profiles in the heat exchanger of the process shown in Fig. 6.3. See Table 6.4 for the composition of the streams.

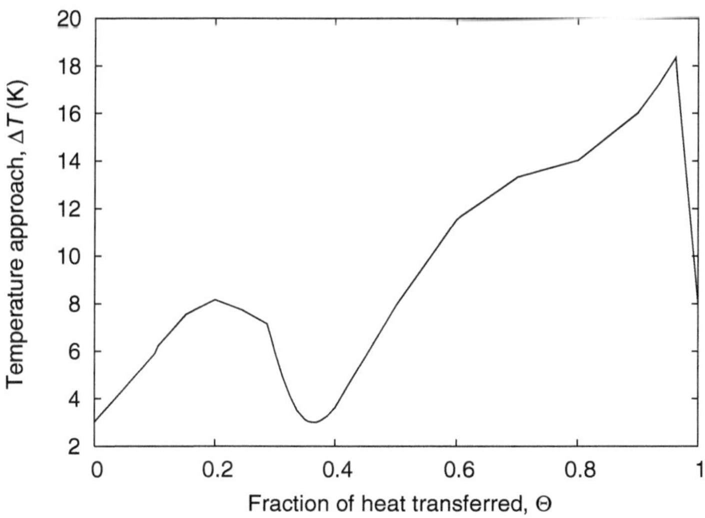

Fig. 6.10. Temperature approach between the hot and cold streams in the heat exchanger of the process shown in Fig. 6.3.

The exergy efficiency of the compressors and aftercooler/condenser has been estimated to be 69.8% (Fig. 6.8), assuming that a two-stage compression process is used (Fig. 6.5). The adiabatic efficiency of each compression stage is assumed to be 80%. A temperature approach of 5 K has been assumed in the aftercooler and condenser. The pressure ratio in each compressor has been chosen so as to minimize

Table 6.6. Temperature approach between streams in the heat exchanger of the single-stage liquefaction process shown in Fig. 6.3 and operating with the refrigerant mixtures shown in Table 6.4

LMTD (ΔT_{lm}), K	7.4
Minimum approach (ΔT_{min}), K	3.0
Warm end approach (ΔT_{we}), K	8.0
Cold end approach (ΔT_{ce}), K	3.0

the overall compressor power. The overall exergy efficiency of the process, excluding the LNG expansion valve, is given by the expression

$$\eta_{ex,o} = \frac{\text{minimum power for liquefaction}}{\text{compressor power input}} = \frac{\dot{n}_7(\text{ex}_7 - \text{ex}_6)}{-\dot{W}_c} \tag{6.3}$$

The overall exergy efficiency of the system is the product of the exergy efficiency of the cold box and that of the compressors and is 37.3% in this example. It can be seen from Fig. 6.8 that the overall exergy loss is about the same in the cold box (32.5%) and in the compressors/aftercoolers (30.2%).

The compressor power input for liquefying the natural gas feed from 40 bar, 300 K to 40 bar, 113 K has been estimated to be 5.57 kW for a natural gas feed flow rate of 0.24 mol/s, or 14.7 kW per Tonne/day of NG feed. The expansion of liquefied natural gas from 40 to 1 bar at constant enthalpy results in a vapor fraction of 6.4%. If it is assumed that the vapor generated during flashing is not used to cool the feed, the power required will be 15.7 kW per Tonne/day of LNG product.

The main disadvantage with the process shown in Fig. 6.3 is the possibility of freezing of high boilers (butanes and pentanes) at low temperatures. Calculations with the TFREEZE routine of the Aspen Plus process simulator show that the freezing of the *n*Butane and *i*Pentane in the refrigerant mixture is likely to occur at 105 K, which is close to the lowest temperature in the system (110 K). As mentioned earlier (see page 103), this result should be accepted with caution. The amount of high boiling point fluids used should be reduced in order to reduce the possibility of freezing. This will, however, result in a drop in the overall exergy efficiency of the system.

It can be seen from Table 6.5 that about 18.3% of the high-pressure refrigerant stream (stream 3) is in a liquid state at the entry of the cold box heat exchanger. This value is close to that specified by Swenson [88]. Because of the large amount of liquid present, it is customary to pass the vapor and liquid phases of the high-pressure refrigerant separately to the heat exchanger to avoid flow maldistribution. A phase separator (separation drum) is normally used in the single-stage liquefaction process, as shown in Fig. 6.11.

The efficiency of the single-stage liquefaction process is also dependent on the ambient temperature. A change in the ambient temperature will result in the following:

- A change in the heat rejected in the condenser/aftercooler, and consequently a change in the vapor fraction of the high-pressure refrigerant entering the heat exchanger.

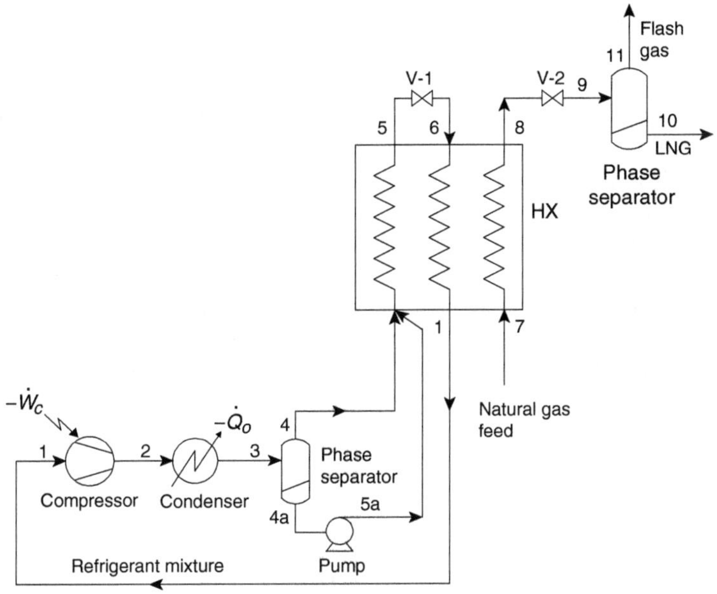

Fig. 6.11. PRICO process with separate entry for liquid and vapor streams at the entry of the heat exchanger [88].

- A change in the liquid level in the phase separator between the condenser and heat exchanger, and consequently a change in the composition of the refrigerant in circulation.

The variation of exergy efficiency and liquefaction rate of the single-stage liquefaction process with ambient temperature depends on the combined effect of the above factors. In general, an increase in ambient temperature results in a lower exergy efficiency or higher power input in all LNG processes. The single-stage liquefaction process has several advantages:

- The operating pressures are low (less than 25 bar on the high-pressure side, and 2–4 bar on the low-pressure side).
- Low-cost plate-fin heat exchangers can be used because of the low operating pressures of the refrigerant mixture.
- The process is simple and ideal for small-scale liquefaction of natural gas.
- The control of the process is relatively simple.

The main disadvantages of the single-stage natural gas liquefaction process are

- higher power consumption compared to other mixed refrigerant processes,
- possibility of freezing of the refrigerant mixture at low temperatures.

6.5 Precooled LNG process without phase separators

The precooled natural gas liquefaction process shown in Fig. 6.12 was one of the first LNG processes to be patented [83] and precedes the single-stage natural gas liquefaction process shown in Fig. 6.3 by many years. The process can be considered a precooled PRICO process. In this process, an external refrigerant is used to cool the process to a temperature of 210 to 245 K. The precooling temperature is dependent on several factors such as (1) separation of propane and other high boilers, (2) dew point temperature of the feed, (3) distribution of work between different compressor drivers, etc. In order to simplify the analysis, the separation of propane and other high boilers has not been considered in the present example. A two-stage as well as a three-stage compression process has been assumed for the precooling refrigerant

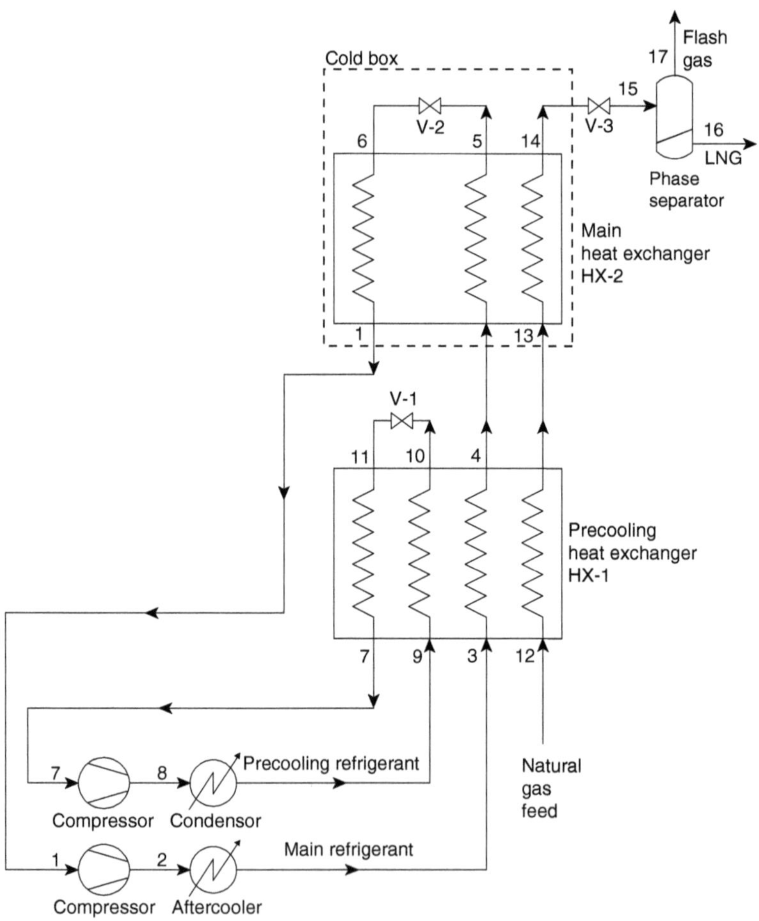

Fig. 6.12. Precooled process without phase separators [83]. (Adapted from Great Britain patent no. 895,094.)

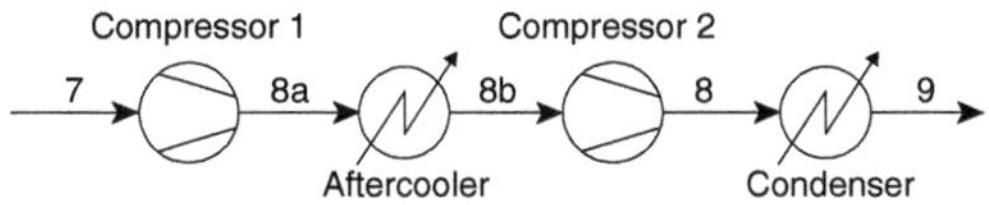

Fig. 6.13. Two-stage compressor required for the compression of the precooling refrigerant of the precooled process shown in Fig. 6.12.

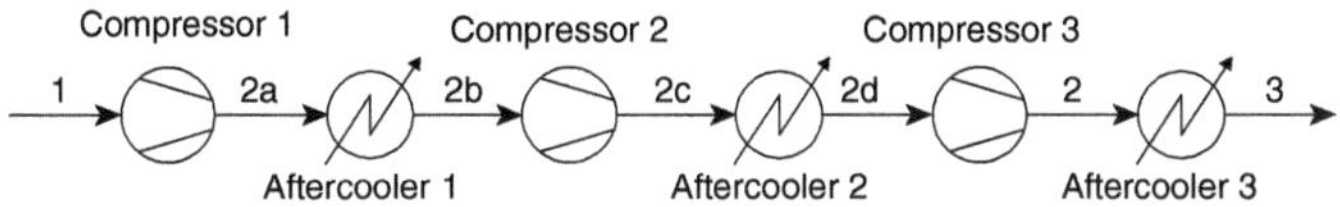

Fig. 6.14. Three-stage compressor required for the compression of the main refrigerant of the precooled process shown in Fig. 6.12.

and the main refrigerant, respectively, as shown in Figs. 6.13 and 6.14. Consider a cold box consisting of only the main heat exchanger and expansion valve as shown in Fig. 6.12.

Table 6.7 shows the optimum composition of the refrigerant for the above example. The composition of the two refrigerants and the operating pressures were estimated by maximizing the exergy efficiency of the cold box and the precooling process, respectively. The precooling refrigerant is completely condensed in the condenser. A temperature approach of 10 K has been assumed at the cold end of the precooling refrigerant condenser.

The temperature, pressure, and vapor fraction at the entry and exit of the main heat exchanger and valve V-2 are shown in Table 6.8, and those in the main- and precooling-compressors are shown in Tables 6.9 and 6.10, respectively. It has been assumed that the natural gas feed and the high-pressure mixed refrigerant are both cooled to 245 K before they enter the main heat exchanger. The natural gas feed (stream 13) is in a superheated vapor state at the exit of the precooler, whereas the main refrigerant (stream 4) is in a two-phase state with a vapor fraction of 57.5%.

The exergy efficiency of the cold box of Fig. 6.12 is given by the expression

$$\eta_{\text{ex, cb}} = \frac{\text{minimum power for liquefaction}}{\text{exergy expenditure}} = \frac{\dot{n}_{14}(\text{ex}_{14} - \text{ex}_{13})}{\dot{n}_{1}(\text{ex}_{4} - \text{ex}_{1})}. \tag{6.4}$$

Consider the cold box of the precooling process, consisting of the precooling-compressors, condenser, and precooling heat exchanger (HX-1). The exergy efficiency of the precooling process is given by the expression

$$\eta_{\text{ex, cb, pre}} = \frac{\text{minimum power for cooling the feed and main refrigerant}}{\text{exergy expenditure}} \tag{6.5}$$

or

$$\eta_{\text{ex, cb, pre}} = \frac{\dot{n}_{12}(\text{ex}_{13} - \text{ex}_{12}) + \dot{n}_{3}(\text{ex}_{4} - \text{ex}_{3})}{\sum -\dot{W}_{c,\ \text{precooler}}}. \tag{6.6}$$

Table 6.7. Composition of the natural gas feed and the refrigerants of the process shown in Fig. 6.12

Component	Natural gas feed (mol%)	Main refrigerant (mol %)	Precooling refrigerant (mol%)
Nitrogen	4.00	10.88	––
Methane	87.50	30.52	––
Ethane	5.50	43.57	31.15
Propane	2.10	15.03	36.02
*n*Butane	0.50	––	32.83
*i*Butane	0.30	––	––
*n*Pentane	––	––	––
*i*Pentane	0.10	––	––

Table 6.8. Temperature, pressure, vapor fraction, and flow rate of different streams of the process shown in Fig. 6.12 and operating with the refrigerant composition shown in Table 6.7

	Stream					
	1	3	4	5	6	7
Temperature, K	240.4	305.0	245.0	113.0	109.2	307.0
Pressure, bar	3.0	24.0	24.0	24.0	3.0	3.1
Vapor fraction	1.000	1.000	0.575	0.000	0.072	1.000
Flow rate, mol/s	1.000	1.000	1.000	1.000	1.000	0.613
	Stream					
	9	10	11	12	13	14
Temperature, K	310.0	245.0	238.1	305.0	245.0	113.0
Pressure, bar	19.6	19.6	3.1	40.0	40.0	40.0
Vapor fraction	0.000	0.000	0.053	1.000	1.000	0.000
Flow rate, mol/s	0.613	0.613	0.613	0.479	0.479	0.479

Table 6.9. Temperature, pressure, vapor fraction, and flow rate of different streams of the main-compressor shown in Fig. 6.13 and operating with the refrigerant composition shown in Table 6.7

	Stream				
	7	8a	8b	8	9
Temperature, K	307.0	344.2	305.0	354.3	310.0
Pressure, bar	3.1	7.1	7.1	19.6	19.6
Vapor fraction	1.000	1.000	1.000	1.000	0.000
Flow rate, mol/s	0.613	0.613	0.613	0.613	0.613

Table 6.10. Temperature, pressure, vapor fraction, and flow rate of different streams of the precooling-compressor shown in Fig. 6.14 and operating with the refrigerant composition shown in Table 6.7

	Stream						
	1	2a	2b	2c	2d	2	3
Temperature, K	240.4	339.8	305.0	326.0	305.0	319.3	305.0
Pressure, bar	3.0	14.4	14.4	19.6	19.6	24.0	24.0
Vapor fraction	1.000	1.000	1.000	1.000	1.000	1.000	1.000
Flow rate, mol/s	1.000	1.000	1.000	1.000	1.000	1.000	1.000

The overall exergy efficiency of the process is given by

$$\eta_{\mathrm{ex},o} = \frac{\text{minimum power for liquefaction}}{\text{total compressor power input}} = \frac{\dot{n}_{12}(\mathrm{ex}_{14} - \mathrm{ex}_{12})}{\sum_i -\dot{W}_{c,\,i}}, \tag{6.7}$$

where $\sum_i -\dot{W}_{c,1}$ refers to the total power input for compressing the main and precooling refrigerants.

Figure 6.15 shows the exergy utilization in the cold box of the precooled liquefaction process shown in Fig. 6.12. At 71.7%, the exergy efficiency of the cold box in the precooled process is much higher than that in the single-stage process (Fig. 6.8) without precooling (53.5%). The higher exergy efficiency of the cold box of the precooled process (Fig. 6.12) is due to the smaller temperature approach between the streams in the heat exchangers of the precooled process.

The exergy efficiency of the precooler consisting of the compressors, condenser of the precooling refrigerant, precooling heat exchanger (HX-1), and expansion valve (V-1) is only 34.4% because of the large exergy loss in the precooling refrigerant condenser.

The overall exergy efficiency of the precooled process (Fig. 6.12) is 45% compared to 37.3% with a single-stage process (Fig. 6.8). The exergy loss in the compressors and condensers/aftercoolers is 32.2% in the precooled process, compared to 30.2% in the single-stage process, while that in the heat exchangers and valves is 22.8% in the precooled process, compared to 32.5% in the single-stage process. The overall exergy efficiency of the precooled process is higher than that of a single-stage liquefaction process without precooling, largely due to the smaller temperature approach in the heat exchangers (Fig. 6.16). A comparison of the refrigerant compositions in Tables 6.4 and 6.7 shows that butanes and pentanes are not used in the precooled process. The chance of the refrigerant freezing at the operating temperatures is, therefore, nearly nonexistent in the precooled case. Different processes can be used for the precooling stage. Some of the possible precooling processes are discussed in the following sections.

As already mentioned, ideally three different mixtures should be used to (1) desuperheat (precooling), (2) condense (liquefy), and (3) subcool the natural gas feed since the $(\partial h/\partial T)$ or c_p is different in the three regions.

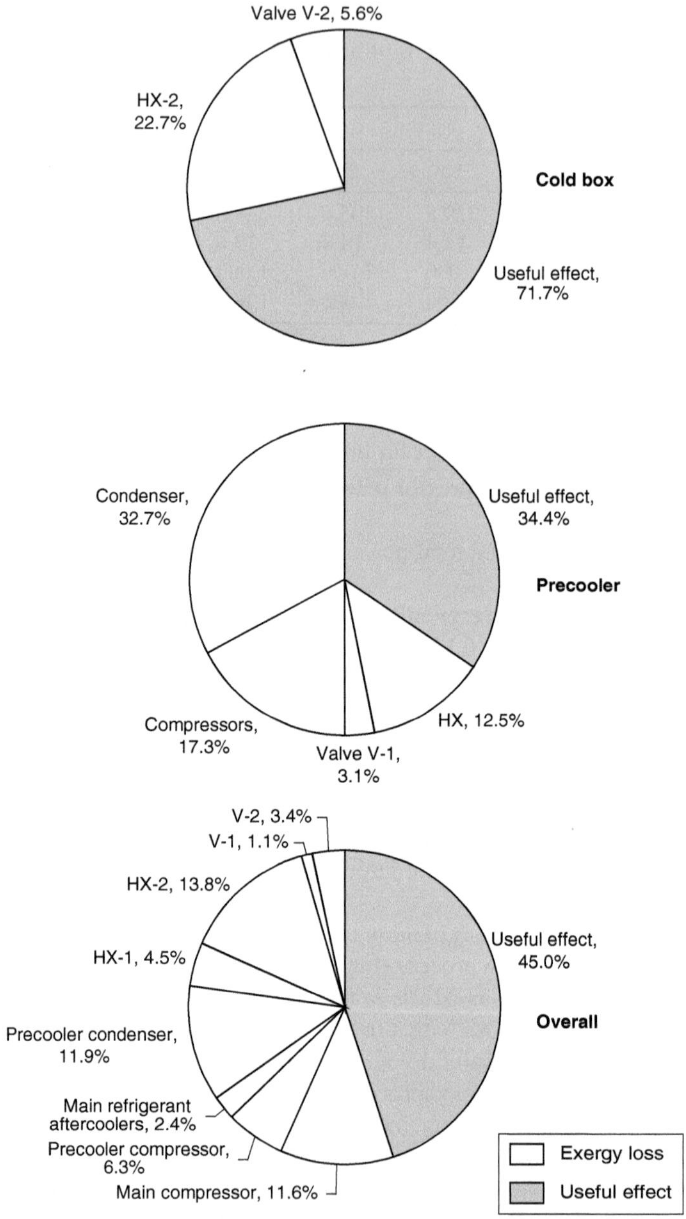

Fig. 6.15. Exergy utilization in the cold box, precooler, and overall process of a precooled mixed refrigerant natural gas liquefaction process shown in Fig. 6.12 with feed and refrigerants shown in Table 6.7.

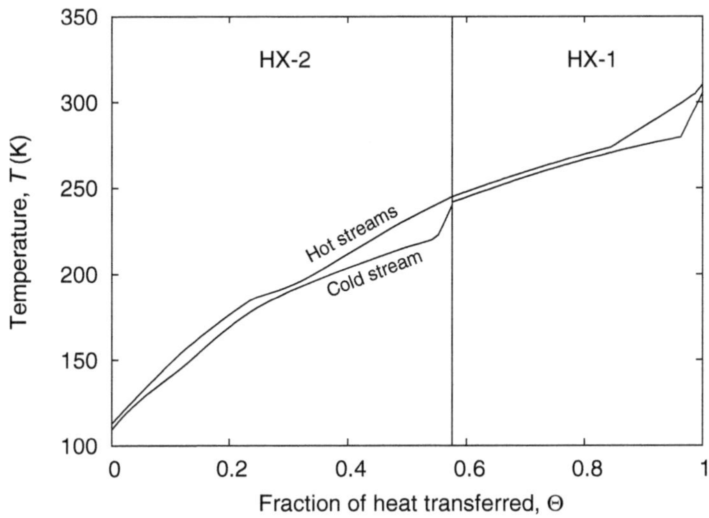

Fig. 6.16. Temperature profiles in the two heat exchangers of the precooled PRICO process (Fig. 6.12).

It can, however, be seen from the present example that the use of two different mixtures, —one for desuperheating (precooling) and another for condensation and subcooling— is better than that in a single-stage liquefaction process where only one refrigerant is used for desuperheating, condensing, and subcooling the natural gas feed. The use of multiple refrigerants, however, makes the liquefaction process complex. Because of the large improvement in the exergy efficiency from 37.3% to 45% (see Fig. 6.15) when precooling is used, however, it makes economic sense to use precooled processes.

Normally, different heat exchanger cores are used for the precooling and main heat exchangers. It is customary to split the vapor and liquid phases at the entry of any heat exchanger, and feed them separately to prevent flow maldistribution to the tubes of a spiral-tube (Giauque–Hampson) heat exchanger, or a plate-fin heat exchanger.

The high-pressure main refrigerant is partially condensed in the precooling heat exchanger. The vapor fraction of the high-pressure main refrigerant stream entering heat exchanger HX-2 (stream 4) is 57.5% in the present example. The natural gas feed (stream 13), on the other hand, is in the superheated vapor state. The liquid and vapor phases of the refrigerant mixture need to be separated in a flash drum (phase separator) and distributed separately to the headers of the main heat exchanger (HX-2) to avoid flow maldistribution, when two separate heat exchanger cores are used for the precooling and main heat exchangers (HX-1 and HX-2).

The need for phase separators can be exploited usefully by designing systems in which the phase separators are an integral part of the process, rather than the heat exchanger flow distribution system. Several phase separator processes have been developed in the last 40 years. Some of the more important phase separator processes are described in the following sections.

The vapor fraction of the main refrigerant at the entry of the main heat exchanger (stream 4) decreases with an increase in the operating pressure (p_4) or a decrease in the precooling temperature (T_4). With an appropriate operating pressure and precooling temperature, the refrigerant entering the main heat exchanger (stream 4) can be either a saturated or a slightly subcooled liquid as in the LIQUEFIN process [35].

6.6 LNG processes with a phase separator

Most of the problems associated with the possibility of freezing of high boilers in the refrigerant of the single-stage process shown in Fig. 6.3 can be easily overcome using phase separators that return high boilers to the compressor at relatively high temperatures (220 to 260 K), much above the freezing point of the high boilers. The simplest of the phase separator processes is shown in Fig. 6.17. The compression system consists of three-stage compressors with aftercoolers/ condensers as shown in Fig. 6.18. In this process, the refrigerant mixture composition and operating pressures are so chosen that partial condensation of the refrigerant occurs in the condenser. The liquid and vapor phases are separated in a phase separator before passing through the precooling heat exchanger (HX-1). The liquid phase, which contains most of the high boilers, is subcooled and expanded to a lower pressure, to provide the refrigeration necessary to cool and partially condense the vapor stream leaving the phase separator to a temperature of 220–260 K. The natural gas is in a superheated state at the exit of the precooling heat exchanger (HX-1). The natural gas feed is condensed (liquefied) and subcooled in the main heat exchanger (HX-2). The phase separator process shown in Fig. 6.17 is similar to the precooled process shown in Fig. 6.12 but with the precooling refrigerant derived from the refrigerant mixture leaving the compressors using the phase separator.

The advantages with a phase separator process can be understood with the help of an example. The design specifications (Table 6.11) are the same as those for the single-stage liquefaction process (Table 6.3). However, the maximum operating pressure of the refrigerant mixture has been increased to 60 bar and that of the natural gas feed

Table 6.11. Design specifications for a natural gas liquefier with a phase separator process (Fig. 6.17)

Minimum temperature approach in condensers	5.0 K
Minimum temperature approach in heat exchangers	3.0 K
Maximum compressor discharge pressure, p_2	60 bar
Minimum compressor suction pressure, p_1	3 bar
Feed (natural gas) operating pressure, p_{13}	65 bar
Pressure drop in the heat exchangers, Δp	0 bar
Adiabatic efficiency of compressors	80%
Adiabatic efficiency of pumps	90%

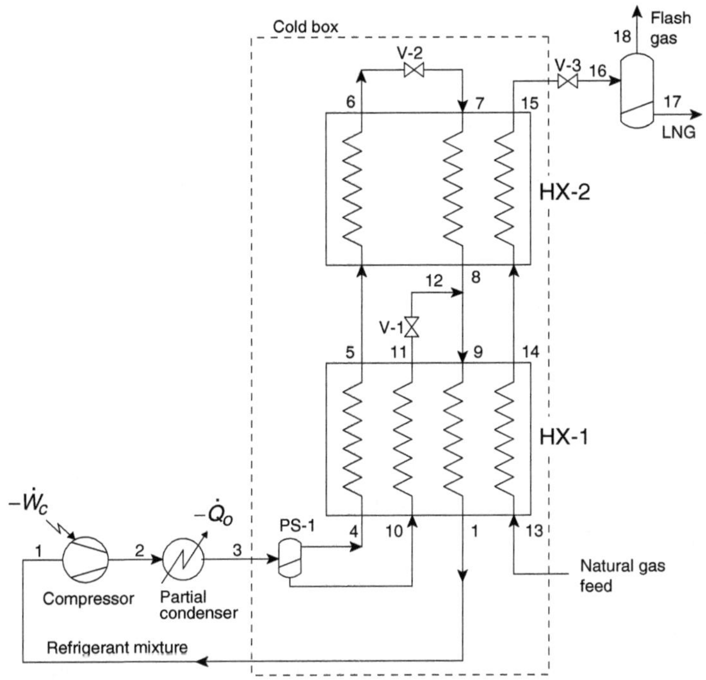

Fig. 6.17. single-stage mixed refrigerant natural gas liquefaction cycle with a phase separator [31]. The dotted line shows the control volume of the cold box defined in this example. (Adapted from U.S. Patent no. 3,932,154.)

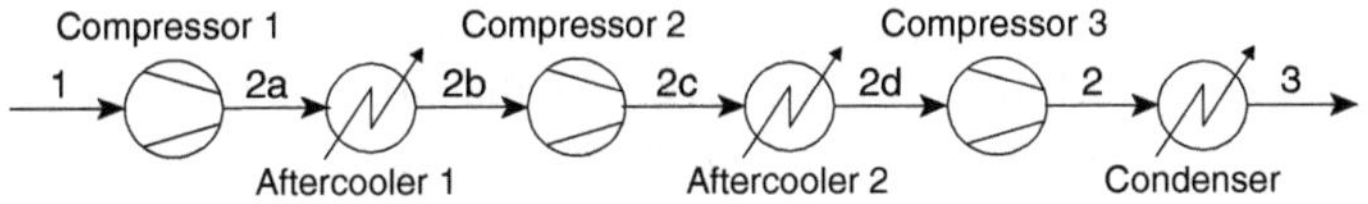

Fig. 6.18. Compressor arrangement for simplified process shown in Fig. 6.17.

to 65 bar, with the assumption that a spiral-tube (Giauque–Hampson) heat exchanger will be used in this example compared to plate-fin heat exchangers in the case of the single-stage LNG process shown in Fig. 6.3. As in the previous example, the LNG expander is not considered part of the cold box. The cold flash vapor is warmed in the heat exchangers and used as a fuel in the plant itself in large plants. In order to simplify the analysis, the refrigeration available with the flash vapor generated at the exit of valve V3 is not used in the liquefaction of the natural gas feed in this example.

The performance of the system has been evaluated for the feed and refrigerant mixture specified in Table 6.12. As in the previous example, the refrigerant composition has been optimized to maximize the exergy efficiency of the cold box subject to the minimum temperature approach between streams being 5 K in the aftercoolers/condensers and 3 K in the heat exchangers in the cold box. The temperature of the high-pressure refrigerant and the natural gas leaving the precooling heat

Table 6.12. Feed and refrigerant composition (mole fraction) for the process shown in Fig. 6.17

Component	NG feed (mol%)	Stream 3[a] (mol%)	Stream 4 (mol%)	Stream 11 (mol%)
Nitrogen	4.0	8.59	12.25	2.16
Methane	87.5	25.97	33.97	11.93
Ethane	5.5	25.42	25.53	25.22
Propane	2.1	39.11	27.84	58.88
*n*Butane	0.5	0.92	0.41	1.81
*i*Butane	0.3	–	–	–
*i*Pentane	0.1	–	–	–

[a] Refrigerant circulating through compressor.

exchanger (HX-1) as well as that of the high-pressure refrigerant leaving the second heat exchanger (HX-2) are considered design variables in the optimization model. The temperature of the natural gas at the end of the precooling heat exchanger (T_{14}) is constrained to prevent partial condensation of the natural gas stream in the first heat exchanger (HX-1).

The problem of freezing of the refrigerant can be minimized in this example by using less than 1% of *n*Butane and no pentanes in the main heat exchanger. The phase separator reduces the content of the *n*Butane to about 0.4% (stream 4). Calculations performed with the TFREEZE routine of the Aspen Plus process simulator showed that freezing may occur with the refrigerant mixture below 85 K.

The details of the temperature, pressure, vapor fraction, and mole flow rate of different streams are shown in Tables 6.13 and 6.14. The power input to the different compressors is shown in Table 6.15. It can be seen that about 36% of the refrigerant mixture is partially condensed in the condenser (stream 3, Fig. 6.17). The liquid phase separated in the phase separator is subcooled in heat exchanger HX-1 before expansion in valve V-1. Subcooling of stream 11 results in lower temperature change across valve V-1 and more balanced heat capacity rates ($\dot{n}c_p$) of the hot and cold fluid streams in the heat exchanger, resulting in a lower exergy loss. The natural gas feed and the high-pressure refrigerant stream (stream 5) are cooled to a temperature of 254.5 K in the precooling heat exchanger (HX-1).

Another important advantage in using a phase separator is the reduction in the flow rate of the refrigerant in the main heat exchanger (HX-2). The feed flow rate is about 63%[1] of that for the refrigerant in the precooling heat exchanger and about 40% in the main heat exchanger (HX-2) of the phase separator process shown in Fig. 6.17, compared to 24% in the single-stage process without phase separators (Fig. 6.3). This results in a significantly smaller main heat exchanger for the phase separator process compared to the single-stage process shown in Fig. 6.3.

[1] $0.40/0.637 = 0.628$.

Table 6.13. Temperature, pressure, vapor fraction, and flow rate of different streams of the natural gas liquefaction process shown in Fig. 6.17 and operating with the refrigerant mixture in Table 6.12

	Stream						
	1	3	4	5	6	7	8
Temperature, K	302.00	305.00	305.00	254.50	113.00	109.00	244.30
Pressure, bar	3.00	46.80	46.80	46.80	46.80	3.00	3.00
Vapor fraction	1.000	0.637	1.000	0.407	0.000	0.094	1.000
Flow rate, mol/s	1.000	1.000	0.637	0.637	0.637	0.637	0.637
	Stream						
	9	10	11	12	13	14	15
Temperature, K	228.20	305.00	254.50	222.30	300.00	254.50	113.00
Pressure, bar	3.00	46.80	46.80	3.00	65.00	65.00	65.00
Vapor fraction	0.765	0.000	0.000	0.287	1.000	1.000	0.000
Flow rate, mol/s	1.000	0.363	0.363	0.363	0.401	0.401	0.401

Table 6.14. Temperature, pressure, and vapor fraction of different streams of the compressors in Fig. 6.18 and operating with the feed and mixture in Table 6.12

	Stream						
	1	2a	2b	2c	2d	2	3
Temperature, K	302.00	355.70	305.00	360.66	305.00	365.10	305.00
Pressure, bar	3.00	7.50	7.50	18.75	18.75	46.75	46.75
Vapor fraction	1.000	1.000	1.000	1.000	1.000	1.000	0.637

Table 6.15. Power input to the three compressors in the process shown in Fig. 6.18 and operating with the mixtures provided in Table 6.12 and a natural gas feed flow rate of 0.401 mol/s

	Compressor		
	C1	C2	C3
Power input, kW	2.99	2.89	2.57
Flow rate, mol/s	1.00	1.00	1.00

The exergy efficiency of the compressors (Fig. 6.18) is given by the expression

$$\eta_{ex,cs} = \frac{\text{minimum power for compression}}{\text{compressor power input}} = \frac{\dot{n}_1(ex_3 - ex_1)}{\text{compressor power input}}. \quad (6.8)$$

The exergy efficiency of the cold box consisting of the heat exchangers, phase separator, and expansion valves is given by the expression

$$\eta_{ex,cs} = \frac{\text{minimum power for liquefaction}}{\text{exergy expenditure}} = \frac{\dot{n}_{16}(ex_{15} - ex_{13})}{\dot{n}_3(ex_3 - ex_1)}. \quad (6.9)$$

The overall exergy efficiency of the process, excluding the LNG expansion valve, is given by the expression

$$\eta_{ex,o} = \frac{\text{minimum power for liquefaction}}{\text{compressor power input}} = \frac{\dot{n}_{16}(ex_{15} - ex_{13})}{\text{compressor power input}} \quad (6.10)$$

or $$\eta_{ex,o} = \eta_{ex,cb}\,\eta_{ex,co}. \quad (6.11)$$

Figure 6.19 shows the exergy utilization in the cold box, the compressors, and the overall process. Only 53.6% of the exergy input to the cold box is used for the liquefaction of natural gas, as shown in Fig. 6.19. The rest of the input exergy is lost predominantly in the heat exchangers and the expansion valves. The exergy loss in the precooling heat exchanger (HX-1) is about double that in the main heat exchanger (HX-2). The exergy loss in the phase separator is zero since the pressure drop in the phase separator has been assumed to be zero.

It can be seen from Figs. 6.20 and 6.21 that the temperature difference between the hot and cold streams is quite large in the first heat exchanger (HX-1), resulting in a large exergy loss. The average LMTD in the HX-1 and HX-2 heat exchangers is 23.6 K and 4.7 K, respectively (see Table 6.16). The NTU of the cold and hot streams of HX-1 is an order of magnitude less than that of the main heat exchanger (HX-2).

One of the disadvantages with any phase separator process is the exergy loss associated with the mixing of streams of different composition (separated in the phase separator). It can be seen from Fig. 6.19 that the exergy loss in the mixer is small and only 0.9% of that in the cold box.

The exergy loss in the compressors and aftercoolers is much more uniformly divided than in the cold box, with the exergy efficiency of compressors and aftercoolers

Table 6.16. Temperature approach between streams in the two heat exchangers of the phase separator process shown in Fig. 6.18 and operating with the mixtures shown in Table 6.12

Temperature approach	HX-1	HX-2
LMTD (ΔT_{lm}), K	23.6	4.7
Minimum approach (ΔT_{min}), K	3.0	3.0
Warm end approach (ΔT_{we}), K	3.0	10.2
Cold end approach (ΔT_{ce}), K	26.3	4.0

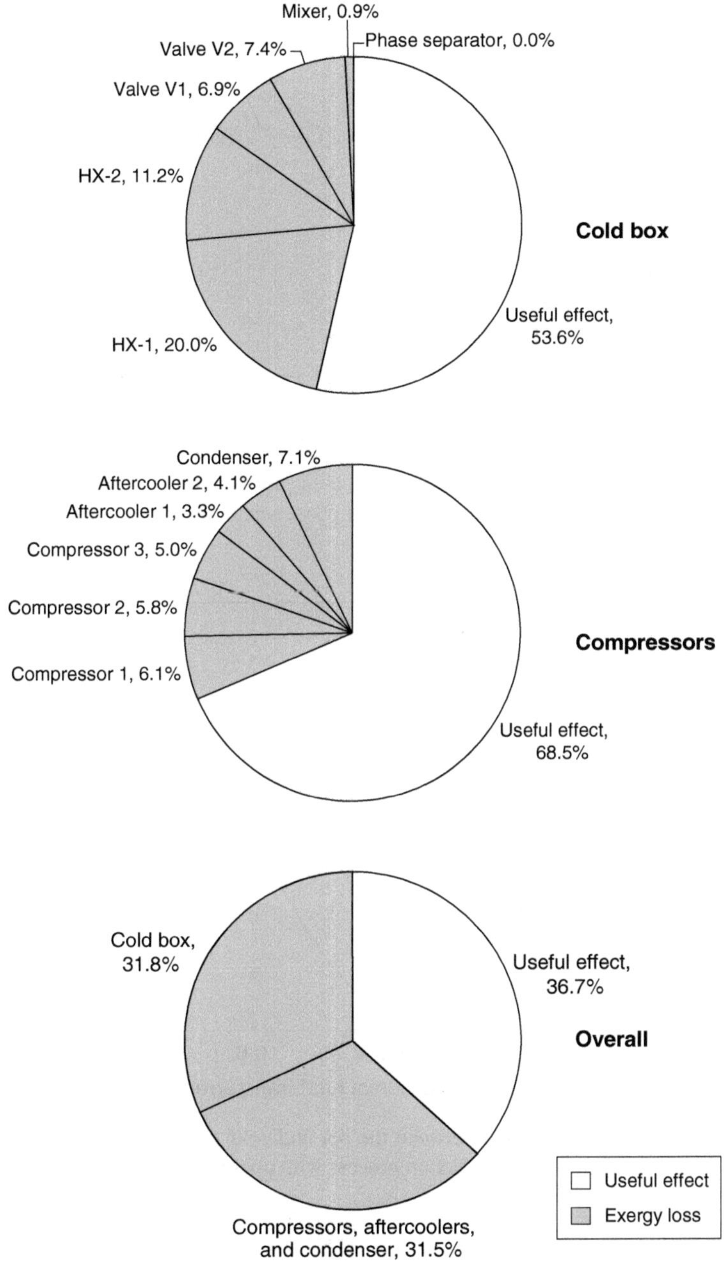

Fig. 6.19. Exergy losses in the cold box and compressors, aftercoolers, and condenser of a mixed refrigerant natural gas liquefaction process with a phase separator shown in Fig. 6.17 and operating with feed and refrigerant mixture shown in Table 6.12.

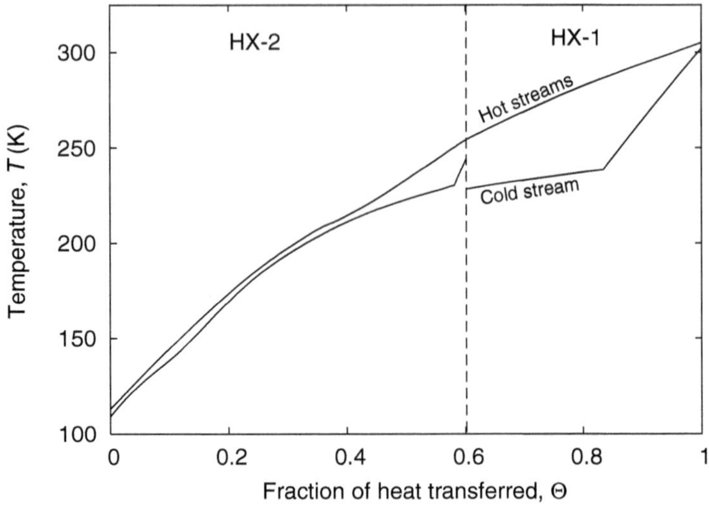

Fig. 6.20. Temperature profiles of the hot and cold fluid streams in the single-stage mixed refrigerant natural gas liquefaction process with phase separator shown in Fig. 6.17.

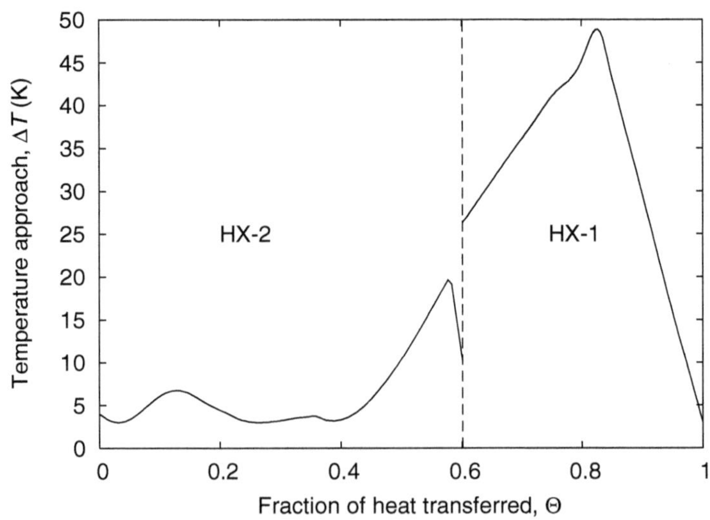

Fig. 6.21. Temperature approach between the hot and cold fluid streams in the single-stage mixed refrigerant natural gas liquefaction process with phase separator shown in Fig. 6.17.

being 68.5%. The overall exergy efficiency of the system is only 36.7%, with the exergy loss being nearly the same in the cold and warm sections of the system. The overall exergy efficiency of the system shown in Fig. 6.17 is therefore about the same as that of the single-stage liquefaction process shown in Fig. 6.3, and much lower than that of the precooled process shown in Fig. 6.12.

The advantages of the phase separator process shown in Fig. 6.17 over the single-stage liquefaction process shown in Fig. 6.3 are

- smaller refrigerant flow rate in the main heat exchanger (HX-2), leading to smaller heat exchanger sizes, and
- no possibility of freezing of the main refrigerant at low temperatures.

The main disadvantage in this process is the high exergy losses in the precooling heat exchanger (HX-1), due to the large temperature approach between the streams in the heat exchanger (Figs. 6.20 and 6.21). A process that uses a separate precooling refrigerant is more efficient than the present example. Most modern base-load plants operate on variants of the phase separator processes shown in Fig. 6.17.

Consider a control volume that includes only the main heat exchanger (HX-2) and valve V-2. The system in the control volume is identical to that in the cold box of the single-stage process shown in Fig. 6.3, except that the temperature of the feed and the high-pressure refrigerant entering the control volume is 254.5 K, instead of 300 K and 305 K, respectively, in the case of the single refrigerant process shown in Fig. 6.3. The exergy efficiency of this control volume ($\eta_{\text{ex, cv}}$) is given by the expression

$$\eta_{\text{ex, cv}} = \frac{\text{minimum power for liquefaction}}{\text{exergy expenditure}} = \frac{\dot{n}_{14}(\text{ex}_{15} - \text{ex}_{14})}{\dot{n}_5(\text{ex}_5 - \text{ex}_8)}. \tag{6.12}$$

When the heat exchanger HX-1 and valve V-1 are excluded, the exergy efficiency of the cold box turns out to be 74.2% and higher than that in the precooled process shown in Fig. 6.12. The overall exergy efficiency of the phase separator process can be improved substantially by decreasing the temperature approach between the streams in the first heat exchanger (HX-1). One way of decreasing the temperature approach between the streams in the first heat exchanger is to add a precooler to the phase separator process shown in Fig. 6.17. Just as precooling with a separate precooling refrigerant improves the exergy efficiency of the single-stage liquefaction process, precooling also improves the exergy efficiency of the phase separator process shown in Fig. 6.17.

Most base-load plants use precooled phase separator processes. The precooling can be done using a separate precooling refrigerant or the precooling refrigerant can be derived from the main refrigerant itself. The different precooled phase separator processes with phase separators are discussed in the next section.

The compressor power required in the single-stage phase separator process can be decreased by using different pressures for the low-pressure refrigerant in the two heat exchangers. Figure 6.22 shows one such patented process [80]. The utilization of input exergy in the process is shown in Fig. 6.23. It can be seen from Figs. 6.19 and 6.23 that the overall exergy efficiency of the process improves by about 3% by using different pressures for the low-pressure refrigerant in the two heat exchangers.

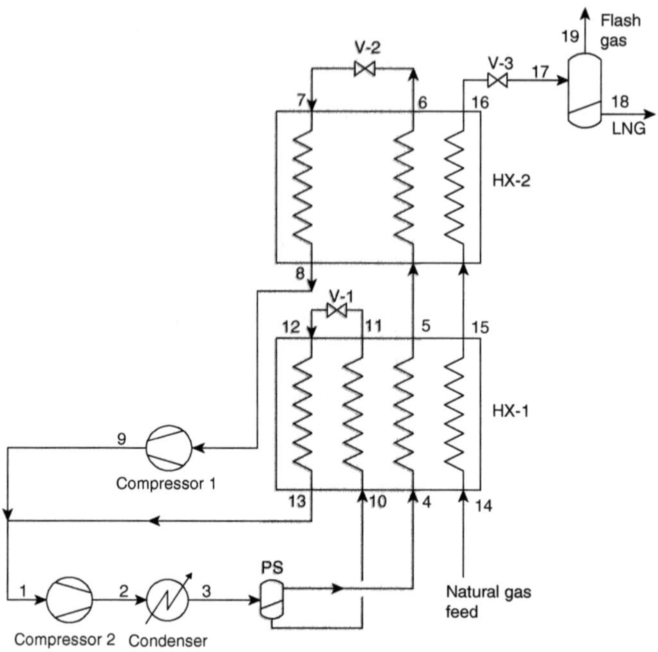

Fig. 6.22. Simplified version of the single-stage mixed refrigerant natural gas liquefaction process with two low-pressure stages presented in Ref. [80]. (Adapted from U.S. Patent no. 6,347,531.)

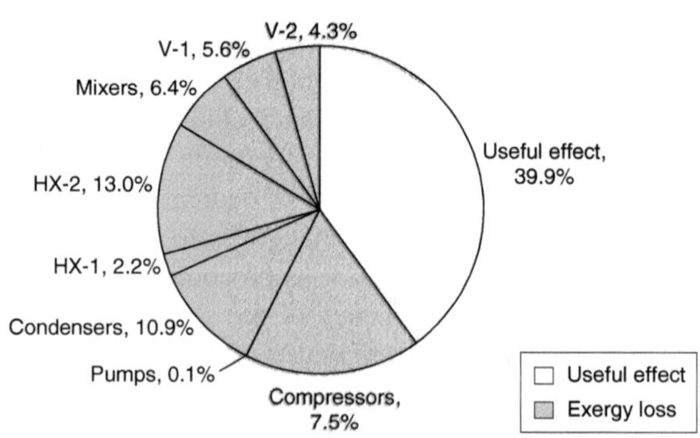

Fig. 6.23. Exergy utilization in a single-stage mixed refrigerant natural gas liquefaction process shown in Fig. 6.22.

6.7 Precooled LNG process with a phase separator

Consider a precooled natural gas liquefaction process with a single phase separator shown in Fig. 6.24. A three-stage compression process is used for the main refrigerant, as shown in Fig. 6.25. The precooling temperature has been assumed to be 240 K.

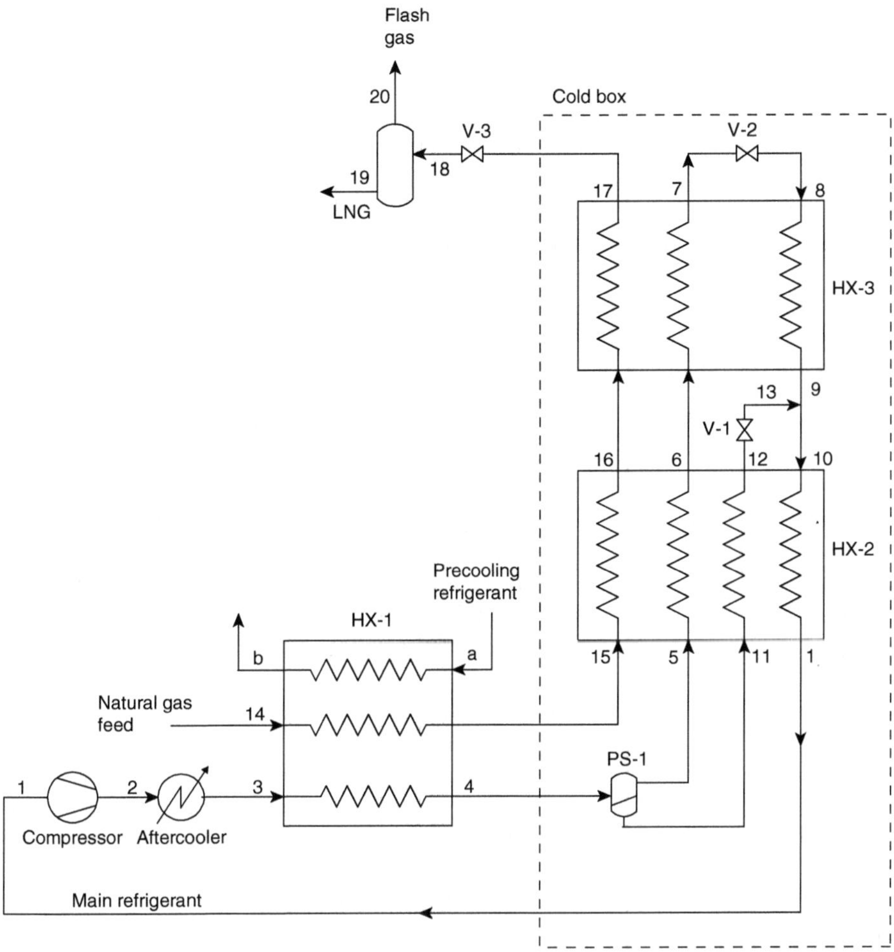

Fig. 6.24. Precooled mixed refrigerant natural gas liquefaction process with a phase separator.

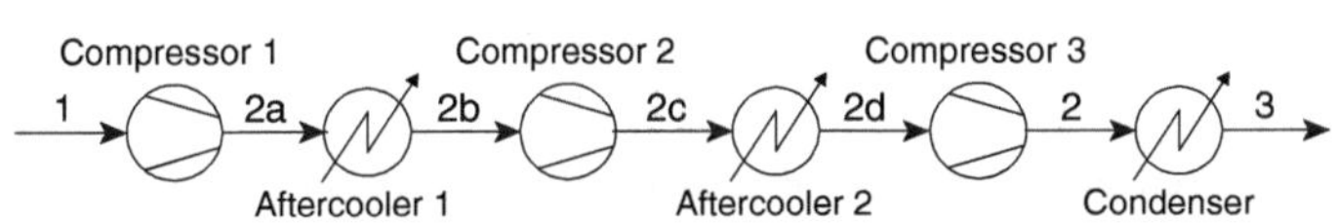

Fig. 6.25. Compressor arrangement for the main refrigerant in the process shown in Fig. 6.24.

The process is similar to the LNG process with a phase separator (Fig. 6.17), except for the additional precooling heat exchanger (HX-1). The performance of the system has been evaluated for the feed and refrigerant specified in Table 6.17.

Three different refrigerants are evaporated in the three heat exchangers to cool the hot fluid streams. However, the refrigerants evaporated in the second and third heat exchangers are derived from a single refrigerant using the phase separator. The three refrigerant compositions can be so chosen that the $(\partial h/\partial T)_p$ of the hot and cold fluid streams satisfy the relationship presented in Eq. (3.10).

Different precooling arrangements can be used and are described in the following section. In this section, the performance of a generic precooled phase separator process is discussed with reference to a cold box that excludes the precooler.

The exergy efficiency of the cold box is given by the expression

$$\eta_{\mathrm{ex,cb}} = \frac{\text{minimum power for liquefaction}}{\text{exergy expenditure}} = \frac{\dot{n}_{17}(\mathrm{ex}_{17} - \mathrm{ex}_{15})}{\dot{n}_4(\mathrm{ex}_4 - \mathrm{ex}_1)}. \quad (6.13)$$

As in previous examples, the refrigerant composition has been optimized to maximize the exergy efficiency of the cold box subject to the minimum temperature approach of 3 K in the heat exchangers in the cold box. The problem of freezing of the refrigerant can be essentially eliminated in this example by not using butanes and pentanes in the main heat exchanger. Propane freezes below 76 K in the present example.

The details of the pressure, temperature, vapor fraction, and flow rate of different streams are shown in Table 6.18. It can be seen that 71% of the refrigerant mixture is partially condensed in the precooling heat exchanger (HX-1) (stream 4, Fig. 6.24). The ratio of mole flow rates for the feed and the low-pressure refrigerant is nearly 75% in the second heat exchanger (HX-2) and about 258% in the third heat exchanger (HX-3). The third heat exchanger (HX-3) can therefore be much smaller in heat load compared to the second heat exchanger (HX-2), as shown in Fig. 6.26. It can be seen from Fig. 6.26 that the temperature profiles of the hot and cold streams are somewhat parallel to a large extent in the second heat exchanger (HX-2) except near the dew point of the low-pressure refrigerant (Fig. 6.27). The maximum temperature approach is limited to about 13 K in HX-2 and slightly over 7 K in HX-3.

Table 6.17. Feed and refrigerant composition for the process shown in Fig. 6.24

Component	Stream 3 (mol%)	Stream 4 (mol%)	Stream 11 (mol%)	NG feed (mol%)
Nitrogen	7.0	17.1	2.9	4.0
Methane	41.8	65.5	32.1	87.5
Ethane	29.9	14.1	36.4	5.5
Propane	21.3	3.3	28.6	2.1
*n*Butane	—	—	—	0.5
*i*Butane	—	—	—	0.3
*i*Pentane	—	—	—	0.1

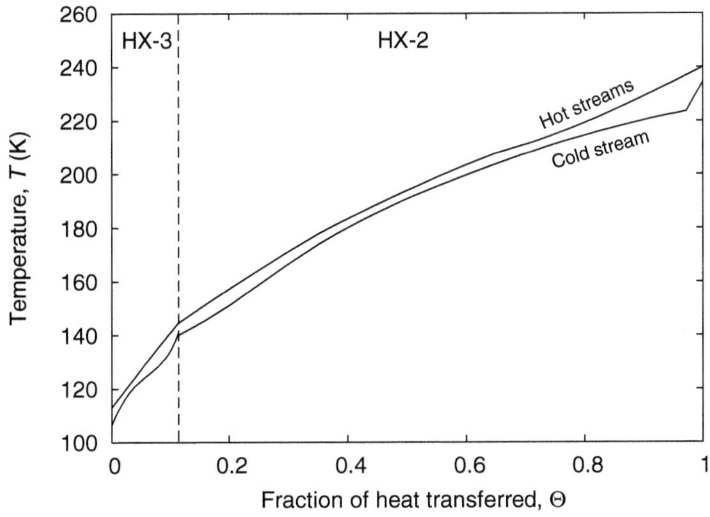

Fig. 6.26. Temperature profiles in the precooled mixed refrigerant natural gas liquefaction process with a phase separator shown in Fig. 6.24.

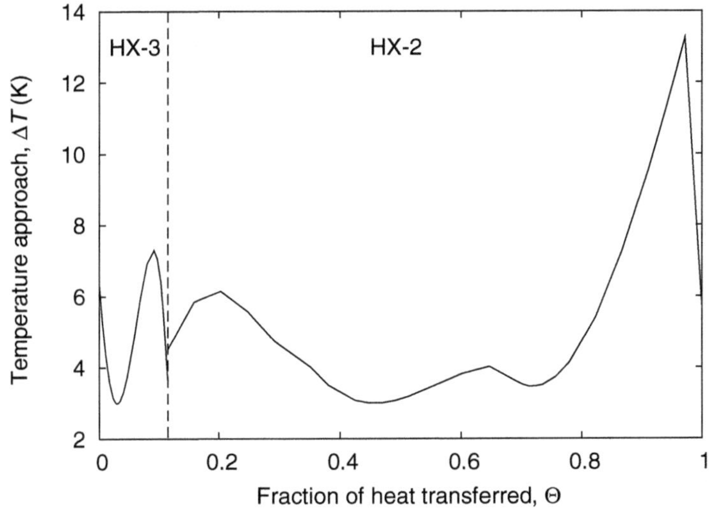

Fig. 6.27. Temperature approach between the hot and cold fluid streams in the precooled mixed refrigerant natural gas liquefaction process with a phase separator shown in Fig. 6.24.

Table 6.18. Temperature, pressure, vapor fraction, and flow rate of different streams of the compressors in Fig. 6.24 and operating with the feed and mixture in Table 6.17

	Stream						
	1	4	5	6	7	8	9
Temperature, K	234.3	240.0	240.0	144.7	113.0	106.7	141.1
Pressure, bar	3.0	48.6	48.6	48.6	48.6	3.0	3.0
Vapor fraction	1.000	0.290	1.000	0.000	0.000	0.095	0.761
Flow rate, mol/s	1.000	1.000	0.290	0.290	0.290	0.290	0.290
	Stream						
	10	11	12	13	15	16	17
Temperature, K	140.2	240.0	144.7	139.2	240.0	144.7	113.0
Pressure, bar	3.0	48.6	48.6	3.0	65.0	65.0	65.0
Vapor fraction	0.276	0.000	0.000	0.084	1.000	0.000	0.000
Flow rate, mol/s	1.000	0.710	0.710	0.710	0.748	0.748	0.748

Figure 6.28 shows the utilization of input exergy for the cold box, which excludes the compressors, the aftercoolers (condensers), and the precooler.

The exergy efficiency of the cold box in this example can be seen from Fig. 6.28 to be 80.3%. The exergy loss in the second expansion valve, V-2, is 5.6% and that in the first is 2.2%. Replacing valve V-2 with a turbine expander will result in a significant gain in this case. The largest exergy loss, at 9.1%, is in the second heat exchanger (HX-2), whereas that in the third heat exchanger (HX-3) is only 2.6%. While the average LMTD of the two heat exchangers is nearly the same in this case (Table 6.19), the temperature approach between the streams is high at the warm end of the second heat exchanger (HX-2) compared to that in the third heat exchanger (HX-3), resulting in a higher exergy loss in the second heat exchanger.

A comparison of the efficiency of the cold box of the LNG process with a phase separator (Fig. 6.17) with that of the current process (Fig. 6.24) shows that the exergy efficiency of the cold box increases from 53.6% (Fig. 6.19) to 80.3% due to precooling (Fig. 6.28).

The phase separator in the process ensures that the high-pressure refrigerant is in the gaseous phase at the entry of the heat exchanger HX-2. The high-pressure

Table 6.19. Temperature approach between streams in different heat exchangers of the phase separator process shown in Fig. 6.24 and operating with mixtures from Table 6.12

Temperature approach	HX-2	HX-3
LMTD (ΔT_{lm}), K	4.5	4.6
Minimum approach (ΔT_{min}), K	3.0	3.0
Warm end approach (ΔT_{we}), K	5.7	3.6
Cold end approach (ΔT_{ce}), K	4.5	6.3

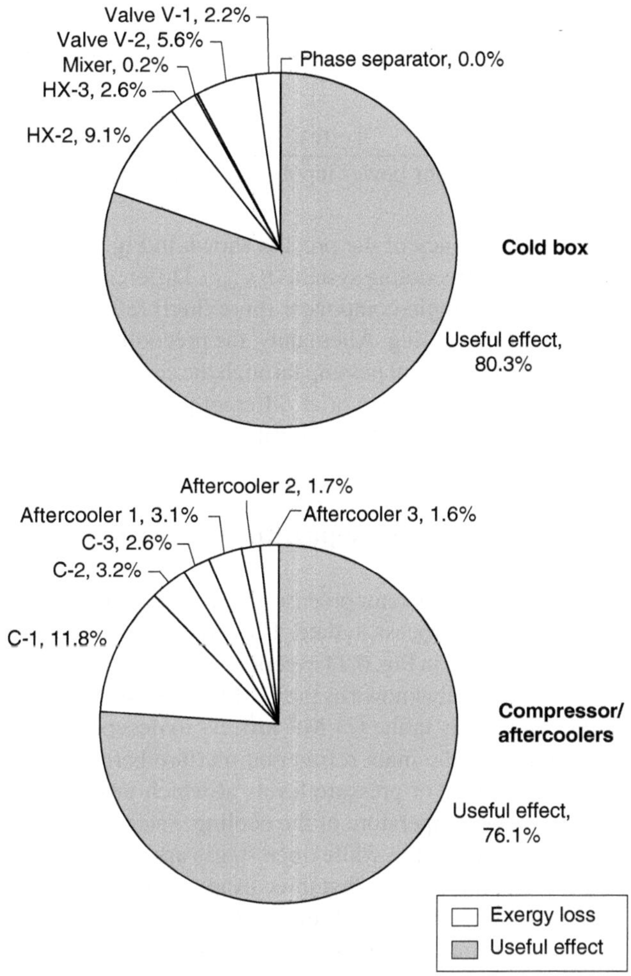

Fig. 6.28. Exergy utilization in the cold box and compressors of a precooled mixed refrigerant natural gas liquefaction process with one phase separator shown in Figs. 6.24 and 6.25 with the feed and refrigerants shown in Table 6.17.

refrigerant entering the third heat exchanger (HX-3) is in a liquid state at the entry. In some processes e.g., Ref. [78] part of the liquid phase leaving the phase separator is mixed with the vapor phase to improve the efficiency of the process.

The exergy efficiency of the compressor section (compressors and aftercoolers) is given by the expression

$$\eta_{\mathrm{ex,cs}} = \frac{\text{compressor power input to a reversible compressor}}{\text{compressor power input}} \tag{6.14}$$

$$= \frac{\dot{n}_1(\mathrm{ex}_3 - \mathrm{ex}_1)}{\sum_{i=1}^{3} -\dot{W}_i}. \tag{6.15}$$

The exergy efficiency of the compressor section of the main refrigerant is 76.1% in the present example (Fig. 6.28). The overall exergy efficiency of the process is given by the expression

$$\eta_{ex,o} = \frac{\dot{n}_{17}(\text{ex}_{17} - \text{ex}_{14})}{\text{compressor power input} + \dot{n}_a(\text{ex}_a - \text{ex}_b)/\eta_{\text{ex,pre}}}. \tag{6.16}$$

The overall exergy efficiency of the process shown in Fig. 6.24 is dependent on the exergy efficiency of the precooling system ($\eta_{\text{ex,pre}}$). Different arrangements can be used for precooling. Either single-component (pure fluid) refrigerants or refrigerant mixtures can be used for precooling. Alternately, the precooling refrigerant can itself be derived from the main refrigerant passing through the compressors. The refrigerant can be evaporated at a single pressure or at different pressures. Different processes result from each of these choices and are described in the following sections.

6.8 Propane precooled phase separator (C3-MR) process

The propane precooled phase separator process, shown in Fig. 6.29, is the most widely used natural gas liquefaction process to date. The process is similar to the precooled phase separator process shown in Fig. 6.24 except for the precooling part. The propane precooled process is also widely known as the C3-MR process. Propane is evaporated at three or four pressure levels in the C3-MR process to desuperheat the natural gas feed and partially condense the main refrigerant mixture before it enters the phase separator (PS-4). The number of pressure levels at which propane is evaporated is normally dependent on the temperature of the cooling water or air used. Four stages are advantageous in warm climates, while three stages are adequate in cold climates. Propane is partially evaporated in the first three evaporators (HX-1–HX-3) and fully evaporated in the final evaporator (HX-4) of Fig. 6.29.

In processes where propane and other high boilers (C_{3+}) are removed from the natural gas feed, the separation of high boilers is carried out at the end of the propane precooling stage. The propane evaporating pressures and flow rate were determined for the main refrigerant mixture shown in Table 6.17. The temperature, pressure, vapor fraction, and flow rate of the main refrigerant are essentially the same as that in Table 6.18.

The exergy efficiency of the precooler ($\eta_{\text{ex, pre}}$) is given by the expression

$$\eta_{\text{ex, pre}} = \frac{\text{minimum power for cooling feed and main refrigerant}}{\text{exergy input}}, \tag{6.17}$$

$$\eta_{\text{ex, pre}} = \frac{\dot{n}_{39}(\text{ex}_{43} - \text{ex}_{39}) + \dot{n}_3(\text{ex}_7 - \text{ex}_3)}{\sum_{i=1}^{4} -W_{c,i,\text{precooler}}}. \tag{6.18}$$

The overall exergy efficiency ($\eta_{\text{ex},o}$) of the C3-MR process shown in Fig. 6.29 is given by the expression

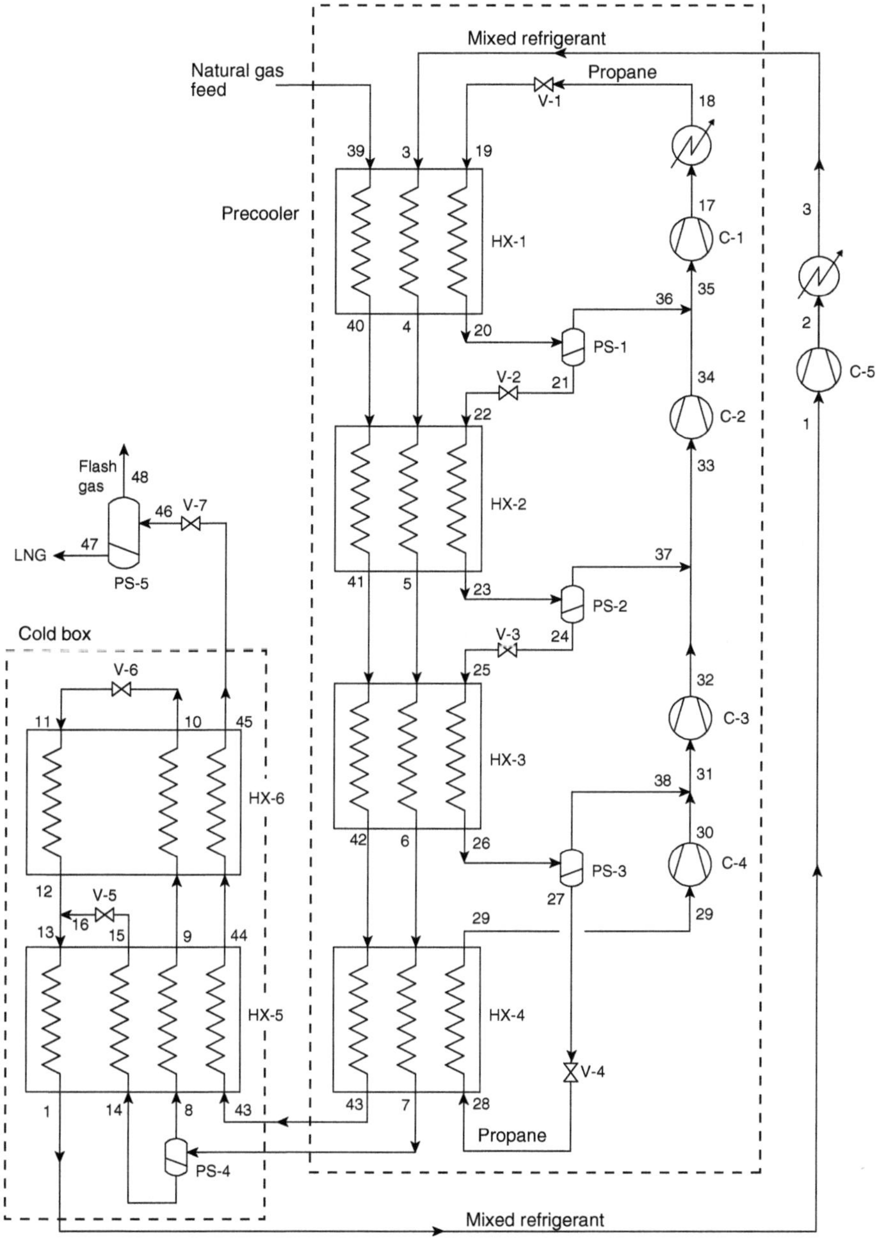

Fig. 6.29. Propane precooled mixed refrigerant natural gas liquefaction process [43]. (Adapted from U.S. Patent no. 3,763,658.)

$$\eta_{\text{ex},o} = \frac{\text{minimum work for cooling feed and main refrigerant}}{\text{total power input}} \tag{6.19}$$

$$= \frac{\dot{n}_{39}(\text{ex}_{45} - \text{ex}_{39})}{\sum_{i=1}^{5} -W_{c,i}}. \tag{6.20}$$

The design specifications for a propane precooled mixed refrigerant natural gas liquefier are shown in Table 6.20. The precooling temperature has been assumed to be 240 K. A condensing temperature of 315 K was assumed for the propane condenser. A small subcooling is usually used in practice. However, the degree of subcooling was assumed to be zero in the present case to simplify the analysis. The evaporating pressures and flow rate of propane were determined using an optimization model in which the exergy efficiency of the propane precooler is maximized subject to a temperature approach of 3 K in all heat exchangers.

The temperature, pressure, vapor fraction, and flow rate at different streams of the precooling section of the propane precooled LNG process (Fig. 6.29) are shown in Table 6.21. The temperature, pressure, vapor fraction, and flow rate of the main refrigeration section (cold box) are essentially the same as those in Table 6.18. The mole flow rate of stream 19 of the precooling refrigerant (1.027 mol/s) is slightly greater than that of stream 3 of the main refrigerant (1.0 mol/s) in the first precooling heat exchanger (HX-1). The precooling refrigerant flow decreases to 0.166 mol/s in the final precooling heat exchanger (HX-4) as most of the refrigerant is returned to the compressors in the three phase separators (PS-1, PS-2, and PS-3).

It can also be observed from Table 6.21 that propane is partially evaporated to a vapor fraction of 35.4% (stream 20) in the first heat exchanger (HX-1). The vapor and liquid propane at the end of the first heat exchanger (HX-1) are separated in a phase separator (PS-1). The vapor phase is mixed with the output of compressor C-2 and sent to compressor C-1, while the liquid phase is expanded in valve V-2 and sent to the second heat exchanger (HX-2). This process is repeated until the propane is completely evaporated in the fourth heat exchanger (HX-4), as shown in Fig. 6.29. Propane is normally in a superheated state at the entry of compressor C-4.

Figure 6.30 shows the exergy utilization in the precooler and the overall process. The exergy efficiency of the optimized precooler is 34.2%, and the overall exergy

Table 6.20. Design specifications for a propane precooled mixed refrigerant process natural gas liquefier (Fig. 6.29)

Temperature of precooling refrigerant leaving condensers	315 K
Temperature of main refrigerant leaving partial condenser	305 K
Minimum temperature approach in heat exchangers	3.0 K
Feed (natural gas) operating pressure, p_{39}	65 bar
Pressure drop in the heat exchangers, Δp	0 bar
Adiabatic efficiency of compressors	80%
Adiabatic efficiency of pumps	90%
Precooling temperature	240 K
Minimum compressor suction pressure	3 bar

Table 6.21. The temperature, pressure, vapor fraction, and flow rate of different streams of the propane precooled natural gas liquefaction process (Fig. 6.29). See Table 6.18 for details of the main refrigeration section (cold box)

	Stream					
	3	4	5	6	7	17
Temperature, K	305.0	290.6	278.8	256.8	240.0	330.3
Pressure, bar	48.6	48.6	48.6	48.6	48.6	14.3
Vapor fraction	1.000	0.885	0.699	0.445	0.290	1.000
Flow rate, mol/s	1.000	1.000	1.000	1.000	1.000	1.027

	Stream					
	18	19	20	21	22	23
Temperature, K	315.0	287.5	287.5	287.5	275.7	275.7
Pressure, bar	14.3	7.2	7.2	7.2	5.1	5.1
Vapor fraction	0.000	0.221	0.354	0.000	0.082	0.312
Flow rate, mol/s	1.027	1.027	1.027	0.663	0.663	0.663

	Stream					
	24	25	26	27	28	29
Temperature, K	275.7	253.8	253.8	253.8	237.0	242.0
Pressure, bar	5.1	2.5	2.5	2.5	1.3	1.3
Vapor fraction	0.000	0.132	0.637	0.000	0.091	1.000
Flow rate, mol/s	0.456	0.456	0.456	0.166	0.166	0.166

	Stream					
	30	31	32	33	34	35
Temperature, K	268.9	259.3	290.5	286.0	301.5	296.6
Pressure, bar	2.5	2.5	5.1	5.1	7.2	7.2
Vapor fraction	1.000	1.000	1.000	1.000	1.000	1.000
Flow rate, mol/s	0.166	0.456	0.456	0.663	0.663	1.027

	Stream					
	36	37	38	39	40	41
Temperature, K	287.5	275.7	253.8	300.0	290.6	278.8
Pressure, bar	7.2	5.1	2.5	65.0	65.0	65.0
Vapor fraction	1.000	1.000	1.000	1.000	1.000	1.000
Flow rate, mol/s	0.363	0.207	0.291	0.748	0.748	0.748

	Stream	
	42	43
Temperature, K	256.8	240.0
Pressure, bar	65.0	65.0
Vapor fraction	1.000	1.000
Flow rate, mol/s	0.748	0.748

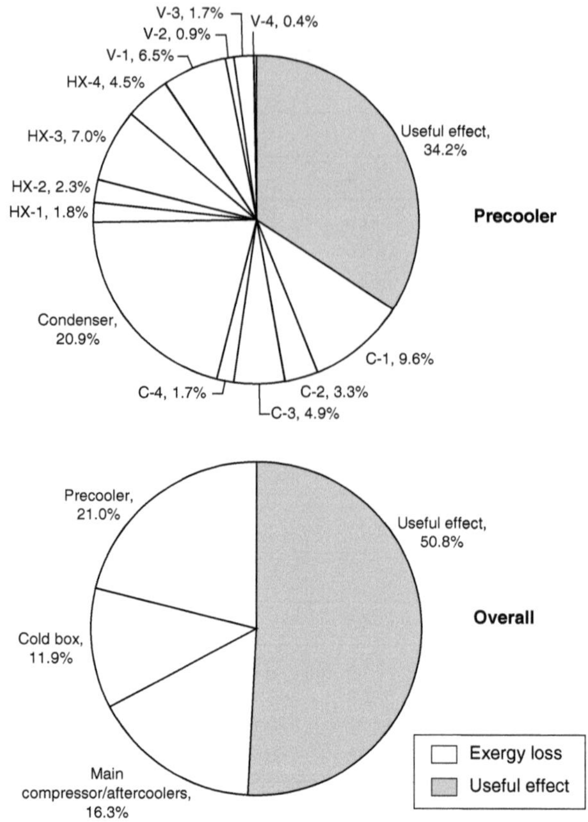

Fig. 6.30. Exergy utilization in the precooler and the overall process of a propane precooled mixed refrigerant natural gas liquefaction process.

efficiency of the liquefaction process is 50.8%. The overall exergy efficiency of the process decreases to less than 50% when the exergy losses due to flash gas and the throttling in valve V-7 are also taken into account. The exergy efficiency of the process can be increased by using smaller temperature approaches in the heat exchangers.

It can be seen from Fig. 6.30 that the maximum exergy loss in the case of the precooler occurs in the propane condenser due to the occurrence of a pinch point in between the two ends of the heat exchanger [see Fig. 6.32(a)]. This loss can be decreased by using a refrigerant mixture instead of propane. The maximum loss among the precooling-compressors occurs in the warmest compressor (C-1) and is nearly the same as that in the other three propane compressors since the flow rate through compressor C-1 and the temperature at the compressor suction are the highest among all the precooling refrigerant compressors. Figure 6.31 shows the temperature profile of the hot streams (main refrigerant mixture, natural gas feed) and the cold stream (propane refrigerant) in the four precooling heat exchangers. The large temperature difference between the hot and cold streams results in a higher exergy loss in the

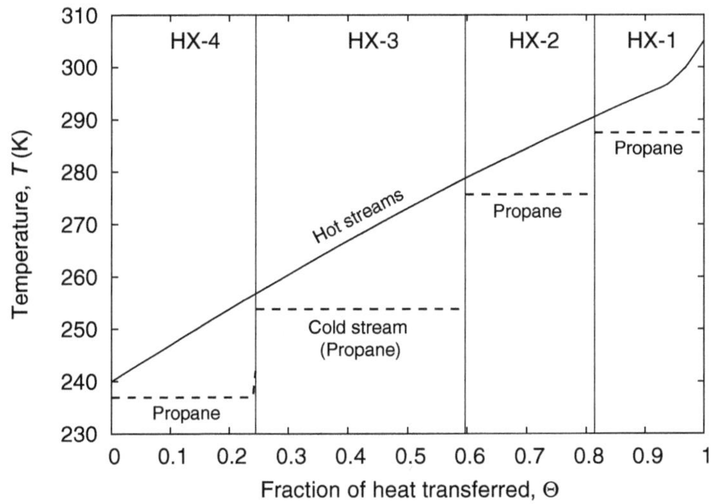

Fig. 6.31. Temperature profiles of the hot and cold (propane) streams in the four precooler heat exchangers of a four-stage precooler of propane precooled mixed refrigerant natural gas liquefaction process shown in Fig. 6.29 and operating with a main refrigerant (stream 3) whose composition is shown in Table 6.18.

propane precooled process compared to the mixed refrigerant precooled process (see Fig. 6.16). When the overall system is considered, the exergy loss in the main section is 28.2% and that in the precooling section is 21%. The exergy loss in the main refrigerant compressor and aftercooler is higher than that in the heat exchangers and valves in the main refrigeration section (cold box).

6.9 Mixed refrigerant precooled phase separator (DMR) processes

Consider the propane precooled mixed refrigerant process with one phase separator shown in Fig. 6.29. The temperature profiles in a water-cooled condenser of the propane precooling refrigerant are shown in Fig. 6.32. The minimum temperature approach between the streams or the pinch point occurs in between the two ends of the condenser due to a change in specific heat of the refrigerant from a finite value in the superheated vapor state to an infinite value during constant-temperature condensation. The temperature of propane leaving the condenser is therefore much higher than that of water or air entering the condenser.

A pinch point will also occur in the condenser at the dew point temperature of the refrigerant when a zeotropic mixture is used as the precooling refrigerant, as shown in Fig. 6.32(b). The temperature of the refrigerant leaving the condenser, however, will be lower than that in the propane case due to refrigerant glide (the difference between the dew and bubble point temperatures) during condensation. The smaller

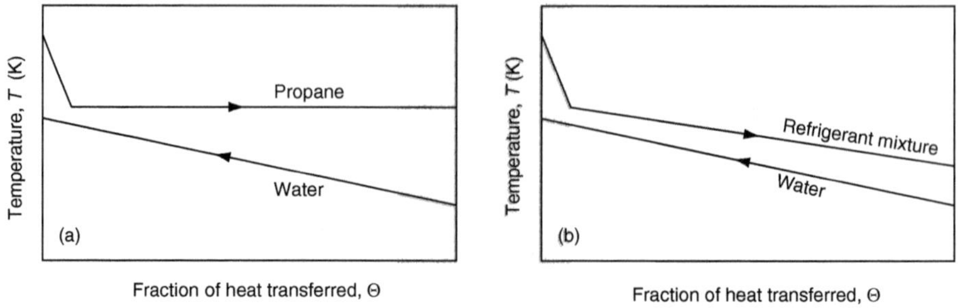

Fig. 6.32. Temperature profiles of the refrigerant and heat transfer fluid (water) in a typical condenser for (a) propane and (b) refrigerant mixture.

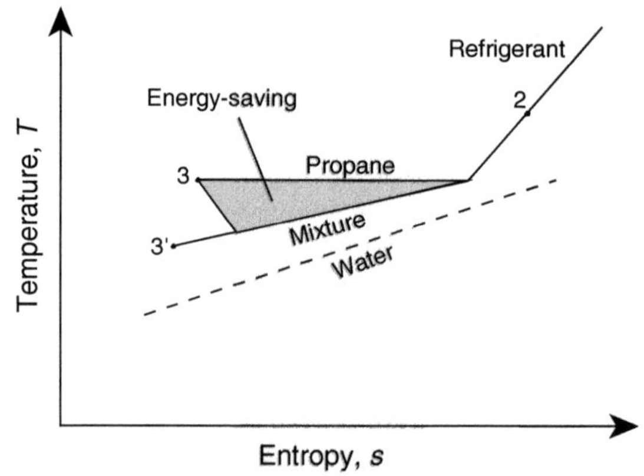

Fig. 6.33. Comparison of condensation process in a refrigerator operating with propane and a mixture.

temperature approach at the cold end leads to lower exergy losses. This results in a saving of compressor power when mixtures are used as the precooling refrigerant instead of pure fluids such as propane, as shown in Fig. 6.33. In the best possible case, the temperature profiles of the refrigerant mixture and water are parallel between the dew and bubble point temperatures of the refrigerant. The use of a refrigerant mixture therefore results in a lower compressor power or higher exergy efficiency.

Consider the precooled mixed refrigerant process with one phase separator shown in Fig. 6.24. Propane is used as the precooling refrigerant in the process shown in Fig. 6.29. A refrigerant mixture can also be used as the precooling refrigerant. A process that uses refrigerant mixtures in the main refrigeration process and precooling process is shown in Fig. 6.34. The process is usually known as the dual mixed refrigerant (DMR) process. The precooling refrigerant is evaporated at a single pressure in the process shown in Fig. 6.34, while it is evaporated at two different pressures in the process shown in Fig. 6.35. The two processes will be termed the DMR-1 process (Fig. 6.34) and the DMR-2 process (Fig. 6.35).

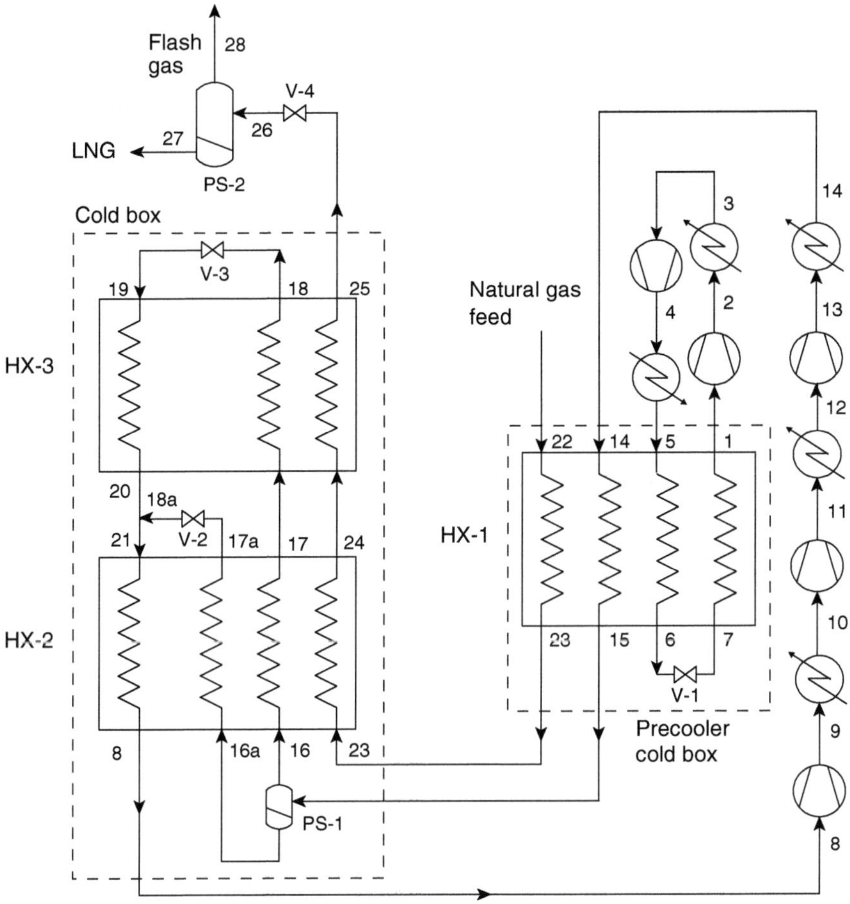

Fig. 6.34. Dual mixed refrigerant process with the precooling refrigerant evaporated at a single pressure (DMR-1).

The design specifications of the dual mixed refrigerant (DMR) process liquefiers are shown in Table 6.22. The temperature of precooling refrigerant leaving the condenser has been assumed to be 310 K, compared to 315 K for the propane precooled mixed refrigerant processor.

Optimum composition and operating pressures of the precooling refrigerant were determined for both DMR processes, by optimizing the exergy efficiency of the cold box of the precooler (excluding the compressor and condensers). The composition and operating pressures of the main (low-temperature) refrigerant and the natural gas feed are the same as those shown in Tables 6.17 and 6.18, respectively. The exergy efficiency of the precooling stage of the two DMR processes is given by the following expression:

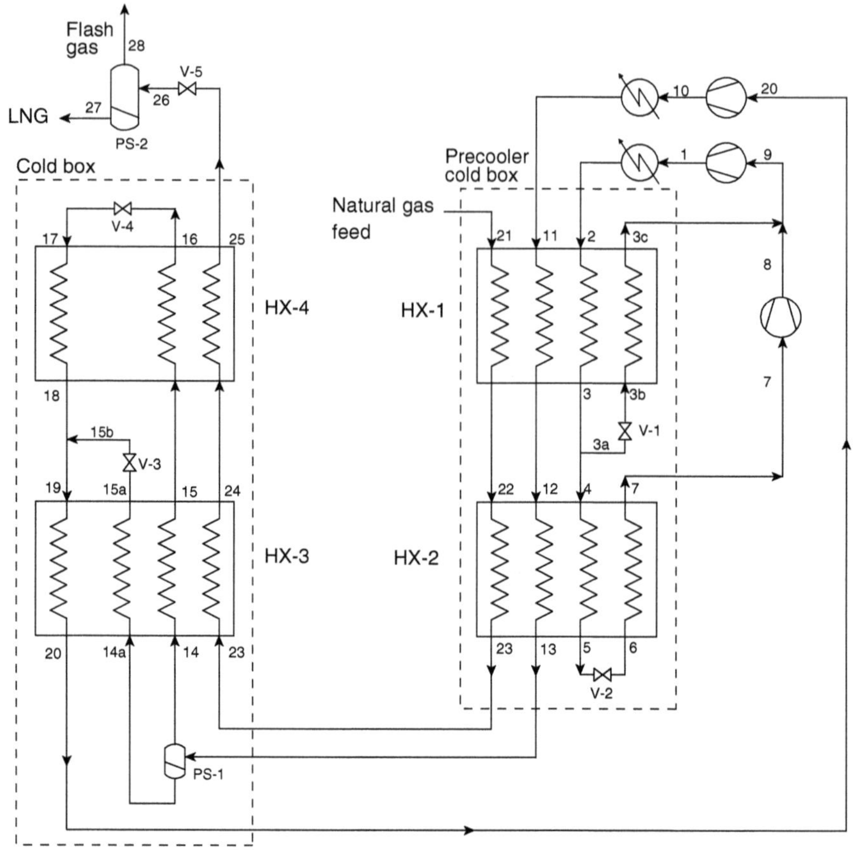

Fig. 6.35. Dual mixed refrigerant process with the precooling refrigerant evaporated at different pressures (DMR-2).

Table 6.22. Design specifications for a mixed refrigerant precooled natural gas liquefier (Figs. 6.34 and 6.35)

Temperature of precooling refrigerant leaving condensers	310 K
Temperature of main refrigerant leaving partial condenser	305 K
Minimum temperature approach in heat exchangers	3.0 K
Feed (natural gas) operating pressure, p_{21}	65 bar
Pressure drop in the heat exchangers, Δp	0 bar
Adiabatic efficiency of compressors	80%
Adiabatic efficiency of pumps	90%
Precooling temperature	240 K
Minimum compressor suction pressure	3 bar

$$\eta_{\text{ex, pre}} = \frac{\text{minimum power for cooling feed and main refrigerant}}{\text{compressor input power}} \tag{6.21}$$

$$= \frac{\dot{n}_{23}(\text{ex}_{23} - \text{ex}_{22}) + \dot{n}_{15}(\text{ex}_{15} - \text{ex}_{14})}{\sum_{i=1}^{2} -\dot{W}_{c,i,\text{precooler}}} \quad \text{(DMR-1)} \tag{6.22}$$

$$= \frac{\dot{n}_{23}(\text{ex}_{23} - \text{ex}_{21}) + \dot{n}_{13}(\text{ex}_{13} - \text{ex}_{11})}{\sum_{i=1}^{2} -\dot{W}_{c,i,\text{precooler}}} \quad \text{(DMR-2).} \tag{6.23}$$

The exergy efficiency of the cold box for the precooler alone is given by the expression

$$\eta_{\text{ex, cb, pre}} = \frac{\text{minimum power for cooling feed and main refrigerant}}{\text{exergy expenditure}} \tag{6.24}$$

$$= \frac{\dot{n}_{23}(\text{ex}_{23} - \text{ex}_{22}) + \dot{n}_{15}(\text{ex}_{15} - \text{ex}_{14})}{\dot{n}_1(\text{ex}_5 - \text{ex}_1)} \quad \text{(DMR-1)} \tag{6.25}$$

$$= \frac{\dot{n}_{13}(\text{ex}_{13} - \text{ex}_{11}) + \dot{n}_{23}(\text{ex}_{23} - \text{ex}_{21})}{\dot{n}_2\text{ex}_2 - \dot{n}_7\text{ex}_7 - \dot{n}_{3c}\text{ex}_{3c}} \quad \text{(DMR-2).} \tag{6.26}$$

The precooling refrigerant composition and operating pressures were determined using the optimization method described in Chapter 5 with the exergy efficiency of the precooler cold box being the objective function. It is also possible to use the exergy efficiency of the entire precooler (including the cold box and the compressors and condensers) as the objective function. However, since the adiabatic efficiency of the compressor is fixed and independent of the mixture composition and operating pressures, the use of exergy efficiency of the precooler cold box as the objective function, in general, will also result in a high overall exergy of the entire precooling process. The optimum composition of the precooling refrigerants for each of the processes is shown in Table 6.23. The major difference between the mixtures used for precooling is in the fraction of propane used. The concentration of propane in the precooling refrigerant is 4.94 mol% for the DMR-1 process and 64.16 mol% for the DMR-2 process.

The performance of the two processes with optimum mixture compositions is presented in Tables 6.24 and 6.25. The composition of the main (low-temperature) refrigerant and the performance of the main cold box are the same as those for the precooled process shown in Fig. 6.24 (see Tables 6.17 and 6.18). It can be seen from

Table 6.23. Optimum composition of precooling refrigerants for the dual mixed refrigerant (DMR) processes shown in Figs. 6.34 and 6.35

Component	DMR-1 (Fig. 6.34) (mol%)	DMR-2 (Fig. 6.35) (mol%)
Ethane	45.47	24.82
Propane	4.94	64.16
*n*Butane	49.59	11.03

Table 6.24. Temperature, pressure, vapor fraction, and flow rate of different streams of the DMR-1 process shown in Fig. 6.34 with mixture composition shown in Table 6.23

	Stream					
	1	2	3	4	5	6
Temperature, K	302.4	328.8	310.0	366.9	310.1	240.0
Pressure, bar	3.7	6.7	6.7	21.8	21.8	21.8
Vapor fraction	1.000	1.000	1.000	1.000	0.000	0.000
Flow rate, mol/s	0.789	0.789	0.789	0.789	0.789	0.789
	Stream					
	7	14	15	22	23	
Temperature, K	235.1	305.0	240.0	300.0	240.0	
Pressure, bar	3.7	48.6	48.6	65.0	65.0	
Vapor fraction	0.042	1.000	0.290	1.000	1.000	
Flow rate, mol/s	0.789	1.000	1.000	0.748	0.748	

Table 6.25. Temperature, pressure, vapor fraction, and flow rate of different streams of the DMR-2 process shown in Fig. 6.35 with the mixture composition shown in Table 6.23

	Stream					
	1	2	3	3a	3b	3c
Temperature, K	355.6	310.0	273.1	273.1	270.0	306.2
Pressure, bar	19.2	19.2	19.2	19.2	7.6	7.6
Vapor fraction	1.000	0.000	0.000	0.000	0.026	1.000
Flow rate, mol/s	0.913	0.913	0.913	0.546	0.546	0.546
	Stream					
	4	5	6	7	8	9
Temperature, K	273.0	240.0	237.0	267.8	312.6	308.8
Pressure, bar	19.2	19.2	2.8	2.8	7.6	7.6
Vapor fraction	0.000	0.000	0.023	1.000	1.000	1.000
Flow rate, mol/s	0.366	0.366	0.366	0.366	0.366	0.913
	Stream					
	21	22	23			
Temperature, K	300.0	273.0	240.0			
Pressure, bar	65.0	65.0	65.0			
Vapor fraction	1.000	1.000	1.000			
Flow rate, mol/s	0.748	0.748	0.748			

Tables 6.24 and 6.25 that the precooling refrigerants are completely condensed in both condensers (stream 5 in Table 6.24 and stream 2 in Table 6.25).

The flow rate of the precooling refrigerant (stream 1) is 0.789 mol/s, corresponding to a natural gas (stream 21) feed flow rate of 0.748 mol/s in the DMR-1 process shown in Fig. 6.34 and 0.913 mol/s (stream 2) in the first precooling heat exchanger (HX-1) and 0.366 mol/s (stream 4) in the second precooling heat exchanger (HX-2) for the DMR-2 process shown in Fig. 6.35. The compression work required is smaller in the case of the DMR-2 process (Fig. 6.35) since the flow rate of precooling refrigerant through the first stage compressor (0.366 mol/s) is smaller than that through the DMR-1 process.

Figure 6.36 shows the exergy utilization in the precooler for the DMR-1 process shown in Fig. 6.34. The temperature of the DMR-1 precooling refrigerant entering the condenser is higher than that in the DMR-2 process by about 10 K. While the precooling refrigerant condenser heat loads are about the same in the two DMR processes, the exergy loss in the DMR-1 precooling condenser is about 1.5 times that of the DMR-2 process. The difference between the dew and bubble point temperatures (glide) of the precooling refrigerant at the condensing pressure is 45 K with the DMR-1 process and 18 K with the DMR-2 process with the mixtures shown in Table 6.23. The large temperature glide of the refrigerant during condensation results in a larger exergy loss for the DMR-1 process. The exergy efficiency of the precooling process therefore increases from 30.2% with the DMR-1 process to 36.3% with the DMR-2 process (Fig. 6.37) when the number of pressures at which the precooling refrigerant is evaporated is increased from one to two.

The overall exergy efficiency of the two DMR processes is given by the expression

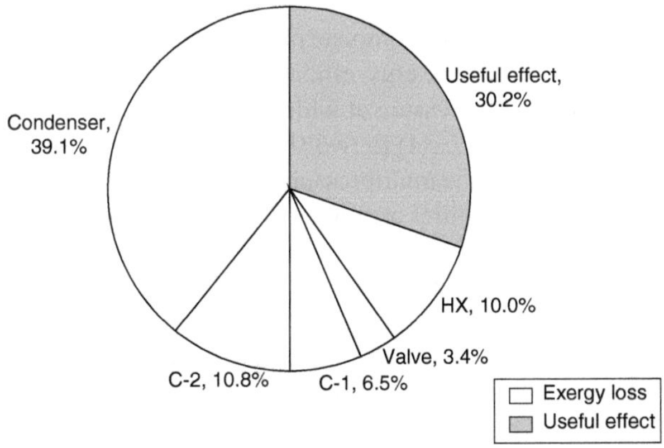

Fig. 6.36. Exergy utilization in the precooler of a DMR-1 process with the precooling refrigerant evaporated at a single pressure (Fig. 6.34).

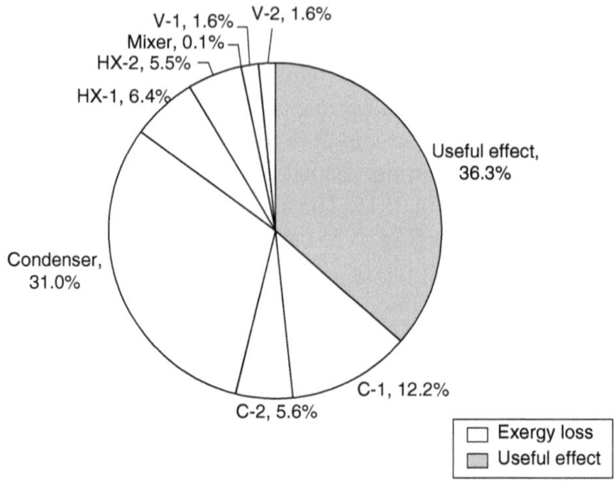

Fig. 6.37. Exergy utilization in the precooler of the DMR-2 process with the precooling refrigerant evaporated at different pressures (Fig. 6.35).

$$\eta_{ex,o} = \frac{\text{minimum power for liquefaction}}{\text{total compressor power input}} \tag{6.27}$$

$$= \frac{\dot{n}_{25}(\text{ex}_{25} - \text{ex}_{22})}{\text{total compressor power input}} \quad \text{(DMR-1)} \tag{6.28}$$

$$= \frac{\dot{n}_{25}(\text{ex}_{25} - \text{ex}_{21})}{\text{total compressor power input}} \quad \text{(DMR-2).} \tag{6.29}$$

The exergy utilization in the complete process is shown for the two DMR processes in Figs. 6.38 and 6.39. The exergy efficiency of the DMR process increases from 48.3% to 51.3% when the pressure at which the precooling refrigerant is evaporated is increased from one (Fig. 6.34) to two (Fig. 6.35). Roberts and Agrawal [78] state that the disadvantage with the multipressure evaporation is that precooling refrigerant streams at different temperatures are mixed in between compression stages, increasing thermodynamic irreversibility and reducing cycle efficiency. For the DMR-2 process (Fig. 6.35), the exergy loss in the mixer (Mixer 2) is only 0.1% of the compressor input power, as shown in Fig. 6.39. Similarly, the overall exergy efficiency of the DMR process with the evaporation of precooling refrigerant at multiple pressure stages is better than that at a single pressure. The exergy efficiency of the DMR process shown in Fig. 6.35 is slightly higher than even the C3-MR process.

Figures 6.40 and 6.41 show the temperature profiles in the precooling heat exchangers in the two DMR processes. The exergy loss in the precooling-compressors of the DMR-2 process is lower than that of the DMR-1 by only 0.5%, whereas the exergy loss in the precooling refrigerant condenser of the DMR-2 process is lower than that of the DMR-1 process by 4%, resulting in a higher overall exergy efficiency in the case of the DMR-2 process.

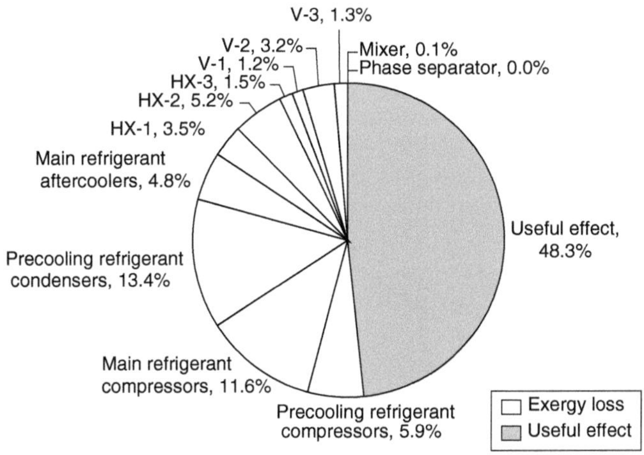

Fig. 6.38. Overall exergy utilization in the DMR-1 process shown in Fig. 6.34.

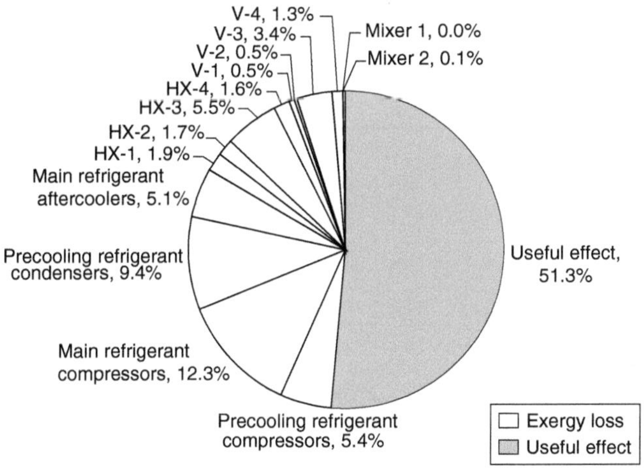

Fig. 6.39. Overall exergy utilization in the DMR-2 process shown in Fig. 6.35.

Garier and Paradowski [42] claim a process similar to the DMR-1 process (Fig. 6.34) except that the natural gas is precooled using the main refrigerant in a separate heat exchanger instead of the precooling heat exchanger (HX-1). Only the main refrigerant is cooled in the precooling heat exchanger. Newton [61] claims a process similar to the DMR-2 process (Fig. 6.35) but in which the precooling refrigerant is evaporated at three different pressures. In addition, Paradowski and Rojey [66] claim the evaporation of precooling refrigerant at multiple pressures. However, this process does not use any phase separator in the process, as in Fig. 6.12.

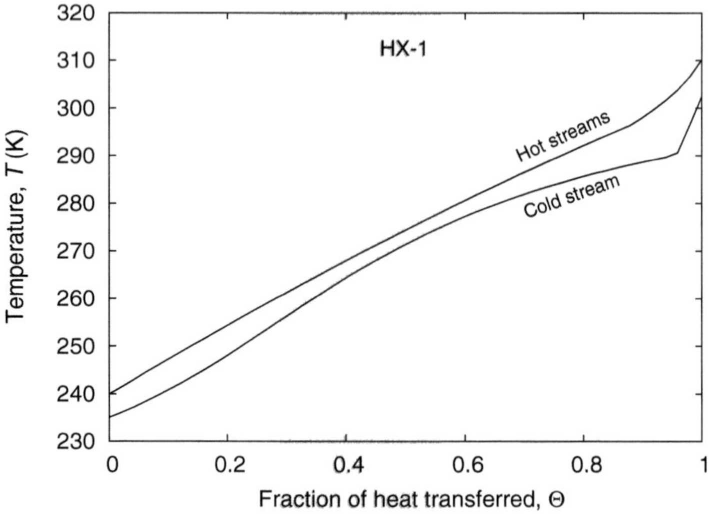

Fig. 6.40. Temperature profiles in the precooling heat exchanger (HX-1) of a DMR-1 process operating with the refrigerant mixture shown in Table 6.23.

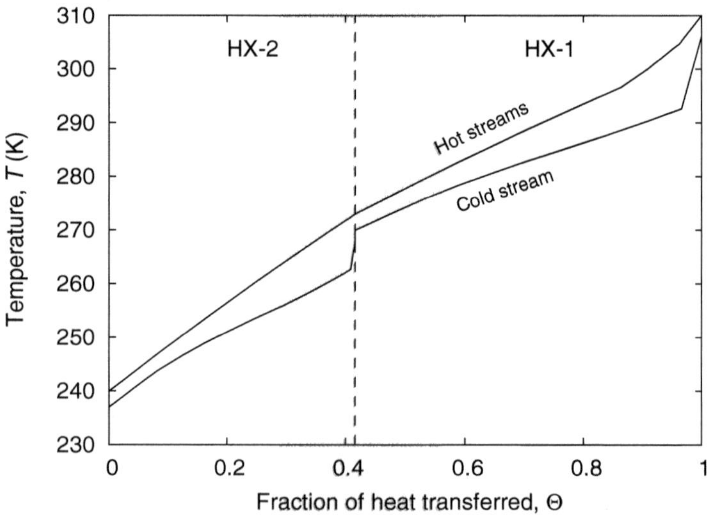

Fig. 6.41. Temperature profiles in the two heat exchangers (HX-1, HX-2) of the precooler of a DMR-2 process operating with refrigerant mixture shown in Table 6.23.

The precooling temperature has been chosen in the present example such that the natural gas is in the single-phase condition but close to its dew point temperature at the exit of the precooling heat exchangers. Alternately, one can choose the precooling system such that the natural gas feed is below the bubble point temperature at the exit of the precooling heat exchangers. Such an approach is adopted by the LIQUEFIN process [66]. When high boilers such as propane are removed from the natural gas feed, the precooling temperature is chosen so as to partially condense the natural gas feed in the precooling heat exchangers. In such cases, the natural gas feed leaving the separation system and entering the main refrigeration process cold box is normally in a saturated vapor state.

6.10 LNG process with multiple phase separators (Kleemenko process)

Figure 6.42 shows a Kleemenko natural gas liquefaction process with two phase separators, presented by Kleemenko in 1959 [50]. The compressor arrangement normally used is shown in Fig. 6.43. The process is similar to the precooled phase separator process shown in Fig. 6.24. While different precooling refrigerants are used in the C3-MR process (Fig. 6.29) and the dual mixed refrigerant processes (Figs. 6.34 and 6.35), the precooling refrigerant is derived from the main refrigerant itself in the Kleemenko process using a phase separator. The main refrigerant is partially condensed in the condenser, and the liquid and vapor phases are separated in the first phase separator (PS-1). The liquid phase, rich in high boilers, is subcooled and expanded in valve V-1 to provide the refrigeration required to partially condense the vapor phase separated in the first phase separator (PS-1). The natural gas is normally desuperheated (precooled) in the first heat exchanger (HX-1). The process is repeated to derive two different refrigerants of different composition to condense (liquefy) the natural gas feed completely in the second heat exchanger (HX-2) and subcool in the third heat exchanger (HX-3).

Kleemenko's invention was special in many respects:

- The Kleemenko process was the first process that used refrigerant mixtures to liquefy natural gas.
- The process used three different refrigerants for the desuperheating (precooling), condensation (liquefaction), and subcooling of natural gas.
- All three refrigerants were derived from a single refrigerant using phase separators.

Kleemenko was probably the first person to identify the advantages in using a zeotropic refrigerant mixture in vapor compression refrigerators that provide refrigeration over a small range of temperatures, typically less than 10 K [50].

Kleemenko's invention [50] led to the development and growth of the LNG industry. A variant of the Kleemenko process commonly known as the TEALARC process was first used in a base-load plant more than 30 years ago and has been replaced by precooled phase separator processes such as the C3-MR process. However, it continues to be used in smaller peak shaving plants [14].

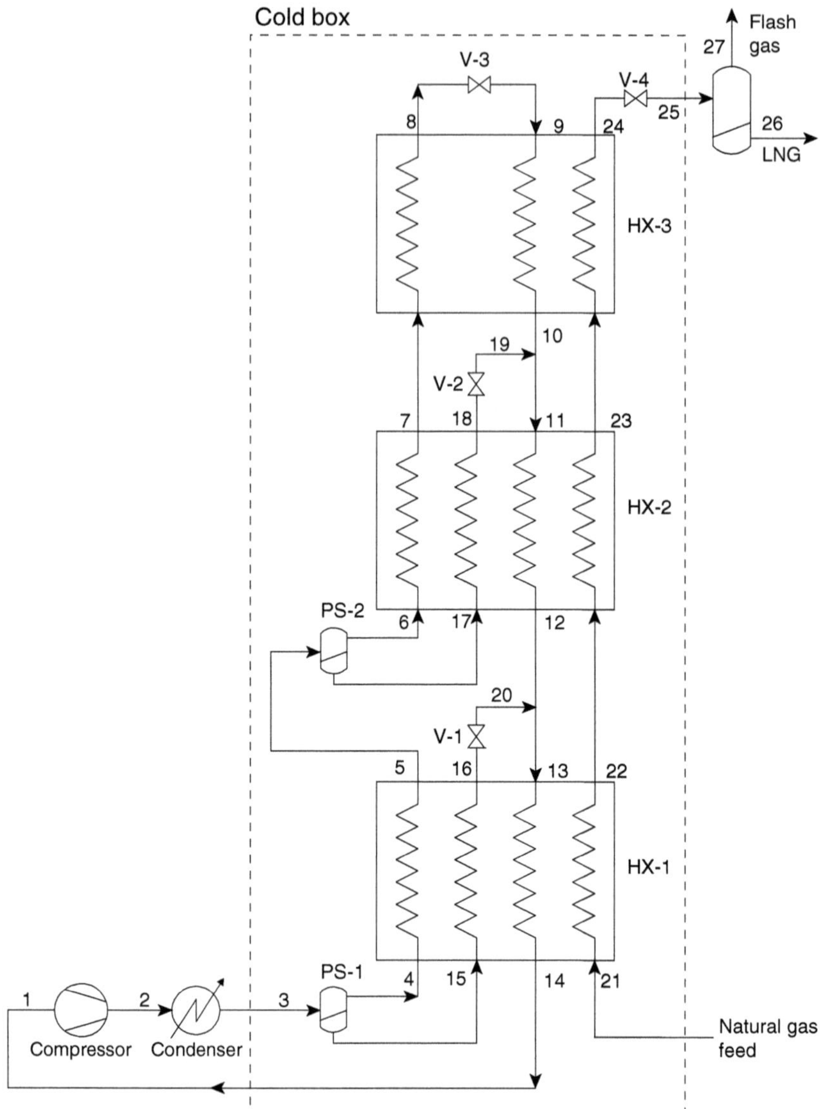

Fig. 6.42. single-stage mixed refrigerant natural gas liquefaction process with two phase separators [50], also known as the Kleemenko process.

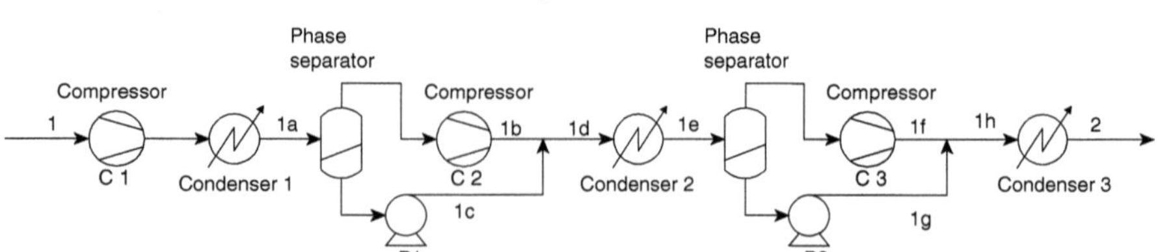

Fig. 6.43. Compressor arrangement for the process shown in Fig. 6.42.

The correct choice of refrigerant composition passing through the compressor and condenser, the operating pressures of the compressor, the temperature of the refrigerants and natural gas feed at the exit of each heat exchanger, and the fraction of liquid at the entry of each of the phase separators are critical to the attainment of a high exergy efficiency with the process. The optimization method used to choose all the above parameters is similar to that presented in Chapter 5 for the cooling/liquefaction of gases. However, additional constraints on the vapor fraction of the refrigerant at the inlet of the phase separators should be used to ensure that a mixture of vapor and liquid enters the phase separators. The exergy efficiency of the cold box is chosen as the objective function for this process also.

Table 6.26 shows the composition of optimum refrigerant at different locations of the Kleemenko process shown in Fig. 6.42. The refrigerant passing through the compressor (stream 3) contains 10.8% of npentane and 4.3% of *n*Butane. Most of the *n*Butane and *n*Pentane are separated in the two separators, and the refrigerant passing through the last heat exchanger (stream 6) consists mostly of nitrogen, methane, and ethane, with a small quantity of propane and negligible quantities of butanes and pentanes. Freezing of the refrigerant mixtures at low temperatures is thus completely avoided despite the large amount of high boilers in the refrigerant mixture.

The pressure, temperature, vapor fraction, and flow rate at the exit and entry of each piece of equipment are shown in Table 6.27. The natural gas feed is in a superheated vapor state at the exit of the first heat exchanger (HX-1) and in a slightly subcooled liquid condition at the exit of the second heat exchanger (HX-2). The deep subcooling of the natural gas feed occurs in the third heat exchanger. It can be observed from Table 6.27 that the vapor fraction of the refrigerant streams entering the warm and cold phase separators (streams 3, 5) is 69.3% and 40.1%, respectively. The flow rate of the refrigerant to the third heat exchanger (HX-3) is therefore much lower than that through the compressor due to the large amount of liquid phase separated in the phase separators. The ratio of flow rates of the low-pressure refrigerant and the natural gas feed decreases from 2.57 in the first heat exchanger (HX-1) to 0.71 in the third heat

Table 6.26. Composition (mol %) of feed and different refrigerant mixture streams of the Kleemenko process shown in Fig. 6.42

	(Mol %)						
Component	NG feed	Stream 3	Stream 4	Stream 15	Stream 5	Stream 6	Stream 17
Nitrogen	4.00	7.52	10.32	1.18	10.32	21.07	3.12
Methane	87.50	28.43	36.68	9.78	36.68	54.72	24.58
Ethane	5.50	43.90	45.03	41.34	45.03	23.41	59.52
Propane	2.10	5.01	3.72	7.91	3.72	0.65	5.78
*n*Butane	0.50	4.34	1.89	9.86	1.89	0.10	3.09
*i*Butane	0.30	—	—	—	—	—	—
nPentane	—	10.81	2.36	29.92	2.36	0.04	3.91
*i*Pentane	0.10	—	—	—	—	—	—

Table 6.27. Temperature, pressure, vapor fraction, and flow rate of different streams of the Kleemenko process shown in Fig. 6.42 with the mixture composition shown in Table 6.26

	Stream					
	3	4	5	6	7	8
Temperature, K	305.0	305.0	239.3	239.3	159.0	113.0
Pressure, bar	42.7	42.7	42.7	42.7	42.7	42.7
Vapor fraction	0.693	1.000	0.401	1.000	0.000	0.000
Flow rate, mol/s	1.000	0.693	0.693	0.278	0.278	0.278
	9	10	11	12	13	14
Temperature, K	108.7	154.3	155.2	234.6	228.9	299.2
Pressure, bar	4.9	4.9	4.9	4.9	4.9	4.9
Vapor fraction	0.074	0.700	0.319	0.949	0.721	1.000
Flow rate, mol/s	0.278	0.278	0.693	0.693	1.000	1.000
	15	16	17	18	19	20
Temperature, K	305.0	239.3	239.3	159.0	154.9	224.4
Pressure, bar	42.7	42.7	42.7	42.7	4.9	4.9
Vapor fraction	0.000	0.000	0.000	0.000	0.069	0.176
Flow rate, mol/s	0.307	0.307	0.415	0.415	0.415	0.307
	21	22	23	24		
Temperature, K	300.0	239.3	159.0	113.0		
Pressure, bar	65.0	65.0	65.0	65.0		
Vapor fraction	1.000	1.000	0.000	0.000		
Flow rate, mol/s	0.389	0.389	0.389	0.389		

exchanger (HX-3). Consequently, the size of the coldest heat exchanger (HX-3) will be much smaller than that of the warmest heat exchanger (HX-1).

The compressor arrangement shown in Fig. 6.43 is assumed for the process. The refrigerant mixture passing through the compressor has a high dew point temperature due to the large concentration of butane and pentane. Hence, part of the refrigerant condenses in the aftercooler between the stages. The liquid and vapor phases are separated in a phase separator and passed through a pump and compressor, respectively, to increase the pressure to the desired level. The power required per mole of refrigerant to raise the pressure of a refrigerant from a pressure p_1 to a pressure p_2 will be higher when the refrigerant is in a gaseous state than in a liquid state since the compression work is inversely proportional to the square of the density of the fluid for the same mass flow rate. The use of pumps in parallel with the compressor is therefore advantageous.

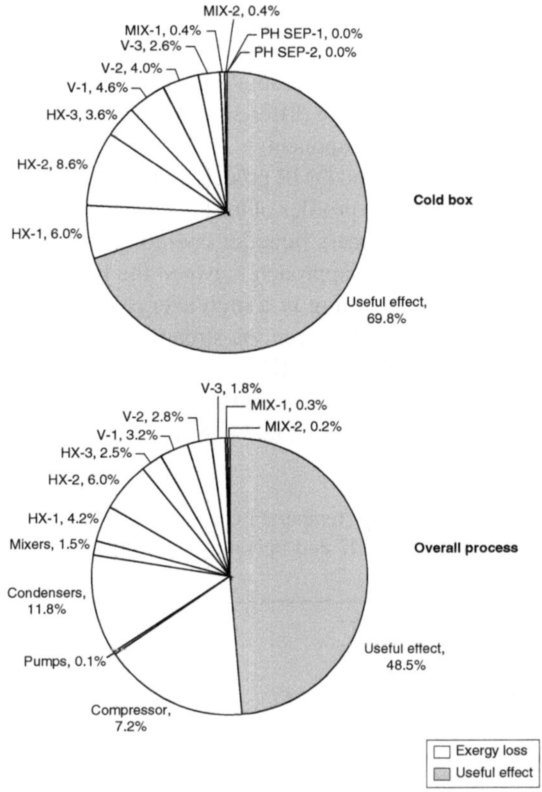

Fig. 6.44. Exergy utilization in the Kleemenko natural gas liquefaction process shown in Fig. 6.42 and operating with a refrigerant from Table 6.26.

The power required by different compressors and pumps is shown in Table 6.28. It can be observed that the power required in the third compressor is smaller than that in the second compressor due to a significant flow rate through the second pump (P2).

Figure 6.44 shows the exergy efficiency of the cold box and the overall process of the Kleemenko process when operating with a refrigerant from Table 6.26. The

Table 6.28. Details of power input to the different compressors in Fig. 6.43 and the Kleemenko process in Fig. 6.42 and operating with mixtures in Table 6.26

	Compressors			Pumps	
	C1	C2	C3	P1	P2
Power, kW	2.99	1.91	1.26	0.011	0.033
Outlet pressure, bar	12.5	24.8	42.7	24.8	42.7
Flow rate, mol/s	1.000	0.927	0.825	0.073	0.175

compressors and condensers have an exergy efficiency of 69.5%, and the cold box has an exergy efficiency of 69.8%. The overall exergy efficiency of the process, which is a product of the cold box and the compressors/condensers, is 48.5%. The exergy loss in the mixers due to mixing of fluids of different composition and temperature is very small compared to that in other components.

As seen earlier, most C3-MR and DMR processes have a slightly higher efficiency. Figure 6.45 shows the temperature profiles of the hot and cold streams in the three heat exchangers of the Kleemenko process liquefier operating with a refrigerant mixture from Table 6.26. The temperature approach between the hot and cold fluid streams is small at low temperatures, resulting in a high exergy efficiency of the cold box. The mean temperature difference between the streams is lower in the case of HX-3 (Table 6.29). The exergy efficiency of the Kleemenko process can be further increased by using a smaller temperature approach in the heat exchangers. Many variants of the Kleemenko process have been patented [52, 80, 81].

Table 6.29. Details of the minimum temperature approach in all heat exchangers of the Kleemenko process shown in Fig. 6.42 and operating with the mixture composition shown in Table 6.26

Temperature approach	HX-1	HX-2	HX-3
LMTD ΔT_{lm}, K	5.2	5.2	4.6
Minimum approach ΔT_{min}, K	3.0	3.0	3.0
Warm end approach ΔT_{we}, K	5.8	4.6	4.7
Cold end approach ΔT_{ce}, K	10.4	3.7	4.3

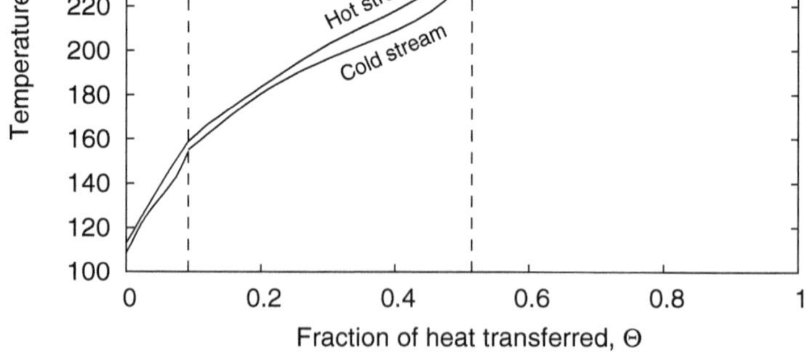

Fig. 6.45. Temperature profiles of the hot and cold fluid streams in a Kleemenko process liquefier operating with refrigerant mixture shown in Table 6.26.

The composition and flow rate of the refrigerant in different heat exchangers are dependent on the vapor fraction of the stream entering the first phase separator. Any change in the ambient temperature affects the composition and refrigerant flow rates. The control of the Kleemenko process is therefore more difficult than the C3-MR or DMR process. The Kleemenko process has been replaced by the other processes with higher exergy efficiency. The Kleemenko process, however, remains an excellent process for the liquefaction of natural gas on a small scale due to its simplicity.

The Kleemenko process is also an excellent process for the liquefaction of other gases such as nitrogen and is discussed in Chapter 7.

6.11 Cascade liquefaction process operating with mixtures

Figure 6.46 shows a patented process similar to the precooled process shown in Fig. 6.12 without any phase separators. The precooled process shown in Fig. 6.12 has two refrigeration stages, one for desuperheating (precooling) and another for condensation (liquefaction) and subcooling. The process shown in Fig. 6.46 has three refrigeration stages with different refrigerants for each stage: one for desuperheating the natural gas feed, the second for condensation, and the third for subcooling. The process can also be considered a cascade liquefaction process operating with refrigerant mixtures. The process is known in the literature as the multifluid cascade (MFC) process and has been used in the Snøhvit project, in Norway [15, 39]. The compressor arrangement for the condensing refrigerant and that for the subcooling refrigerant are shown in Figs. 6.47 and 6.48, respectively. Two-stage compression is used for the precooling refrigerant (stream 16), as shown in Fig. 6.46.

Consider the conventional cascade liquefaction process shown in Fig. 6.2. The large number of phase separators and heat exchangers that need to be used makes the system quite complex. Additionally, the main disadvantage with the conventional cascade process operating with pure fluid (single-component) refrigerants is that the refrigeration is provided at constant temperature at discrete temperature levels. On the other hand, mixed refrigerant processes provide refrigeration over a range of temperatures. The process designer's job essentially consists of determining the optimum composition of the refrigerant, the operating pressures, the temperatures in each of the three cooling stages, and the distribution of the precooling (first-stage) refrigerant between the first two heat exchangers (HX-1, HX-2) that will lead to the maximum exergy efficiency. The optimization method described in Chapter 5 has been used to determine the above parameters. The exergy efficiency of the cold box is chosen as the objective function. Apart from the constraint on the minimum temperature approach in each heat exchanger, additional constraints should be used to prevent liquid at the entry of the compressors. The design specifications adopted for this process are the same as those for the DMR processes discussed earlier.

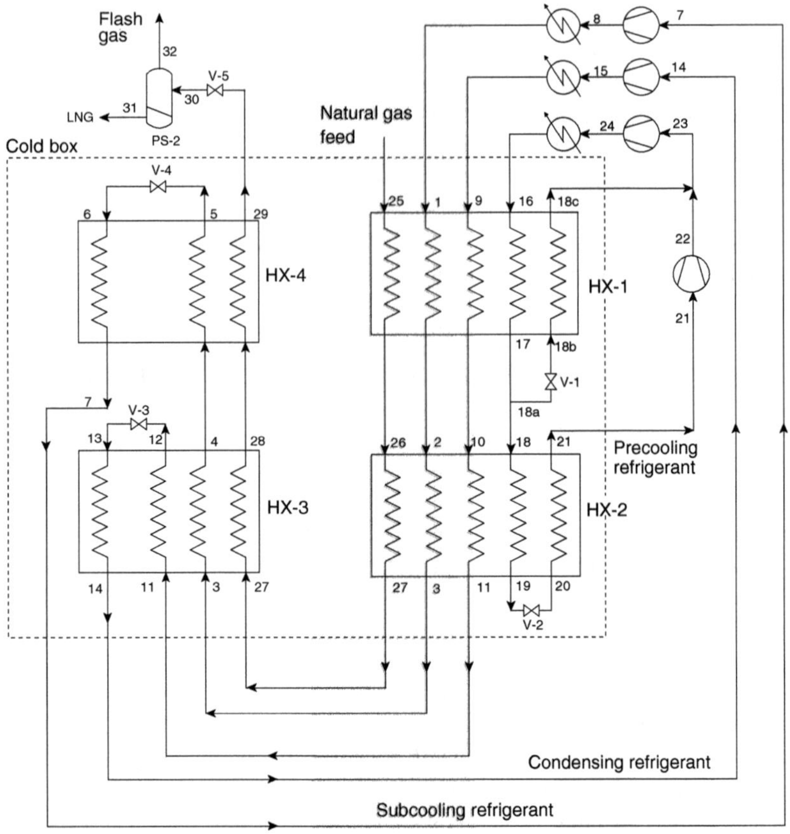

Fig. 6.46. Cascade process operating with mixtures [86]. (Adapted from U.S. Patent no. 6,253,574.)

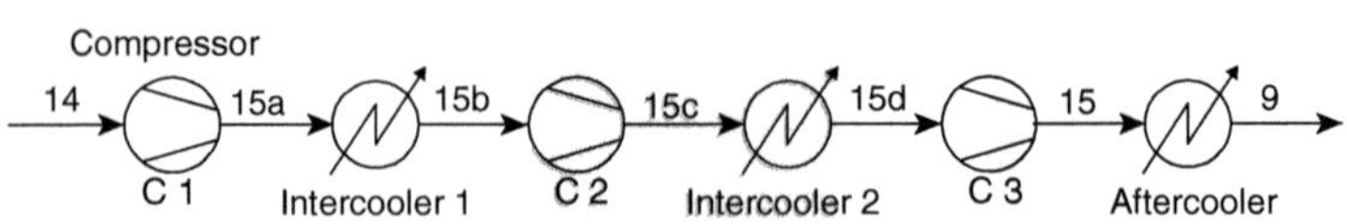

Fig. 6.47. Compressor arrangement for the condensing refrigerant (stream 9).

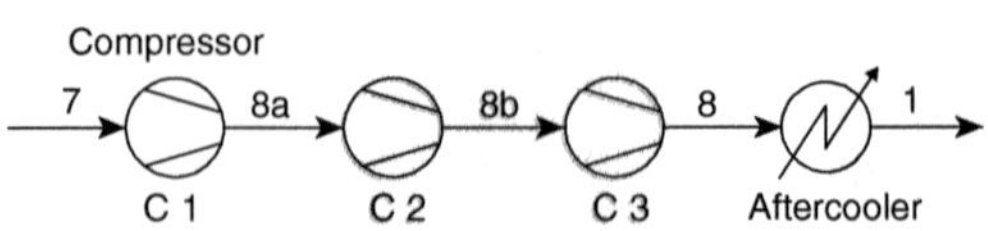

Fig. 6.48. Compressor arrangement for the subcooling refrigerant (stream 1).

The exergy efficiency of the cold box ($\eta_{ex,\,cb}$) is given by the following expression:

$$\eta_{ex,\,cb} = \frac{\text{minimum power for liquefaction}}{\text{exergy expenditure}}, \tag{6.30}$$

$$= \frac{\dot{n}_{25}(\mathrm{ex}_{29} - \mathrm{ex}_{25})}{\dot{n}_1(\mathrm{ex}_1 - \mathrm{ex}_7) + \dot{n}_9(\mathrm{ex}_9 - \mathrm{ex}_{14}) + \dot{n}_{16}\mathrm{ex}_{16} - \dot{n}_{18c}\mathrm{ex}_{18c} - \dot{n}_{21}\mathrm{ex}_{21}}. \tag{6.31}$$

The overall exergy efficiency of the process ($\eta_{ex,\,o}$) is given by the following expression:

$$\eta_{ex,\,o} = \frac{\text{minimum power for liquefaction}}{\text{total compressor power input}} = \frac{\dot{n}_{25}(\mathrm{ex}_{29} - \mathrm{ex}_{25})}{\sum_{i=1}^{3} -\dot{W}_{c,\,i}}, \tag{6.32}$$

where $-\dot{W}_{c,\,i}$ refers to the total power input to the compressors of the three refrigerants.

Table 6.30 shows the optimum mixture composition for the three refrigeration stages; precooling, condensation, and subcooling. The precooling refrigerant requires ethylene, propane, and *n*Butane, the condensation refrigerant uses methane, ethane, ethylene, and propane, and the subcooling refrigerant contains nitrogen, methane, and ethylene. The exergy efficiency will be slightly lower if ethane is used completely instead of ethylene.

Table 6.31 shows the temperature, pressure, vapor fraction, and flow rate of different streams of the process. The precooling refrigerant (stream 16) is completely condensed in the condenser. All the other streams entering the first heat exchanger (HX-1) at the warm end are in a superheated state. The high-pressure condensation refrigerant (stream 9) is condensed partially in the first heat exchanger and leaves the second heat exchanger (HX-2) as stream 11 in a subcooled state. On the other hand, the subcooling refrigerant (stream 1) leaves the third heat exchanger (HX-3) as

Table 6.30. Composition of the natural gas feed and the refrigerants of the multifluid cascade natural gas liquefaction process (Fig. 6.46)

	(mol%)			
Component	Stream 25 Natural gas feed	Stream 1 Subcooling refrigerant	Stream 9 Condensation refrigerant	Stream 16 Precooling refrigerant
Nitrogen	4.00	17.31	—	
Methane	87.50	42.45	12.65	—
Ethane	5.50	—	32.92	0.01
Ethylene	—	40.24	27.77	11.29
Propane	2.10	—	26.65	73.56
*n*Butane	0.50	—	—	15.13
*i*Butane	0.30	—	—	—
*i*Pentane	0.10	—	—	—

Table 6.31. Temperature, pressure, vapor fraction, and flow rate of different streams of the multifluid cascade natural gas liquefaction process (Fig. 6.46) with the mixture composition as in Table 6.30

	Stream					
	1	2	3	4	5	6
Temperature, K	310.0	276.2	247.9	186.7	113.9	106.6
Pressure, bar	33.9	33.9	33.9	33.9	33.9	3.5
Vapor fraction	1.000	1.000	1.000	0.297	0.000	0.109
Flow rate, mol/s	0.721	0.721	0.721	0.721	0.721	0.721

	Stream					
	7	8	9	10	11	12
Temperature, K	181.1	310.0	310.0	276.2	247.7	191.5
Pressure, bar	3.5	33.9	27.9	27.9	27.9	27.9
Vapor fraction	1.000	1.000	1.000	0.395	0.000	0.000
Flow rate, mol/s	0.721	0.721	1.023	1.023	1.023	1.023

	Stream					
	13	14	15	16	17	18
Temperature, K	180.9	242.7	326.1	310.0	282.0	282.0
Pressure, bar	3.1	3.1	27.9	16.9	16.9	16.9
Vapor fraction	0.098	1.000	1.000	0.000	0.000	0.000
Flow rate, mol/s	1.023	1.023	1.023	1.369	1.369	0.513

	Stream					
	18a	18b	18c	19	20	21
Temperature, K	282.0	272.8	305.2	251.0	243.8	275.9
Pressure, bar	16.9	6.7	6.7	16.9	3.0	3.0
Vapor fraction	0.000	0.077	1.000	0.000	0.054	1.000
Flow rate, mol/s	0.855	0.855	0.855	0.513	0.513	0.513

	Stream					
	22	23	24	25	26	27
Temperature, K	311.1	307.5	352.2	300.0	276.2	247.9
Pressure, bar	6.7	6.7	16.9	65.0	65.0	65.0
Vapor fraction	1.000	1.000	1.000	1.000	1.000	1.000
Flow rate, mol/s	0.513	1.369	1.369	1.000	1.000	1.000

	Stream	
	28	29
Temperature, K	186.8	113.0
Pressure, bar	65.0	65.0
Vapor fraction	0.000	0.000
Flow rate, mol/s	1.000	1.000

stream 4 in a partially condensed state. All the low-pressure refrigerant streams enter the different compressors in a superheated condition.

Figure 6.49 shows the temperature of the hot and cold fluid streams in the four heat exchangers. It can be seen that the temperature approach between the streams is small and nearly uniform throughout the length, except at temperatures close to the dew point temperature of the low-pressure refrigerant (cold) stream. This results in a smaller log mean temperature difference, or LMTD (see Table 6.32).

The small temperature difference between the streams (Table 6.32), particularly at low temperatures, results in a high exergy efficiency of the cold box and the overall process. Figures 6.50 and 6.51 show the utilization of exergy in the cold box and the overall process. It can be seen that the exergy efficiency of the cold box is 76.8%, while the overall exergy efficiency is 52.1% and nearly the same as the exergy efficiency of the DMR-2 process (Fig. 6.39).

Table 6.33 shows the distribution of the input power across different compressors. The number of compression stages and aftercoolers/condensers is also shown. No

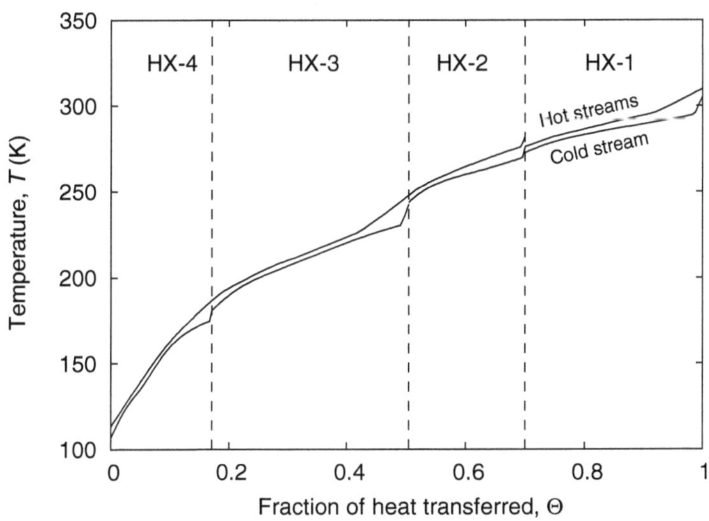

Fig. 6.49. Temperature profiles in a mixed fluid cascade LNG process (Fig. 6.46) operating with feed and mixtures shown in Table 6.30.

Table 6.32. Temperature approach between the streams in the different heat exchangers of the cascade LNG process shown in Fig. 6.46 and operating with the feed and mixtures shown in Table 6.30

Temperature approach	HX-1	HX-2	HX-3	HX-4
LMTD ΔT_{lm}, K	4.3	4.4	3.9	4.2
Minimum approach ΔT_{min}, K	3.0	3.0	3.0	3.0
Warm end approach ΔT_{we}, K	4.8	6.2	5.3	5.8
Cold end approach ΔT_{ce}, K	3.4	3.9	5.8	6.4

Table 6.33. Power input to different compressors of the cascade LNG process (Fig. 6.46) operating with the feed and mixtures shown in Table 6.30

	Compressor			
	C-1	C-2	C-3	C-4
Input power, kW	1.13	3.63	6.14	3.93
Outlet pressure, bar	6.7	16.9	27.9	33.9
Inlet pressure, bar	3.0	6.7	3.1	3.5
No. of stages	1	1	3	3

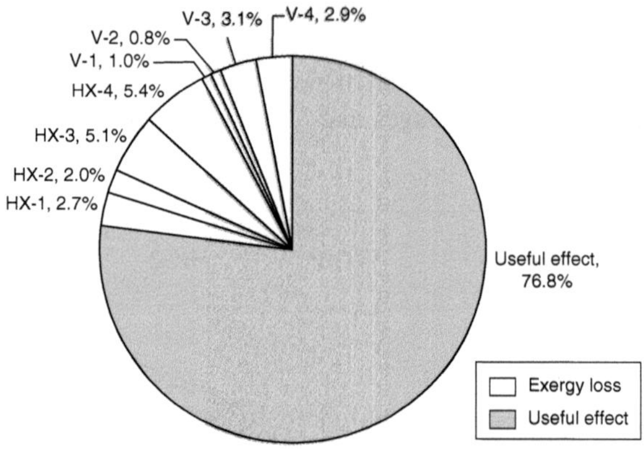

Fig. 6.50. Exergy utilization in the cold box of the mixed fluid cascade LNG process (Fig. 6.46) operating with the feed and mixtures shown in Table 6.30.

intercooling is necessary for the subcooling and precooling refrigerant compressors, while two intercoolers are required for the condensation refrigeration compressors, as shown in Fig. 6.48.

Figure 6.52 shows a patented process [77] similar to the MFC process shown in Fig. 6.46. In this process, the subcooling refrigerant is cooled and partially condensed separately in heat exchanger HX-3. The refrigeration required to partially condense the subcooling refrigerant is obtained from the liquid separated in the phase separator similar to the phase separator processes discussed earlier (Figs. 6.24, 6.29, 6.34, 6.35, and 6.42). Other processes can also be used to partially condense the subcooling refrigerant (see Ref. [77]). Similarly, other variants such as evaporation of the precooling refrigerant at two or more pressure levels as in Fig. 6.35 can improve the exergy efficiency as in the case of the DMR processes.

One of the main advantages of the MFC processes is the possibility to control the flow rates of the different refrigerants independently to accommodate the variation in ambient temperatures and feed composition.

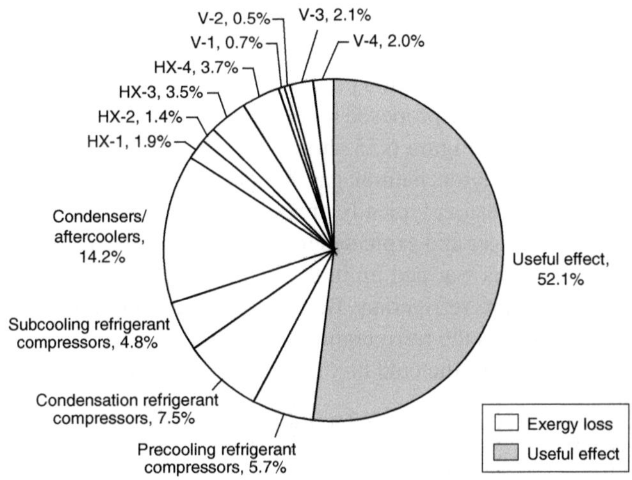

Fig. 6.51. Exergy utilization in the overall process of the mixed fluid cascade LNG process (Fig. 6.46) operating with the feed and mixtures shown in Table 6.30.

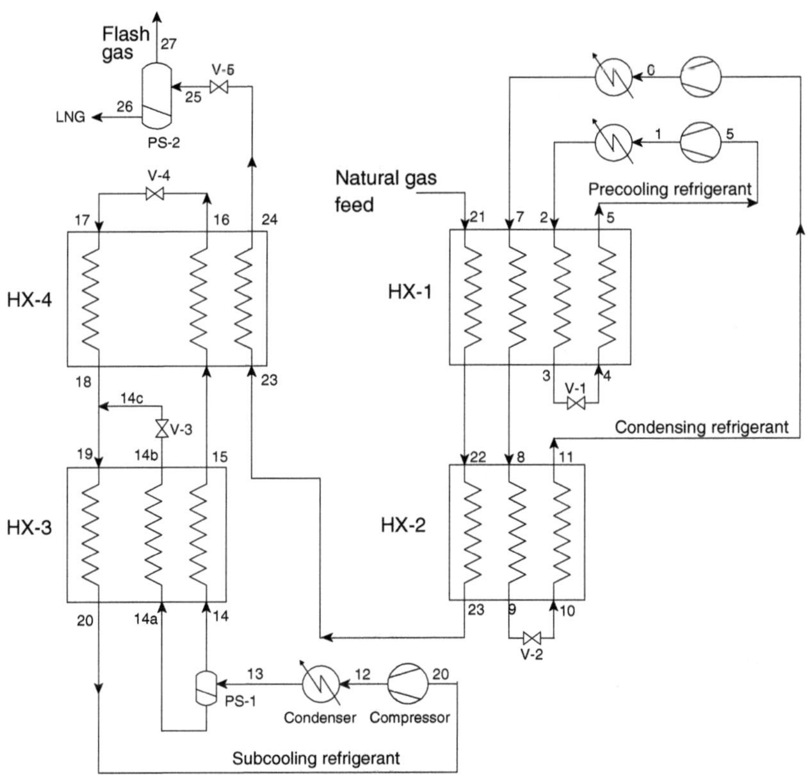

Fig. 6.52. Cascade process operating with mixtures but with subcooling refrigerant compressed at near ambient temperatures [77]. (Adapted from U.S. Patent no. 7,086,251.)

6.12 LNG processes with turbines

Turbine-based natural gas liquefaction processes have been widely used in peak shaving plants. They have also been proposed for offshore plants [12] and for reliquefaction plants on LNG ships [44]. Figure 6.53 shows the simplest of the turbine processes, known as the reverse Brayton natural gas liquefaction process. The refrigerant is compressed to a high-pressure, typically greater than 100 bar. The refrigerant is precooled in the heat exchanger and expanded to low temperatures in a turbine. The cold, low-pressure refrigerant is warmed up in the heat exchanger to cool the natural gas feed and the high-pressure refrigerant. Both nitrogen and mixtures of nitrogen and methane have been used as the refrigerant in this process.

The exergy efficiency of the cold box (Fig. 6.53) is given by the expression

$$\eta_{\text{ex, cb}} = \frac{\text{minimum power for liquefaction}}{\text{exergy expenditure}} \tag{6.33}$$

$$= \frac{\dot{n}_6(\text{ex}_7 - \text{ex}_6)}{\dot{n}_1(\text{ex}_3 - \text{ex}_1) - \dot{W}_e} = \frac{\dot{n}_6(\text{ex}_7 - \text{ex}_6)}{\dot{n}_1\{(\text{ex}_3 - \text{ex}_1) - (\text{ex}_4 - \text{ex}_5)/\eta_{\text{ex},t}\}}, \tag{6.34}$$

where $\eta_{\text{ex},t}$ refers to the exergy efficiency of the turbine (see Table 1.2). The ratio of the flow rate of natural gas feed and the refrigerant is obtained from an energy balance over the cold box as follows:

$$\frac{\dot{n}_6}{\dot{n}_1} = \left(\frac{h_1 - h_3}{h_6 - h_7}\right) + \frac{\dot{W}_e}{\dot{n}_1(h_6 - h_7)}, \tag{6.35}$$

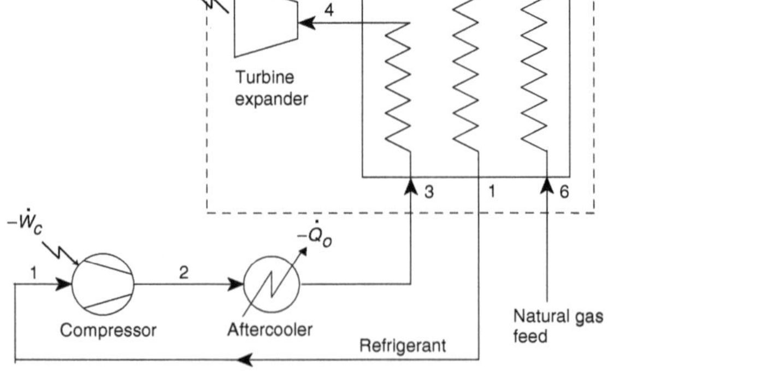

Fig. 6.53. Reverse Brayton process for the liquefaction of natural gas

The term $(h_1 - h_3)$ is proportional to the Joule–Thomson coefficient (μ_{J-T}) and the pressure difference $(p_2 - p_1)$ as shown in Eq. (3.2) is higher for a mixture of nitrogen-methane than for pure nitrogen (Fig. 3.7). The flow rate of the refrigerant will therefore be lower when a mixture of nitrogen and methane is used as the refrigerant instead of pure nitrogen.

The overall exergy efficiency of the reverse Brayton liquefier (Fig. 6.53) is given by the expression

$$\eta_{\text{ex},o} = \frac{\text{minimum power for liquefaction}}{\text{net power input}} \tag{6.36}$$

$$= \frac{\dot{n}_6(\text{ex}_7 - \text{ex}_6)}{-\dot{W}_c - \dot{W}_e} = \frac{\dot{n}_6(\text{ex}_7 - \text{ex}_6)}{\dot{n}_1\{(\text{ex}_2 - \text{ex}_1)/\eta_{\text{ex},c} - (\text{ex}_4 - \text{ex}_5)/\eta_{\text{ex},t}\}}, \tag{6.37}$$

where the above expression, $\eta_{\text{ex},t}$ and $\eta_{\text{ex},c}$ refer to the exergy efficiency of the turbine and compressor, respectively.

Table 6.34 shows the design specifications for a reverse Brayton natural gas liquefaction process. An adiabatic efficiency of 80% has been assumed for the turbine expander, and the maximum operating pressure is limited to 120 bar. The methods presented in Chapter 5 can also be used for turbine processes operating with both pure fluids such as nitrogen as well as mixtures. The operating pressure and the temperature at the entry of the turbine, as well as the composition of the mixture, have been determined using an optimization method with the exergy efficiency of the cold box chosen as the objective function. The minimum temperature approach between the streams in the heat exchanger (HX) has been assumed to be 3 K. The composition of the natural gas feed is the same as that in Table 6.30.

Tables 6.35 and 6.36 show the temperature, pressure, vapor fraction, and flow rate of different streams in the cold box of a reverse Brayton process with nitrogen and a

Table 6.34. Design specifications for reverse Brayton natural gas liquefier

Minimum temperature approach in heat exchangers	3.0 K
Pressure drop in the heat exchangers, Δp	0 bar
Natural gas feed pressure, p_6	65 bar
Maximum refrigerant pressure, p_2	120 bar
Adiabatic efficiency of turbine	80%

Table 6.35. Temperature, pressure, vapor fraction, and flow rate of different streams of the reverse Brayton process operating with nitrogen as the refrigerant

Stream	3	4	5	1	6	7
Temperature, K	300.0	228.6	109.8	296.2	300.0	113.0
Pressure, bar	120.0	120.0	6.3	6.3	65.0	65.0
Vapor fraction	1.00	1.00	1.00	1.00	1.00	0.00
Flow rate, mol/s	4.690	4.690	4.690	4.690	1.000	1.000

Table 6.36. Temperature, pressure, vapor fraction, and flow rate of different streams of the reverse Brayton process operating with a nitrogen-methane (76.9/23.1 mol%) mixture as the refrigerant

Stream	3	4	5	1	6	7
Temperature, K	300.0	246.5	109.8	297.0	300.0	113.0
Pressure, bar	106.5	106.5	3.5	3.5	65.0	65.0
Vapor fraction	1.00	1.00	1.00	1.00	1.00	0.00
Flow rate, mol/s	3.767	3.767	3.767	3.767	1.000	1.000

mixture of nitrogen and methane (76.9/23.1 mol%), respectively. The pressure ratio across the compressor is higher when a mixture of nitrogen and methane is used than when nitrogen alone is used. The flow rate of the refrigerant, on the other hand, is much higher when nitrogen alone is used. This results in a lower power requirement when a mixture of nitrogen and methane is used as the refrigerant, resulting in a higher exergy efficiency of the cold box (Fig. 6.54).

It can be observed from Fig. 6.54 that the exergy loss in the heat exchanger is much higher than that in most natural gas liquefaction processes operating with refrigerant mixtures. The large exergy loss in the heat exchanger is essentially due to a large temperature approach between the hot and cold fluid streams, as shown in Figs. 6.55 and 6.56. A small temperature approach all along the length of the heat exchanger can never be achieved when a single-component refrigerant is used for precooling, condensation and subcooling of the natural gas feed since the specific heat c_p or $(\partial h/\partial T)_p$ is not the same in all three regions (see Fig. 6.7). The exergy loss in the heat exchanger is lower when a mixture of nitrogen and methane is used due to a slightly closer temperature approach in the heat exchanger at low temperatures compared to that when nitrogen alone is used as the refrigerant (Figs. 6.55 and 6.56).

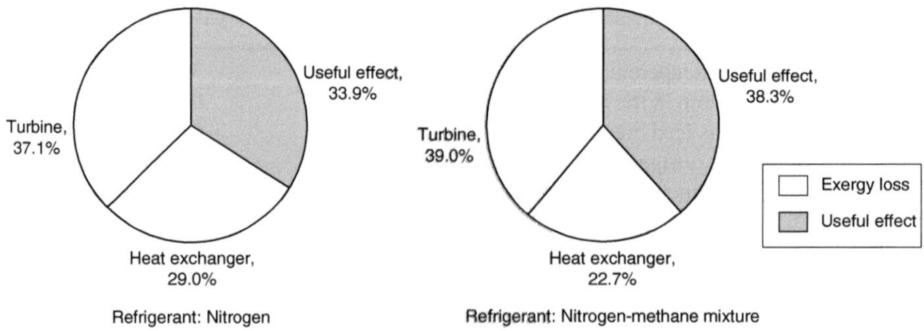

Fig. 6.54. Comparison of exergy utilization in the cold box of a reverse Brayton natural gas liquefaction process operating with nitrogen and a nitrogen-methane mixture (76.9/23.1 mol%) as the refrigerant.

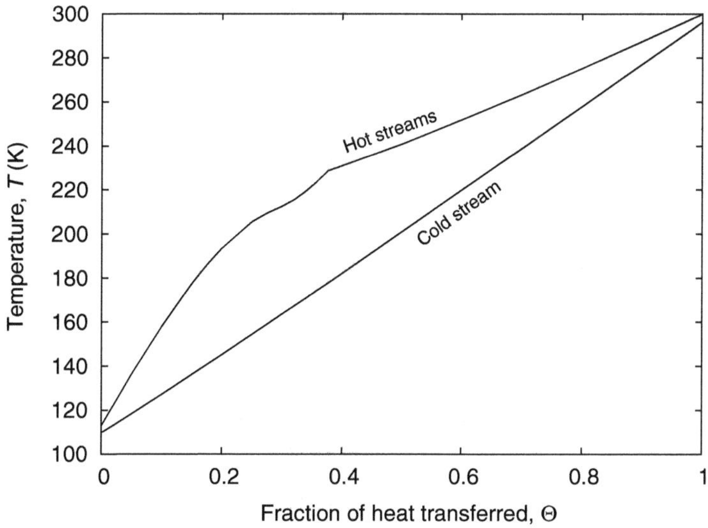

Fig. 6.55. Temperature profiles in the heat exchanger of a reverse Brayton natural gas liquefier operating with nitrogen as the refrigerant and at the operating pressures and temperatures shown in Table 6.35.

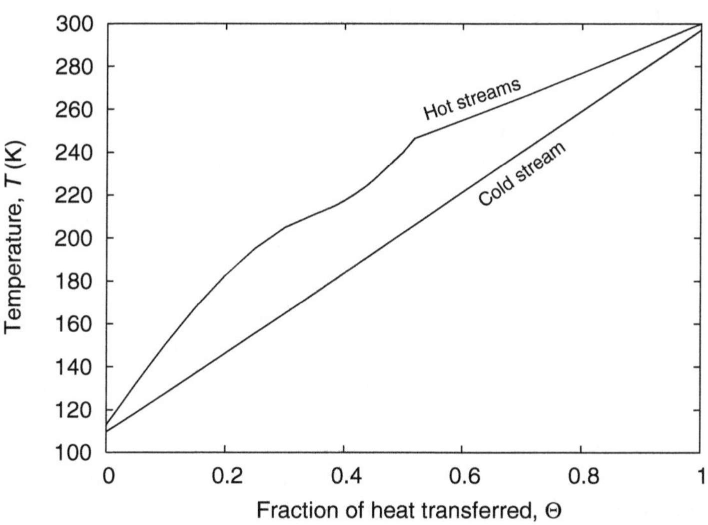

Fig. 6.56. Temperature profiles in the heat exchanger of a reverse Brayton natural gas liquefier operating with a nitrogen-methane mixture (76.9/23.1 mol%) as the refrigerant and at the operating pressures and temperatures shown in Table 6.36.

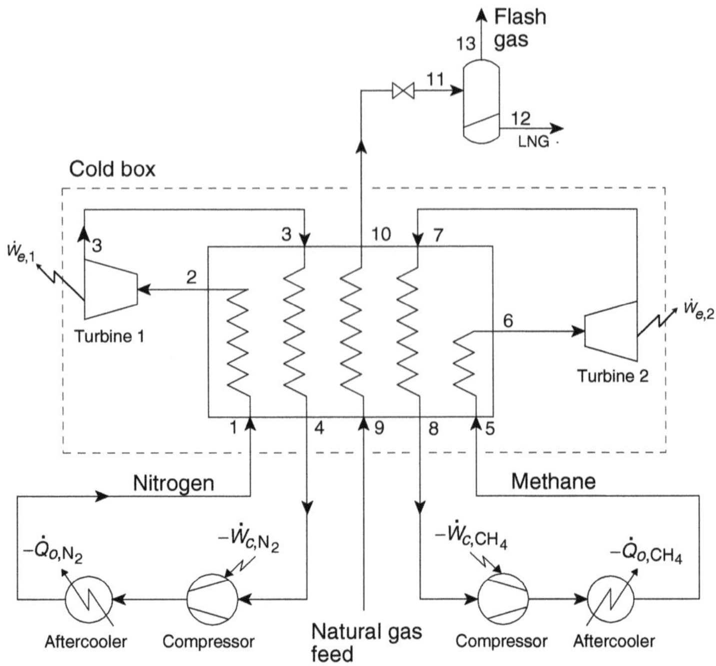

Fig. 6.57. Dual expander natural gas liquefaction process, adapted from Ref. [37]. (Adapted from U.S. Patent no. 6,412,302.)

It can be observed from Fig. 6.54 that the exergy loss in the turbine is higher than that in the heat exchanger in both cases, in spite of the turbine adiabatic efficiency being 80%.

Precooling improves the exergy efficiency of the reverse Brayton liquefiers, as in the case of Linde–Hampson liquefiers (Fig. 1.25). Any of the different processes discussed in the previous sections for precooling mixed refrigerant processes can also be used for precooling the reverse Brayton process. It is also possible to use a second turbine to provide the necessary precooling. Figure 6.57 shows one such patented process, known as the dual expander natural gas liquefaction process [37]. Methane is used as the refrigerant in the precooling process as shown in Fig. 6.57, while nitrogen is used in the main process.

The exergy efficiency of the cold box of a dual expander natural gas liquefier shown in Fig. 6.57 is given by the expression

$$\eta_{\text{ex, cb}} = \frac{\text{minimum power for liquefaction}}{\text{exergy expenditure}} \tag{6.38}$$

$$= \frac{\dot{n}_{10}(\text{ex}_{10} - \text{ex}_9)}{\dot{n}_1(\text{ex}_1 - \text{ex}_4) + \dot{n}_5(\text{ex}_5 - \text{ex}_8) - \dot{W}_{e,1} - \dot{W}_{e,2}}. \tag{6.39}$$

Table 6.37 shows the operating pressure, temperature, vapor fraction, and flow rate of the nitrogen and methane streams determined using an optimization method that maximizes the exergy efficiency of the cold box. The adiabatic efficiency of both turbines was assumed to be 80%.

Figure 6.58 shows the exergy utilization in the cold box of the dual expander process. The exergy efficiency of the cold box is 52.7% compared to 38.3% observed in the case of a reverse Brayton process operating with a mixture of nitrogen and methane.

It can be seen that most of the exergy loss occurs in the two turbines, even in the dual turbine process. The exergy loss in the heat exchanger is very small in this case compared to a reverse Brayton process (see Fig. 6.54) due to closer temperature approaches between the hot and cold streams of the heat exchanger, as shown in

Table 6.37. Temperature, pressure, vapor fraction, and flow rate of the different streams of the dual turbine natural gas liquefaction process shown in Fig. 6.57

	Streams				
	1	2	3	4	5
Temperature, K	300.0	223.2	109.8	291.2	300.0
Pressure, bar	116.4	116.4	6.9	6.9	100.2
Vapor fraction	1.000	1.000	1.000	1.000	1.000
Flow rate, mol/s	2.121	2.121	2.121	2.121	2.804
	Streams				
	6	7	8	9	10
Temperature, K	278.6	192.3	297.0	300.0	113.0
Pressure, bar	100.2	21.9	21.9	65.0	65.0
Vapor fraction	1.000	1.000	1.000	1.000	0.000
Flow rate, mol/s	2.804	2.804	2.804	1.000	1.000

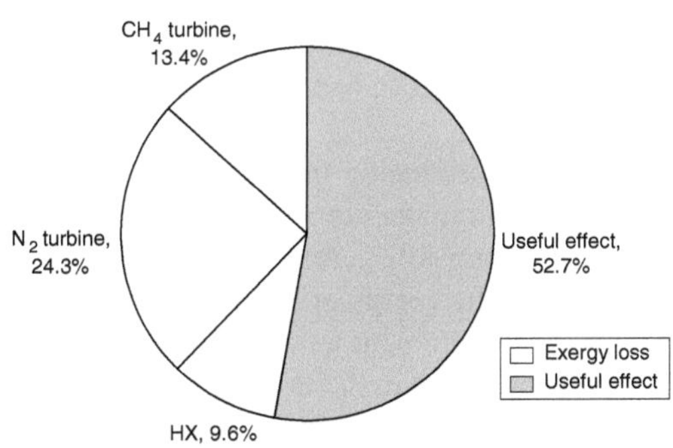

Fig. 6.58. Exergy utilization in the dual expander natural gas liquefaction process shown in Fig. 6.57 operating with nitrogen and methane refrigerants.

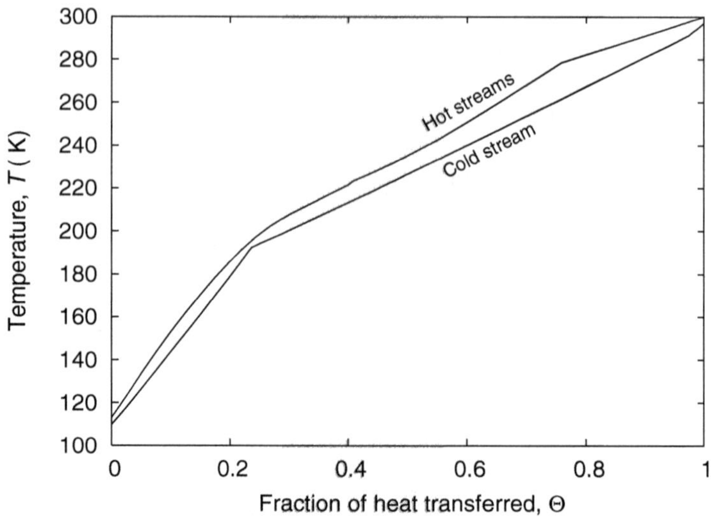

Fig. 6.59. Temperature profiles in the dual expander natural gas liquefaction process operating with nitrogen and methane refrigerants.

Fig. 6.59, all along the heat exchanger length, particularly at low temperatures. The exergy efficiency of the cold box varies between 70 and 80% in the case of the mixed refrigerant processes discussed in the previous sections, while the exergy efficiency of the dual expander process shown in Fig. 6.57 is only 52.7%. The overall exergy efficiency of turbine-based processes, taking into account the exergy loss in the aftercooler and compressor, will also be much lower than that of mixed refrigerant processes. Turbine-based processes, however, are used in small plants due to their simplicity and quick startup time. They are also ideal for offshore platforms where a small hydrocarbon inventory is preferable from the safety point of view.

Table 6.37 shows that the temperature of the methane (stream 6) at the entry of the turbine is 278.6 K. If the methane refrigerant (stream 6) is introduced into Turbine 2 at 300 K, then the exergy efficiency of the cold box of a dual turbine process shown in Fig. 6.57 would decrease to 52.2%, or translating into a drop in exergy efficiency of 0.5%.

The main advantage for introducing the methane refrigerant (stream 6) directly into the turbine is the reduction in the number of streams in the heat exchanger from five to four. When the methane refrigerant is replaced by a mixture of 98 mol% methane and 2 mol% ethane, the exergy efficiency improves to 53.7% in spite of the mixture entering the turbine at ambient temperature (300 K). The use of refrigerant mixtures is thus advantageous even in the dual turbine process.

Precooling the dual turbine process will result in a further improvement in the exergy efficiency. The process will then be similar to many mixed refrigerant processes with three different refrigerants, one each for precooling, condensation, and subcooling.

It is also possible to use recompression processes or multiple turbines to provide the necessary refrigeration to liquefy a natural gas feed. One such patented processes is shown in Fig. 6.60.

6.13 Summary

A number of mixed refrigerant processes have been discussed in this chapter. It is evident from the examples provided that processes that use different refrigerants for precooling (desuperheating), condensation (liquefaction), and subcooling operate at high exergy efficiency. The three refrigerants can be derived from a single refrigerant, as in the case of a Kleemenko process (Fig. 6.42) using phase separators. Alternately, the condensation and subcooling refrigerants can be derived from a single refrigerant using a phase separator, and a separate precooling refrigerant used in the case of the C3-MR (Fig. 6.29) and DMR processes (Figs. 6.34 and 6.35). On the other hand, three separate refrigerants, one each for precooling, condensation, and subcooling, are used in cascade refrigerators operating with refrigerant mixtures (Figs. 6.46 and 6.52). There are also attempts to use a nitrogen expander process for subcooling the natural gas feed, while two separate refrigerants are used for precooling and condensation [79].

The exergy efficiency of most precooled liquefaction processes described in this chapter is nearly the same. The choice of a process also depends on other criteria such as the size of heat exchangers, the cost and availability of equipment, ambient temperatures, etc.

The exergy efficiency of practical large LNG plants is higher than that shown in the examples due to the use of smaller minimum temperature approaches in the heat exchangers (typically 1.8 to 2 K), as well as the use of dense fluid and two-phase expansion turbines instead of expansion valves.

Most of the processes described in this chapter can be used in small natural gas liquefiers being proposed for liquefying stranded wells, biogas from landfills, municipal wastes, etc. Processes with phase separators such as the Kleemenko process or the PRICO process are ideal for such applications because of their simplicity. Precooled liquefaction processes are preferable for large liquefaction systems.

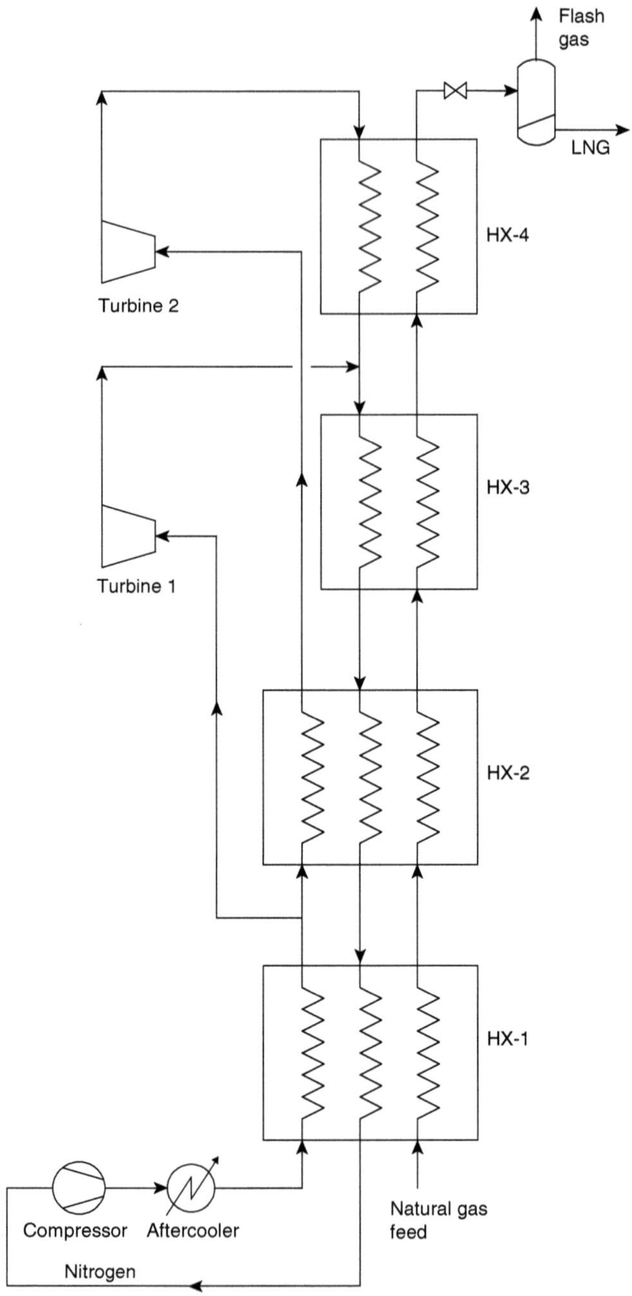

Fig. 6.60. Nitrogen expander process for the liquefaction of natural gas [33]. (Adapted from U.S. Patent no. 6,250,244.)

7

Cooling and liquefaction of air and its constituents

Mixed refrigerant processes were originally developed for the liquefaction of natural gas. There has, however, been considerable interest in academia as well as in industry in developing mixed refrigerant processes for the liquefaction of nitrogen and air [1, 2, 22–24, 29, 30, 45–47, 84]. Mixed refrigerant processes have also been developed to replace expansion turbines in air separation processes [25].

There are several differences between the liquefaction of natural gas and pure fluids such as nitrogen:

- Nitrogen condenses at a much lower temperature than natural gas. Some of the high boilers in the refrigerant mixture can freeze at those temperatures. Decreasing their concentration to prevent freezing results in a lower exergy efficiency.
- Natural gas is subcooled by 80 to 100 K to reduce the fraction of flash gas generated after expansion of the liquid. A few degrees of subcooling (typically 5–15 K) are used in the case of pure fluids such as nitrogen and air. Therefore, only two refrigerants, one for precooling (desuperheating) and another for condensation, are sufficient to liquefy nitrogen efficiently. In the case of natural gas liquefaction, it is advantageous to use three separate refrigerants: for desuperheating (precooling), condensation (liquefaction), and subcooling.
- The latent heat of vaporization is about a third of the total enthalpy change between room temperature and liquefied natural gas temperature before expansion (100 to 120 K). On the other hand, the latent heat of vaporization of nitrogen is about half of the total enthalpy change between room temperature and saturated liquid nitrogen.
- Condensation occurs at constant temperature in the case of pure fluids, whereas the temperature of natural gas varies during condensation. This leads to considerable exergy loss in the condenser in the case of nitrogen or other pure fluids.

These and other differences make the liquefaction of pure fluids somewhat different from that of natural gas, even when a similar process is used. Some of the important processes that can be used for the liquefaction of pure fluids such as nitrogen are described in this chapter.

G. Venkatarathnam, *Cryogenic Mixed Refrigerant Processes*,
DOI: 10.1007/978-0-387-78514-1_7,

7.1 Single-stage processes for the sensible cooling of a pure fluid such as nitrogen

Figure 7.1 shows a single-stage process for the sensible cooling of a pure fluid such as nitrogen to temperatures above its boiling point. The process is identical to the single-stage natural process for the liquefaction of natural gas (Fig. 6.3). Consider the example of cooling nitrogen from 300 K to 130 K at a constant pressure of 10 bar. The design specifications of the process are shown in Table 7.1. A limit of 20 bar has been imposed on the compressor discharge pressure to allow traditional refrigeration compressors to be used. The nitrogen operating pressure has been chosen as 10 bar. The minimum temperature approach between the hot and cold streams of the heat exchanger has been assumed to be 3 K.

The refrigerant mixture composition and operating pressures were determined using the optimization procedure described in Chapter 5. Table 7.2 shows the composition of the optimum refrigerant mixture. It can be seen that a large fraction of propane and *i* Pentane is used in the mixture. However, the high boilers in the mixture are likely to freeze only at temperatures below 90 K. The use of a large fraction of high boilers in the refrigerant mixture results in a large refrigeration effect (Joule–Thomson coefficient at room temperature). Table 7.3 shows the performance of the process with the mixture given in Table 7.2. It can be seen that 20% of the refrigerant mixture (stream 3) condenses in the condenser due to the large fraction of high boilers in the mixture. The minimum temperature approach between the hot and cold streams of the heat exchanger occurs at the warm end of the heat exchanger. The high-pressure

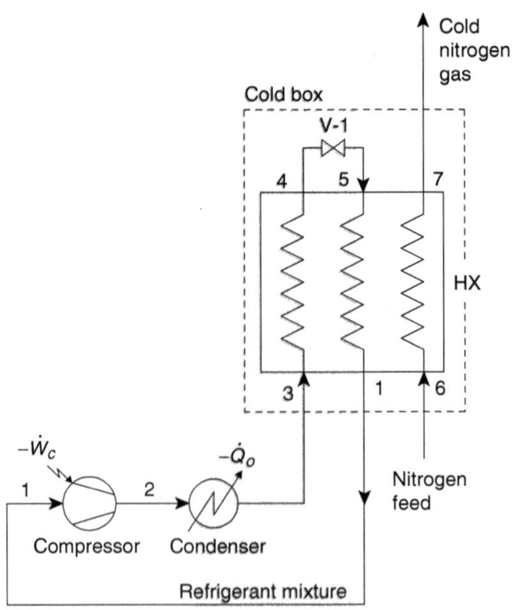

Fig. 7.1. Single-stage mixed refrigerant process for sensible cooling of a gas such as nitrogen.

Table 7.1. Design specifications for a single-stage mixed refrigerant process for the sensible cooling of a nitrogen stream from 300 K to 130 K at a constant pressure of 10 bar

Exergy efficiency of compressor and aftercooler, $\eta_{ex, cs}$	100%
Volumetric efficiency of the compressor, η_v	100%
Maximum compressor discharge pressure, p_2	20 bar
Minimum compressor suction pressure, p_1	3 bar
Minimum temperature approach in the heat exchanger, ΔT_{min}	3 K
Pressure drop in the heat exchangers, Δp	0
Nitrogen feed pressure, p_6	10 bar
Entry temperature of nitrogen, T_6	300 K
Exit temperature of nitrogen, T_7	130 K

Table 7.2. Mixture composition for cooling a nitrogen stream from 300 K to 130 K at a constant pressure of 10 bar using the process shown in Fig. 7.1

Component	Concentration (mol%)
Nitrogen	8.6
Methane	28.9
Ethylene	25.5
Propane	23.7
*i*Pentane	13.2

Table 7.3. Temperature, pressure, vapor fraction, and flow rate of different streams of the process shown in Fig. 7.1 and operating with the mixture shown in Table 7.2

	Stream					
	1	3	4	5	6	7
Temperature, K	297.0	300.0	130.2	126.6	300.0	130.0
Pressure, bar	5.3	20.0	20.0	5.3	10.0	10.0
Vapor fraction	1.000	0.800	0.000	0.072	1.000	1.000
Flow rate, mol/s	1.000	1.000	1.000	1.000	0.771	0.771

refrigerant is in a subcooled liquid state at the entry of the expansion valve (state 4). The low-pressure refrigerant is evaporated completely and superheated in the heat exchanger to provide the refrigeration required to cool the feed gas as well as the high-pressure refrigerant.

The exergy efficiency of the cold box is given by the expression

$$\eta_{ex, cb} = \frac{\text{minimum power for liquefaction}}{\text{exergy expenditure}} = \frac{\dot{n}_7 (ex_7 - ex_6)}{\dot{n}_3 (ex_3 - ex_1)}. \quad (7.1)$$

The overall exergy efficiency of the process is given by the expression

$$\eta_{\mathrm{ex},o} = \frac{\text{minimum power for liquefaction}}{\text{compressor power input}} = \frac{\dot{n}_7\,(\mathrm{ex}_7 - \mathrm{ex}_6)}{-\dot{W}_c}. \quad (7.2)$$

The exergy efficiency of the cold box can also be written as follows:

$$\eta_{\mathrm{ex,cb}} = 1 - \frac{\text{exergy loss in heat exchanger + exergy loss in valve}}{\text{exergy expenditure}}. \quad (7.3)$$

The exergy loss in the valve is minimized by ensuring that the temperature change during expansion is small, or the degree of subcooling of the liquid refrigerant (stream 4) is high. Figures 7.2 and 7.3 show the process on temperature-enthalpy and temperature-entropy diagrams for the refrigerant mixture. The refrigerant chosen is in a subcooled liquid state at the entry of the expansion valve.

The exergy loss in the heat exchanger is minimized when the mixture chosen satisfies the following condition as closely as possible over the entire heat exchanger length [Eq. (3.10)]:

$$\left(\frac{\partial h}{\partial T}\right)_{p,6} - \left(\frac{\partial h}{\partial T}\right)_{p,3} = \frac{\dot{n}_7}{\dot{n}_3}\, c_{p,6}, \quad (7.4)$$

where the numerals in the subscripts refer to the stream numbers in Fig. 7.1. The temperature profiles of the hot and cold streams will be nearly parallel if the above condition is satisfied over the length of the heat exchanger (see Section 3.4). It can be seen from Fig. 7.2 that the variation of enthalpy of the mixture with temperature at the operating pressures (20 and 5.3 bar) is such that it results in a very close temperature approach between the streams (Fig. 7.4), and thereby results in an efficient cooling of a gas. It can be seen from Fig. 7.5 that the exergy efficiency of the cold box (heat exchanger and valve) is quite high and close to 70%.

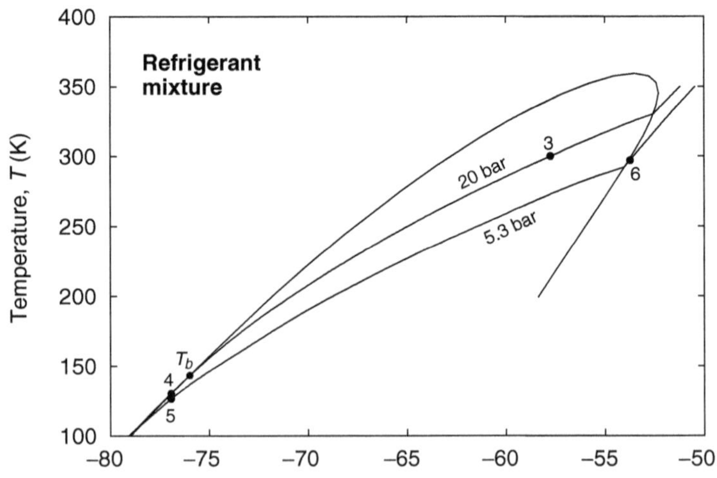

Fig. 7.2. Single-stage mixed refrigerant process for the cooling of nitrogen.

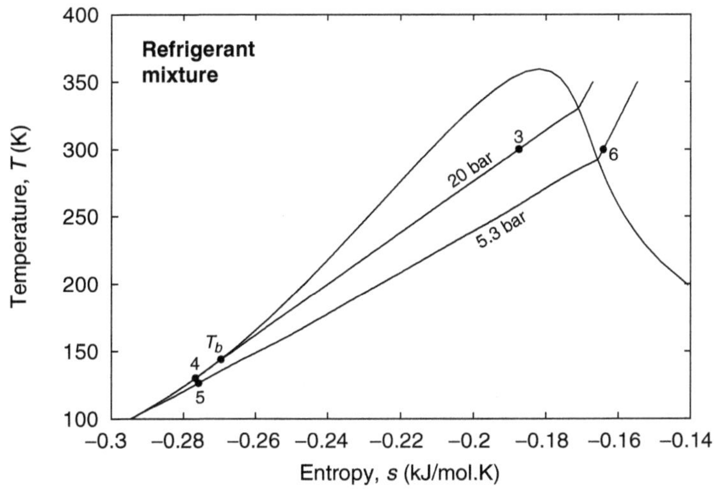

Fig. 7.3. Single-stage mixed refrigerant process for the cooling of nitrogen.

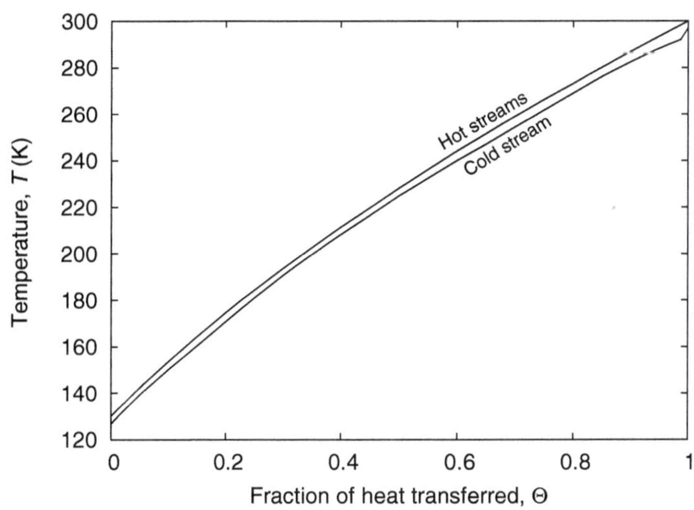

Fig. 7.4. Temperature profiles of the hot and cold streams of the heat exchanger in Fig. 7.1 operating with the mixture shown in Table 7.2.

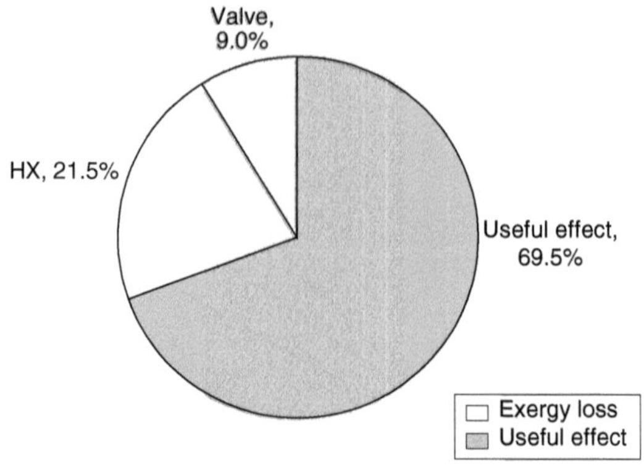

Fig. 7.5. Exergy utilization in the cold box of the process shown in Fig. 7.1 and operating with the mixture shown in Table 7.2.

A comparison of the exergy efficiency of a mixed refrigerant nitrogen cooler (Fig. 7.1) with that of an ideal gas cooler (Figs. 1.5 and 1.14) shows that an exergy efficiency of 70% can be attained in a turbine-based ideal gas cooler only when the operating pressure is on the order of 325 bar and the turbine adiabatic efficiency is higher than 90%. Adiabatic efficiencies higher than 80% are somewhat hard to attain in practical expanders. Hence, cold box exergy efficiency of more than 70% is nearly impossible to attain with an ideal gas cooler operating with the expander shown in Fig. 1.15. It is evident from the above comparison that the mixed refrigerant process shown in Fig. 7.1 is superior to a turbine-based process for gas-cooling applications. The overall exergy efficiency of both turbine-based and mixed refrigerant processes will be much lower due to the exergy losses in the compressor and after-cooler/condenser.

The exergy efficiency of the mixed refrigerant nitrogen cooler shown in Fig. 7.1 can be improved further by using a smaller temperature approach between the streams in the heat exchanger or using an efficient compressor. A single-stage compressor can be used in the case of a mixed refrigerant process shown in Fig. 7.1 since the compression ratio is only 3.8, compared to 25 in the case of a turbine-based ideal gas cooler shown in Fig. 1.5. The exergy efficiency of the compressor/aftercooler is dependent on the adiabatic efficiency of the compressor used. The adiabatic efficiency of fractional horsepower compressors is quite small, typically about 50 to 60% [68].

Figure 7.6 shows the variation of exergy efficiency of the compressor and after-cooler with the adiabatic efficiency of the compressor. The overall exergy efficiency of the system is a product of the exergy efficiency of the cold box and the compressor/aftercooler sections. When the exergy loss in the condenser/aftercooler is also taken into account, the exergy efficiency of the compressor section (compressor and condenser) of small mixed refrigerant systems will typically be 30 to 45%. An overall exergy efficiency of about 20% is therefore attainable even in small mixed refrigerant nitrogen gas coolers.

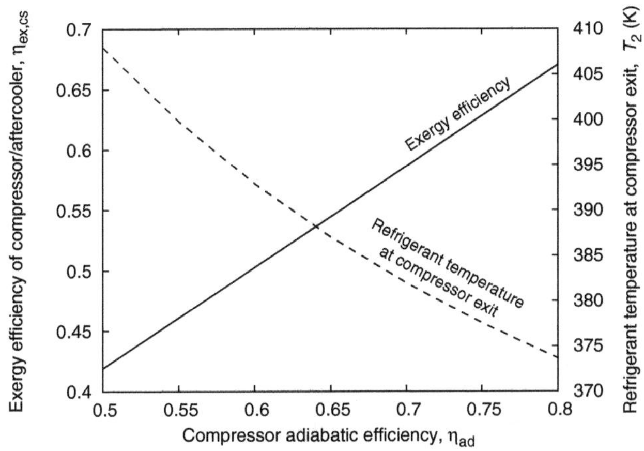

Fig. 7.6. Variation of exergy efficiency of the compressor and aftercooler of the single-stage mixed refrigerant process for the cooling of nitrogen shown in Fig. 7.1 with compressor adiabatic efficiency. The composition of the mixture is presented in Table 7.2.

7.2 Single-stage process for the liquefaction of pure fluids such as nitrogen

Figure 7.7 shows the single-stage process for the liquefaction of pure fluids such as nitrogen. The process is similar to the mixed refrigerant process for the cooling of a pure fluid (Fig. 7.1). Nitrogen feed pressure has been chosen as 40 bar, and the minimum temperature approach between the hot and cold streams has been chosen as 3 K. The design specifications of the process are shown in Table 7.4.

The optimum composition of the mixture was determined for the process using the methods described in Chapter 5. Table 7.5 shows the optimum composition of the refrigerant for the above design specifications. The temperature, pressure, vapor fraction, and flow rate of the refrigerant and nitrogen streams are shown in Table 7.6.

It can be seen from Table 7.5 that a six-component mixture is required for the liquefaction of nitrogen using the process shown in Fig. 7.7. The dew point temperature

Table 7.4. Design specifications for a single-stage mixed refrigerant process liquefier for the liquefaction of nitrogen (Fig. 7.7)

Exergy efficiency of compressor and aftercooler, $\eta_{ex,cs}$	100%
Volumetric efficiency of the compressor, η_v	100%
Maximum compressor discharge pressure, p_2	20 bar
Minimum compressor pressure, p_1	3 bar
Minimum temperature approach in the heat exchanger, ΔT_{min}	3 K
Pressure drop in the heat exchangers, Δp	0
Nitrogen pressure, p_6	40 bar

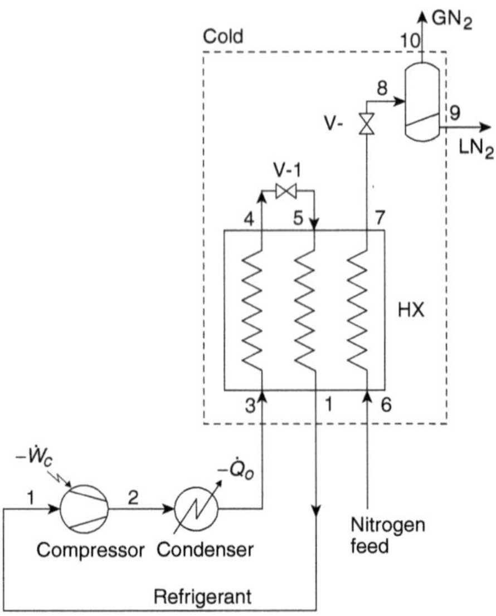

Fig. 7.7. Single-stage mixed refrigerant process for the liquefaction of nitrogen.

Table 7.5. Refrigerant composition for the single-stage nitrogen liquefier shown in Fig. 7.7

Component	Concentration (mol%)
Nitrogen	12.11
Methane	38.19
Ethylene	19.50
Propane	13.34
*i*Butane	9.13
*i*Pentane	7.73

is high due to the large fraction of propane, *i*Butane, and *i*Pentane in the refrigerant mixture. About 11.3% of the refrigerant (stream 3) condenses in the condenser. Both *i*Butane and *i*Pentane can freeze at temperatures below 120 K depending on their concentration in the mixture. The amount of high boilers has been limited to prevent freezing of *i*Butane and *i*Pentane at temperatures lower than 110 K. The molar flow rate of nitrogen through the heat exchanger (stream 6) is 28.8% of that of the refrigerant (stream 3), while that for the liquefied nitrogen (stream 9) is only 10.5% of that of the refrigerant (stream 3).

The exergy efficiency of the cold box is given by the expression

$$\eta_{\text{ex, cb}} = \frac{\text{minimum power liquefaction}}{\text{exergy expenditure}}. \tag{7.5}$$

Since only a fraction of the feed nitrogen is liquefied, the minimum power for liquefaction should be based on the liquid nitrogen flow rate (stream 9). Similarly, the

Table 7.6. Temperature, pressure, vapor fraction, and flow rate of different streams of the single-stage mixed refrigerant nitrogen liquefier shown in Fig. 7.7 and operating with the refrigerant mixture shown in Table 7.5

	Stream					
	3	4	5	1	6	7
Temperature, K	300.0	129.8	122.8	295.8	300.0	125.8
Pressure, bar	18.7	18.7	5.1	5.1	40.0	40.0
Vapor fraction	0.887	0.000	0.116	1.000	1.000	1.000
Flow rate, mol/s	1.000	1.000	1.000	1.000	0.288	0.288

	Stream		
	8	9	10
Temperature, K	77.4	77.4	77.4
Pressure, bar	1.0	1.0	1.0
Vapor fraction	0.634	0.000	1.000
Flow rate, mol/s	0.288	0.105	0.182

exergy input into the cold box is the sum of the difference of exergy of the refrigerant streams entering and leaving the cold box (streams 3 and 1) and the difference between the exergy of the unliquefied portion of the nitrogen feed (stream 6) and the flash nitrogen leaving the phase separator (stream 10).

The exergy efficiency of the cold box can be written as follows:

$$\eta_{\text{ex, cb}} = \frac{\text{minimum power of liquefaction}}{\text{exergy expenditure}} \tag{7.6}$$

$$= \frac{\dot{n}_9\left(\text{ex}_9 - \text{ex}_6\right)}{\dot{n}_3\left(\text{ex}_3 - \text{ex}_1\right) + \dot{n}_{10}\left(\text{ex}_6 - \text{ex}_{10}\right)}. \tag{7.7}$$

The overall exergy efficiency of the process is given by the expression

$$\eta_{\text{ex}, o} = \frac{\text{minimum power for liquefaction}}{\text{compressor power input + net exergy expenditure of nitrogen streams}}$$

$$= \frac{\dot{n}_9\left(\text{ex}_9 - \text{ex}_6\right)}{-\dot{W}_c + \dot{n}_{10}\left(\text{ex}_6 - \text{ex}_{10}\right)}. \tag{7.8}$$

The exergy efficiency of the cold box is quite low — only 36.2%— due to the large exergy loss in the nitrogen expansion valve (V-2) as well as that in the heat exchanger. The higher exergy loss in the heat exchanger in the present case (Fig. 7.8) compared to the previous case (Fig. 7.5) is due to the larger temperature approach between the streams in the heat exchanger (Fig. 7.9).

Thirty-six percent of the nitrogen liquefied in the heat exchanger (stream 7) is lost due to flashing of liquid nitrogen (stream 7) in the expansion valve. The exergy efficiency of the process decreases significantly due to the large fraction of flash

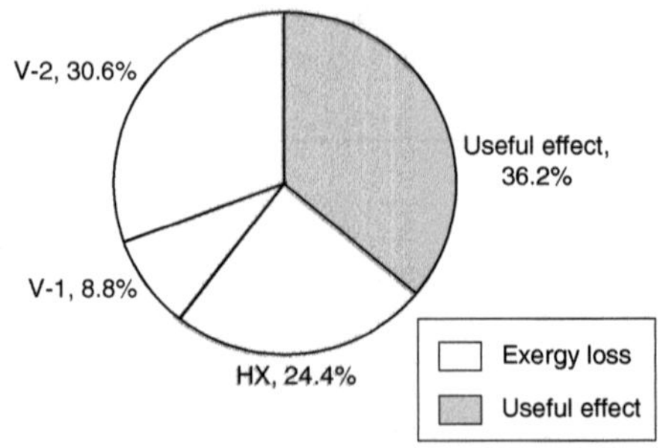

Fig. 7.8. Exergy utilization in the cold box of a single-stage mixed refrigerant process for the liquefaction of nitrogen and operating with the mixture shown in Table 7.5.

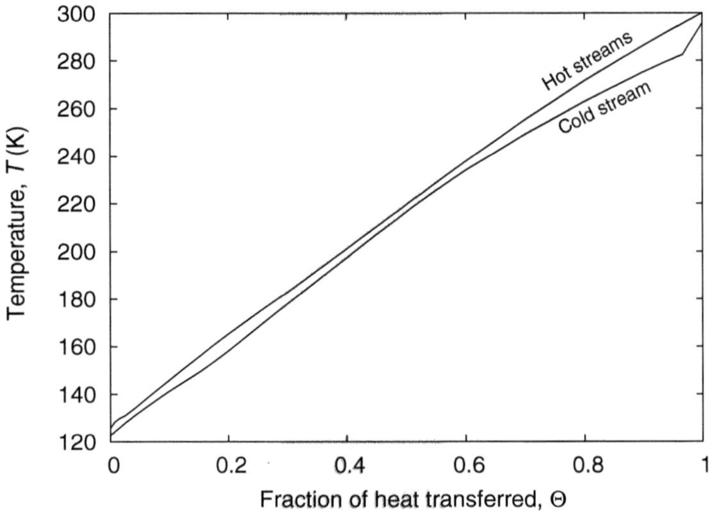

Fig. 7.9. Temperature profiles of the hot and cold streams of the heat exchanger in the mixed refrigerant nitrogen cooling process shown in Fig. 7.7 and operating with the mixture presented in Table 7.5.

gas generated in the nitrogen expansion valve (V-2). Table 7.6 shows that the flow rate of the flash gas (stream 10) generated in the nitrogen expansion valve (V-2) is much higher than the liquid nitrogen product (stream 9). This is the most important disadvantage of the single-stage process shown in Fig. 7.7. The exergy loss in the nitrogen expansion of nitrogen in the expansion valve (V-2) is also quite high due to a large pressure change across the expansion valve.

The overall exergy efficiency of the system that includes the nitrogen compressor power input would therefore be quite low. One way of reducing the work required to liquefy nitrogen is to use the flash nitrogen vapor to cool the high-pressure nitrogen stream as in a Linde–Hampson liquefier (Fig. 1.18). Such a system is shown in Fig. 7.10.

7.3 Mixed refrigerant precooled Linde–Hampson liquefaction process

Figure 7.10 shows a mixed refrigerant precooled Linde–Hampson liquefaction process. The system is similar to the precooled Linde–Hampson liquefier shown in Fig. 1.25, except for the subcooling of the precooling refrigerant before expansion. The design specifications of the system are the same as that in Table 7.4.

The performance of the system has been evaluated with a refrigerant mixture from Table 7.7. As in the previous examples, the refrigerant composition has been optimized to maximize the exergy efficiency of the cold box subject to the minimum temperature approach between streams to be 3 K in the heat exchangers.

The exergy efficiency of the cold box is given by the expression

$$\eta_{\mathrm{ex,\,cb}} = \frac{\text{minimum power for liquefaction}}{\text{exergy expenditure}} = \frac{\dot{n}_{11}\,(\mathrm{ex}_{11} - \mathrm{ex}_{14})}{\dot{n}_8\,(\mathrm{ex}_8 - \mathrm{ex}_6) + \dot{n}_3\,(\mathrm{ex}_3 - \mathrm{ex}_1)}. \tag{7.9}$$

The overall exergy efficiency of the process is given by the expression

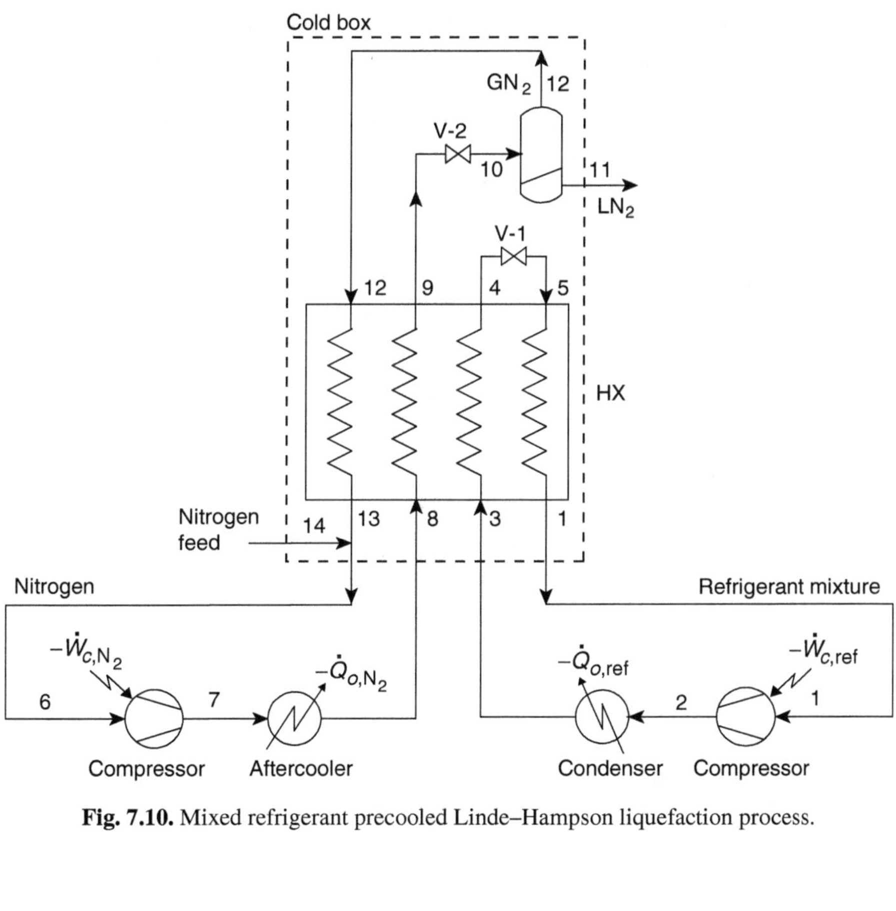

Fig. 7.10. Mixed refrigerant precooled Linde–Hampson liquefaction process.

Table 7.7. Refrigerant composition for a mixed refrigerant precooled Linde–Hampson nitrogen liquefaction process from Fig. 7.10

Component	Concentration (mol%)
Nitrogen	14.27
Methane	39.41
Ethylene	17.76
Propane	14.48
i Butane	7.21
i Pentane	6.87

$$\eta_{\mathrm{ex},\,o} = \frac{\text{minimum power for liquefaction}}{\text{exergy expenditure}} = \frac{\dot{n}_{11}\,(\mathrm{ex}_{11} - \mathrm{ex}_{14})}{(-\dot{W}_{c,\mathrm{ref}}) + (-\dot{W}_{c,\mathrm{N_2}})}, \tag{7.10}$$

where $-\dot{W}_{c,\mathrm{ref}}$ and $-\dot{W}_{c,\mathrm{N}_2}$ refer to the power input to the mixed refrigerant compressor and the nitrogen compressor, respectively.

Consider the cold box that excludes only the compressors and condenser or aftercoolers shown in Fig. 7.10. Figure 7.11 shows the exergy utilization in the cold box of the process. An exergy efficiency of 59.3% can be obtained with the process shown in Fig. 7.10 and operating with the mixtures shown in Table 7.7. The exergy efficiency of the cold box of the process increases from 36.2% when the flash nitrogen is not used, as in Fig. 7.8, to 59.3% when the cold nitrogen flash vapor is used to cool the hot streams in the heat exchanger. It can be seen from Fig. 7.11 that in the present case, the largest exergy loss occurs in the heat exchanger and the nitrogen expansion valve. The nitrogen expansion valve losses can be reduced by decreasing the operating pressure of nitrogen from the 40 bar adopted in this example to a smaller value or by using a turbine expander instead of an expansion valve.

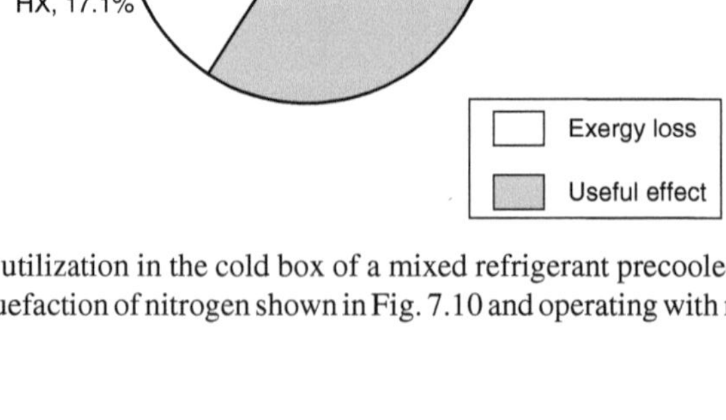

Fig. 7.11. Exergy utilization in the cold box of a mixed refrigerant precooled Linde–Hampson process for the liquefaction of nitrogen shown in Fig. 7.10 and operating with mixtures presented in Table 7.7.

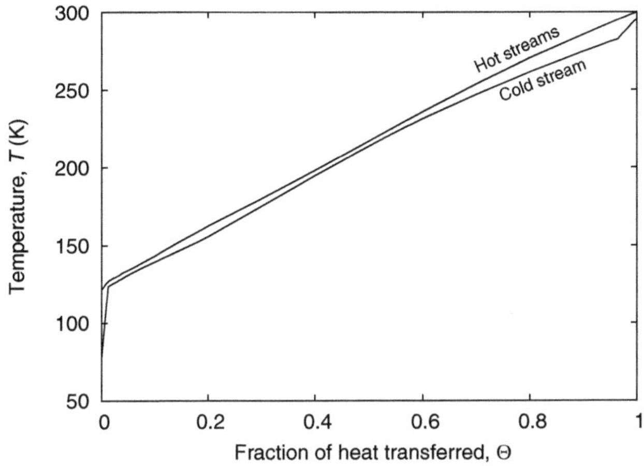

Fig. 7.12. Temperature profiles in the mixed refrigerant precooled Linde–Hampson liquefaction process shown in Fig. 7.10 and operating with the mixture shown in Table 7.7 and Fig. 7.10.

Figure 7.12 shows the temperature profiles in the heat exchanger. The temperature difference between the high-pressure nitrogen (hot stream) and the low-pressure flash nitrogen stream (cold stream) is large (44 K) at the cold end of the heat exchanger. The large temperature difference, however, exists only across a small length of the heat exchanger. On the other hand, the temperature difference across the refrigerant expansion valve (V-1) is only 8.5 K.

Table 7.8 shows the temperature, pressure, vapor fraction, and mole flow rate of different streams in the process. A comparison of the results of the single-stage liquefaction process without flash nitrogen recycle (Table 7.6) with the current process (Table 7.8) shows that the liquid nitrogen flow rate per mole of refrigerant flow increases from 0.105 to 0.168 due to better utilization of the cold in the mixed refrigerant precooled Linde–Hampson liquefier (Fig. 7.10). Similarly, the liquid yield increases from 36.4% to 46.6% due to the utilization of sensible cooling available with the flash nitrogen stream. This will result in a higher exergy efficiency for the current process. The liquid yield in a conventional Linde–Hampson liquefaction process shown in Fig. 1.18 (without precooling) and with a minimum temperature approach of 3 K would have been only 1.2% at the same nitrogen operating pressures (40/1 bar). Similarly, the exergy efficiency of such a system would be only 0.5% compared to 59.3% when a mixed refrigerant precooling is used. The use of mixed refrigerant precooling therefore improves the liquid yield and exergy efficiency of a Linde–Hampson liquefier by many fold.

Table 7.8 shows that the temperature of the high-pressure refrigerant at the exit of the heat exchanger is 132.1 K and is higher than the high-pressure nitrogen at the exit of the heat exchanger (121.4 K). Similarly, the low-pressure refrigerant enters the heat exchanger at a temperature of 123.6 K, again higher than that for the low-pressure nitrogen, which enters the heat exchanger at 77.4 K. Thus, the four-stream

Table 7.8. Temperature, pressure, vapor fraction, and flow rate of different streams of the mixed refrigerant precooled Linde–Hampson liquefier (Fig. 7.10) and operating with the mixture presented in Table 7.7

	Stream					
	3	4	5	1	8	9
Temperature, K	300.0	132.1	123.6	295.6	300.0	121.4
Pressure, bar	19.980	19.980	5.847	5.847	40.000	40.000
Vapor fraction	0.910	0.000	0.137	1.000	1.000	0.000
Flow rate, mol/s	1.000	1.000	1.000	1.000	0.360	0.360
	Stream					
	10	11	12	13		
Temperature, K	77.4	77.4	77.4	297		
Pressure, bar	1.000	1.000	1.000	1.000		
Vapor fraction	0.534	0.000	1.000	1.000		
Flow rate, mol/s	0.360	0.168	0.192	0.192		

heat exchanger essentially acts as a two-stream heat exchanger at the cold end. The heat exchanger shown in Fig. 7.10 is a combination of a four- and a two-stream heat exchanger. Boiarskii et al. [24] studied a process similar to that shown in Fig. 7.10. In their process, shown here in Fig. 7.13, they used a two- and a three-stream heat exchanger combination instead of the four-stream heat exchanger shown in Fig. 7.10. The cost of a single heat exchanger core that incorporates both the four-stream and the two-stream heat exchangers is normally lower than many separate heat exchanger cores.

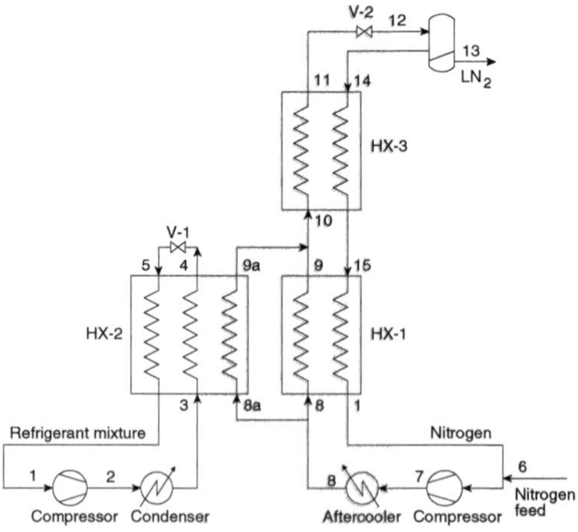

Fig. 7.13. Mixed refrigerant precooled nitrogen liquefaction process with three heat exchangers. (Adapted from Ref. [24] with kind permission of Springer Science and Business Media.)

7.4 Mixed refrigerant precooled Kapitza liquefaction process

Figure 7.14 shows a mixed refrigerant precooled Kapitza liquefaction process. The design specifications that are shown in Table 7.9 are the same as those in Table 7.4 except for the adiabatic efficiency of the turbine, which has been assumed to be 80%. The mixture composition, operating pressures, and temperature were determined by optimizing the exergy efficiency of the cold box. Unlike the natural gas liquefaction processes, the nitrogen precooling temperature is also considered a design variable in the optimization of all nitrogen liquefaction processes. Table 7.10 shows the temperature, pressure, vapor fraction, and flow rate of different streams operating with the mixture shown in Table 7.11.

The high-pressure refrigerant leaves the first heat exchanger (HX-1) at a temperature of 238.9 K and the low-pressure refrigerant enters the first heat exchanger at

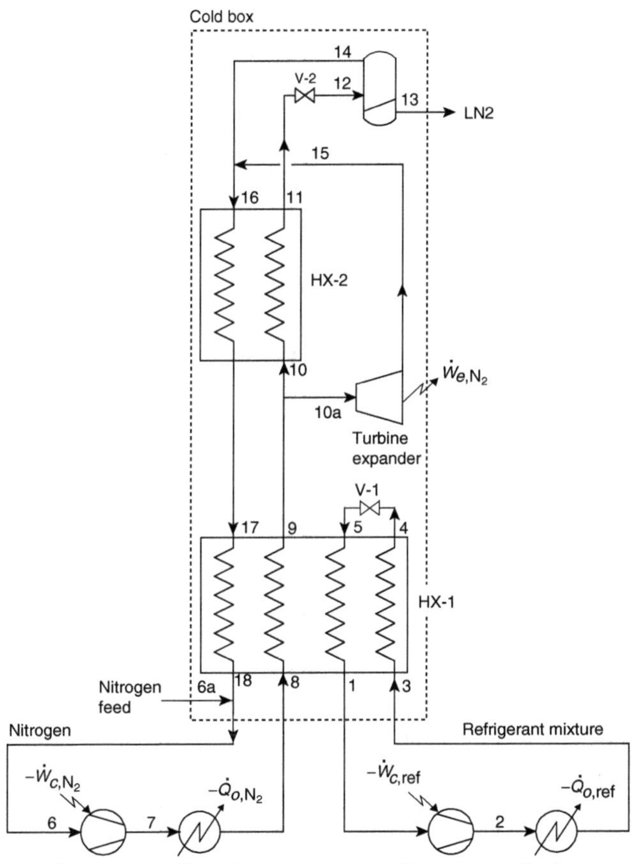

Fig. 7.14. Mixed refrigerant precooled Kapitza liquefaction process [64]. (Adapted from U.S. Patent no. 6,041,620.)

Table 7.9. Design specifications for a mixed refrigerant precooled Kapitza nitrogen liquefier

Exergy efficiency of compressor and aftercooler, $\eta_{ex,cs}$	100%
Volumetric efficiency of the compressor, η_v	100%
Nitrogen feed pressure, p_{6a}	40 bar
Maximum compressor discharge pressure, p_2	20 bar
Minimum compressor suction pressure, p_1	3 bar
Minimum temperature approach in the heat exchanger, ΔT_{min}	3 K
Pressure drop in the heat exchangers, Δp	0
Adiabatic efficiency of the turbine, $\eta_{ad,t}$	80%

Table 7.10. Temperature, pressure, vapor fraction, and flow rate of different streams of the mixed refrigerant precooled Kapitza liquefier (Fig. 7.14) and operating with the mixture presented in Table 7.11

	Stream					
	1	3	4	5	8	9
Temperature, K	291.2	300	238.9	215.7	300.0	184.2
Pressure, bar	3.037	18.230	18.230	3.037	40.000	40.000
Vapor fraction	1.000	0.316	0.000	0.196	1.000	1.000
Flow rate, mol/s	1.000	1.000	1.000	1.000	14.391	14.391

	Stream					
	10	10a	11	12	13	14
Temperature, K	184.2	184.2	97.2	77.4	77.4	77.4
Pressure, bar	40.000	40.000	40.000	1.000	1.000	1.000
Vapor fraction	1.000	1.000	0.000	0.212	0.000	1.000
Flow rate, mol/s	4.108	10.283	4.108	4.108	3.237	0.871

	Stream			
	15	16	17	18
Temperature, K	77.4	77.4	162.4	297
Pressure, bar	1.000	1.000	1.000	1.000
Vapor fraction	1.000	1.000	1.000	1.000
Flow rate, mol/s	10.283	11.154	11.154	11.154

a temperature of 215.7 K. Since the operating temperatures of the precooling refrigerant are high, very little methane is required in the refrigerant mixture, as shown in Table 7.11.

The exergy efficiency of the cold box of the mixed refrigerant precooled Kapitza liquefaction process shown in Fig.7.14 is given by the expression

Table 7.11. Refrigerant composition for the nitrogen liquefaction process shown in Fig. 7.14

Component	Concentration (mol%)
Methane	0.53
Ethane	24.31
Ethylene	24.28
*n*Butane	50.88

$$\eta_{\text{ex, cb}} = \frac{\text{minimum power for liquefaction}}{\text{exergy expenditure}} = \frac{\dot{n}_{13}(\text{ex}_{13} - \text{ex}_{6\text{a}})}{\dot{n}_3(\text{ex}_3 - \text{ex}_1) + \dot{n}_8\,(\text{ex}_8 - \text{ex}_6) - \dot{W}_{e,\text{N}_2}}. \tag{7.11}$$

The exergy of the high-pressure refrigerant and that of the high-pressure nitrogen entering the cold box decrease on expansion. In the present example, the work of expansion of the turbine is assumed to be reused in the plant itself. Therefore, the exergy expenditure will be the sum of exergy change of the refrigerant and nitrogen between the inlet and outlet of the cold box minus the work of expansion, as shown in Eq. (7.11).

The expression of the overall exergy of the process is as follows:

$$\eta_{\text{ex, o}} = \frac{\text{minimum power for liquefaction}}{\text{total compressor power input} - \text{expander power output}} \tag{7.12}$$

$$= \frac{\dot{n}_{13}(\text{ex}_{13} - \text{ex}_6)}{(-\dot{W}_{c,\text{ref}}) + (-\dot{W}_{c,\text{N}_2}) - \dot{W}_{e,\text{N}_2}} \tag{7.13}$$

where $-\dot{W}_{c,\text{ref}}$ and $-\dot{W}_{c,\text{N}_2}$ refer to the power input to the mixed refrigerant compressor and the nitrogen compressor, respectively. The exergy utilization in the cold box of the mixed refrigerant precooled Kapitza liquefaction process is shown in Fig. 7.15. The maximum exergy loss occurs in the turbine, as is the case for a conventional Kapitza process (Fig. 1.45). The exergy loss in the second heat exchanger is somewhat high due to the occurrence of a pinch point (see Fig. 7.16). The pinch point occurs due to a large change in the specific heat of nitrogen at 40 bar, close to the critical temperature as shown in Fig. 1.52.

Consider a conventional Kapitza nitrogen liquefaction process operating at 40/1 bar and with 82.5% of the compressor flow diverted through the turbine. The exergy efficiency of the cold box of an optimized conventional Kaptiza process would be 50.4%, whereas with mixed refrigerant precooling, the exergy efficiency of the cold box increases to 63.4% (Fig. 7.15). Similarly, the liquid yield with the conventional Kaptiza process and mixed refrigerant precooled Kapitza process would be 16.8% compared to 22.5% with a mixed refrigerant precooled Kapitza process. Thus, mixed refrigerant precooling improves the performance of a Kapitza liquefaction process. Since the flow rate of the mixed refrigerant is only 6.9% of that through the nitrogen

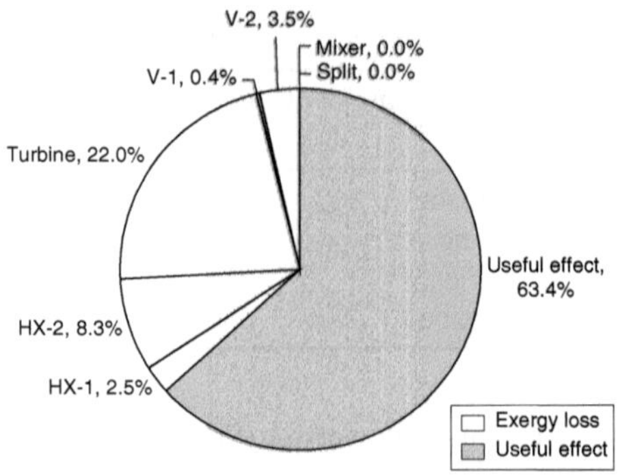

Fig. 7.15. Exergy utilization in the cold box of a mixed refrigerant precooled Kapitza process for the liquefaction of nitrogen (Fig. 7.14).

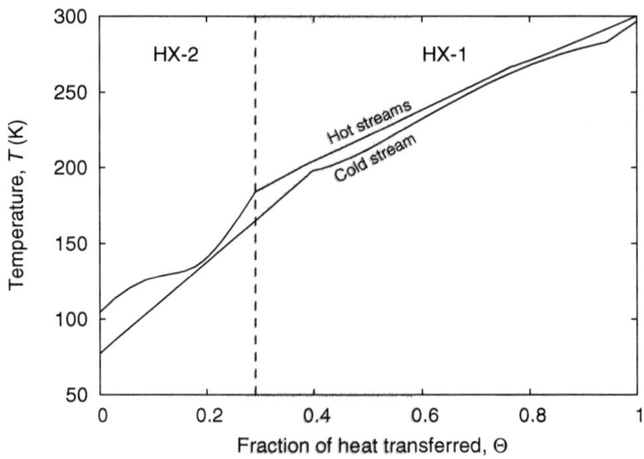

Fig. 7.16. Temperature profiles in the mixed refrigerant precooled Linde liquefaction process (Fig. 7.14) operating with the mixture shown in Table 7.11.

compressor, the additional surface area needed in the first heat exchanger due to precooling is small. The turbine inlet temperature has been chosen to ensure that no liquefaction occurs during expansion of the nitrogen in the turbine.

The performance of a mixed refrigerant precooled Linde–Hampson process and that of a Kapitza liquefaction process are compared in Table 7.12. It can be seen that the exergy efficiency of the mixed refrigerant precooled Kaptiza process is about 4% higher than that of the mixed refrigerant precooled Linde–Hampson process. However, the liquid yield of the precooled Kaptiza process is about half that of the mixed refrigerant precooled Linde–Hampson process. The exergy efficiency of the mixed refrigerant precooled Kapitza process is lower than that of the Linde–Hampson process due to the work recovered in the expansion turbine, which reduces the total power input to the compressors.

Table 7.12. Comparison of the performances of mixed refrigerant precooled Linde–Hampson liquefier (Fig. 7.10) and Kapitza liquefier (Fig. 7.14) operating with optimum refrigerant mixtures (see Tables 7.8 and 7.10)

	Mixed refrigerant precooled	
Performance parameter	Linde–Hampson liquefier	Kapitza liquefier
Exergy efficiency of cold box ($\eta_{ex,cb}$)	59.3%	63.4%
Liquid yield (Y)	46.6%	22.5%

A mixed refrigerant precooled Linde–Hampson liquefier is preferable in the case of small liquefiers since the initial cost of the system will be much lower without the turbine.

The precooling provided by the mixed refrigerant process in the precooled Kapitza process shown in Fig. 7.14 can also be provided by a second turbine. The process would be similar to that of a Collins liquefaction process (Fig. 1.48). The coldest heat exchanger (HX-5) used in the Collins liquefier is not necessary in the dual turbine nitrogen liquefier. If the same operating pressures (40/1 bar) are assumed for nitrogen, the optimum exergy efficiency of such a process would be 45%, compared to 63.4% in the case of a mixed refrigerant precooled nitrogen liquefaction process, with a turbine adiabatic efficiency of 80%.

Figure 7.17 shows a patented dual pressure Kapitza process with mixed refrigerant precooling. In this process, nitrogen is expanded in two stages to reduce the exergy losses during expansion as in a dual pressure Linde–Hampson process [13, 90] since the nitrogen operating pressures are above 70 bar. Precooling is provided using a mixed refrigerant process.

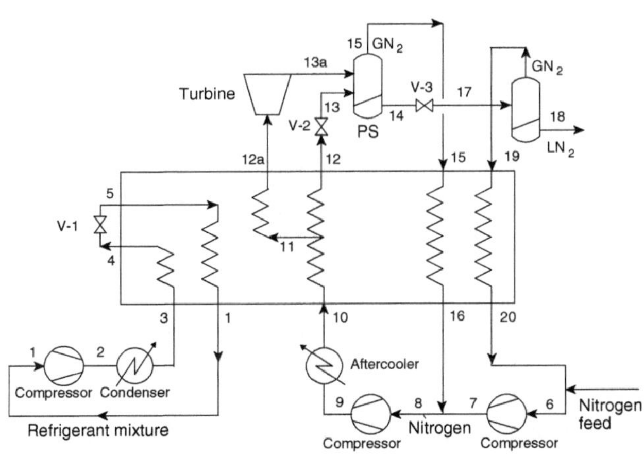

Fig. 7.17. Mixed refrigerant precooled Kapitza process with two-stage nitrogen expansion [29]. (Adapted from U.S. Patent no. 6,298,688.)

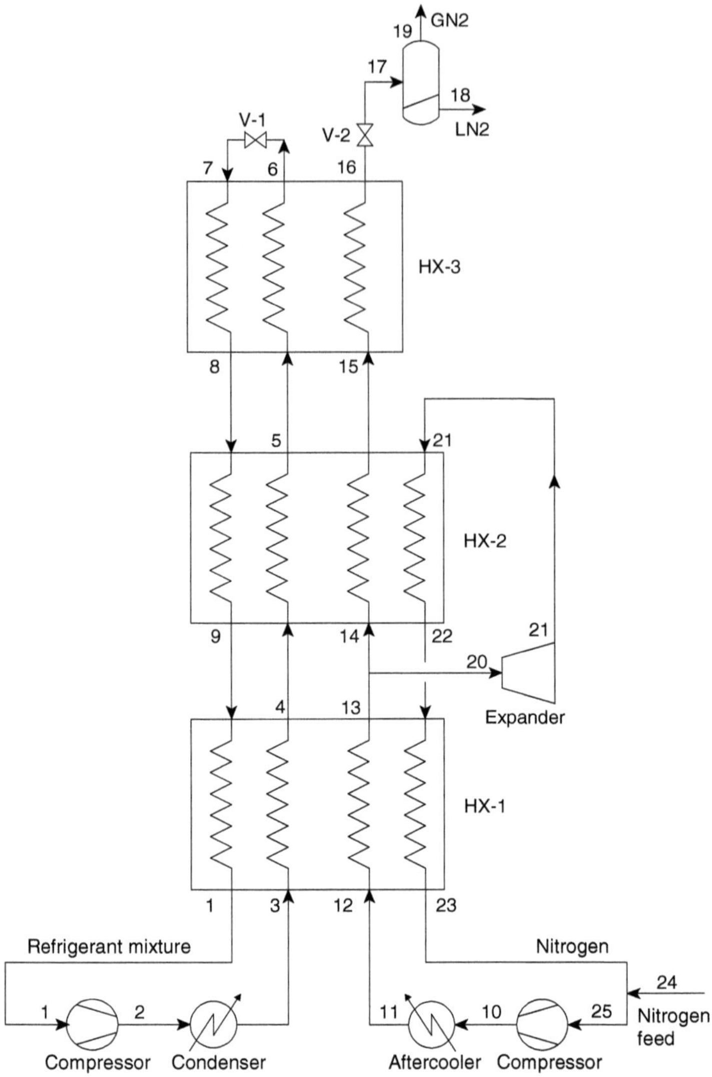

Fig. 7.18. Precooled Kapitza process for liquefaction of industrial gases with very low temperature precooling [64]. (Adapted from U.S. Patent no. 6,041,620.)

It is also possible to extend the mixed refrigerant precooling to very low temperatures as in the patented process shown in Fig. 7.18. The very low temperature precooling ensures that nitrogen is subcooled to a larger extent and that very little flash vapor is generated on expansion of the subcooled liquid nitrogen.

7.5 Liquefaction of nitrogen using the Kleemenko process

Kleemenko [50] was the first to liquefy natural gas using a mixed refrigerant process. The Kleemenko process is still used in small LNG plants. The Kleemenko process (Fig. 7.19) can also be used for the liquefaction of nitrogen, air, or other gases. The design specifications of the process are the same as those for a single-stage liquefaction process (Table 7.4). The procedure for the optimization of mixture composition, operating temperatures, and pressures of the refrigerant is identical to that for the liquefaction of natural gas using the Kleemenko processs. The exergy efficiency of the cold box is considered the objective function, as in the case of the liquefaction of nitrogen. Consider the cold box of the mixed refrigerant precooled Kleemenko process (Fig. 7.19) that excludes the compressor and condenser of the mixed refrigerant. The exergy efficiency of the cold box for the Kleemenko process shown in Fig. 7.19 is given by the expression

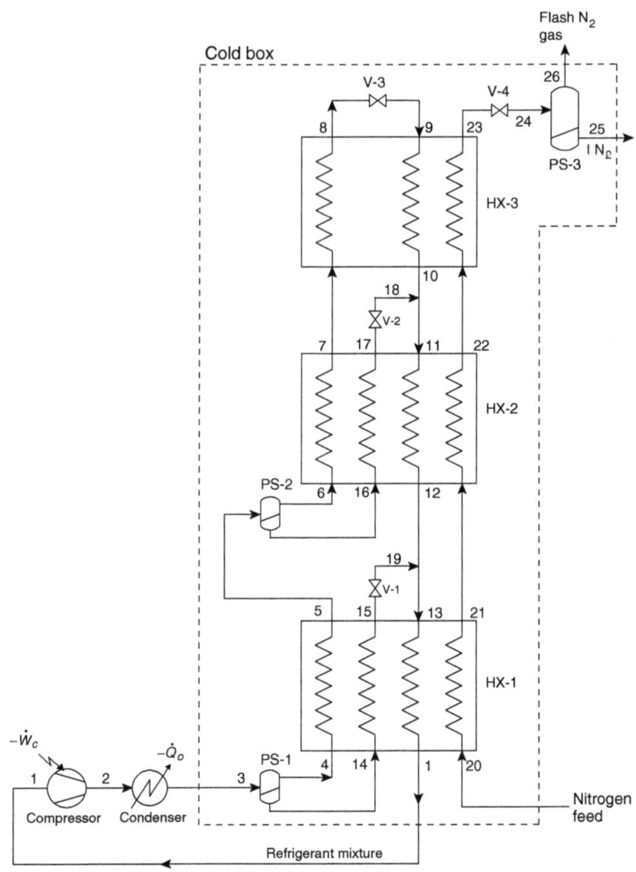

Fig. 7.19. Kleemenko process for the liquefaction of nitrogen, air, and other gases [50].

$$\eta_{\mathrm{ex,cb}} = \frac{\text{minimum power for liquefaction}}{\text{exergy expenditure}} \tag{7.14}$$

$$= \frac{\dot{n}_{25}(\mathrm{ex}_{25} - \mathrm{ex}_{20})}{\dot{n}_3(\mathrm{ex}_3 - \mathrm{ex}_1) + \dot{n}_{26}(\mathrm{ex}_{20} - \mathrm{ex}_{26})}. \tag{7.15}$$

The overall exergy efficiency is given by the expression

$$\eta_{\mathrm{ex},o} = \frac{\text{minimum power for liquefaction}}{\text{compressor power input}} = \frac{\dot{n}_{25}(\mathrm{ex}_{25} - \mathrm{ex}_{20})}{-\dot{W}_{c,\mathrm{ref}}}. \tag{7.16}$$

Table 7.13 shows the optimum composition of the mixture. A small amount of neon is used in the mixture. The addition of neon helps in several ways, as described in Section 4.5. Though the concentration of npentane in the mixture passing through the compressor is 8.9% and that of *n*Butane is 9.5%, their concentration in the refrigerant mixture circulating in the final heat exchanger is quite small and the high boilers in the mixture do not freeze at low temperatures. The refrigerant mixture passing through the third heat exchanger essentially consists of nitrogen, methane, and ethane along with very small fractions of other fluids, including neon.

Table 7.14 shows the temperature, pressure, vapor fraction, and flow rate of different streams. About 14.4% of the refrigerant circulating through the compressor is condensed in the condenser (stream 3). Most of the pentane and about half of the butane in the refrigerant mixture passing through the compressor are separated in the first phase separator (PS-1) in the liquid phase. The liquid phase is subcooled and expanded in valve V-1 to provide the necessary refrigeration to partially condense refrigerant stream 5 in the first heat exchanger (HX-1). The nitrogen feed and refrigerant stream 5 are cooled to a temperature of 241.6 K in the first heat exchanger (HX-1). About 26% of refrigerant stream 5 is condensed in the first heat exchanger. The liquid phase separated in the second phase separator (PS-2) contains almost all of the pentane and most of the butanes and propane. The liquid stream is subcooled and expanded in valve V-2 to provide the refrigeration to cool refrigerant stream 6 and nitrogen stream 22 to a temperature of 142.6 K in the second heat exchanger. Most

Table 7.13. Composition of high-pressure refrigerant at different locations of the Kapitza nitrogen liquefaction process shown in Fig. 7.19

	Mol %				
Component	Stream 3	Stream 4	Stream 14	Stream 6	Stream 17
Nitrogen	7.4	8.6	0.3	11.4	0.5
Methane	37.7	43.2	4.6	54.8	9.9
Ethane	33.4	35.9	18.4	31.4	48.9
Propane	2.7	2.5	3.7	1.0	7.0
*n*Butane	9.5	6.2	29.0	0.7	22.0
*n*Pentane	8.9	3.1	43.9	0.1	11.6
Neon	0.4	0.5	0.0	0.7	0.0

Table 7.14. Temperature, pressure, vapor fraction, and mole flow rate of different streams of the Kleemenko process nitrogen liquefaction process shown in Fig. 7.19 and operating with the refrigerant mixtures shown in Table 7.13

	Stream						
	1	3	4	5	6	7	8
Temperature, K	290.6	300.0	300.0	241.6	241.6	142.6	118.8
Pressure, bar	3.5	18.1	18.1	18.1	18.1	18.1	18.1
Vapor fraction	1.000	0.856	1.000	0.741	1.000	0.025	0.000
Flow rate, mol/s	1.000	1.000	0.856	0.856	0.635	0.635	0.635

	Stream						
	9	10	11	12	13	14	15
Temperature, K	113.0	134.7	138.9	237.0	237.7	300.0	241.6
Pressure, bar	3.5	3.5	3.5	3.5	3.5	18.1	18.1
Vapor fraction	0.084	0.380	0.273	0.923	0.799	0.000	0.000
Flow rate, mol/s	0.635	0.635	0.856	0.856	1.000	0.144	0.144

	Stream						
	16	17	18	19	20	21	22
Temperature, K	241.6	142.6	143.5	237.4	300.0	241.6	142.6
Pressure, bar	18.1	18.1	3.5	3.5	40.0	40.0	40.0
Vapor fraction	0.000	0.000	0.000	0.062	1.000	1.000	1.000
Flow rate, mol/s	0.222	0.222	0.222	0.144	0.310	0.310	0.310

	Stream			
	23	24	25	26
Temperature, K	116.4	77.4	77.4	77.4
Pressure, bar	40.0	1.0	1.0	1.0
Vapor fraction	1.000	0.450	0.000	1.000
Flow rate, mol/s	0.310	0.310	0.170	0.139

of the latent heat of the refrigerant in stream 6 is also removed in the second heat exchanger such that the vapor fraction of stream 7 reduces to 2.5%. The temperatures at the exit of HX-2 can be modified such that refrigerant stream 7 is completely condensed in HX-2. This will be helpful if separate heat exchanger cores are used for the second and third heat exchangers. The nitrogen stream is cooled to a temperature of 116.4 K in the final heat exchanger before expansion in expansion valve V-4. It can be seen from Table 7.14 that the vapor fraction at the exit of the nitrogen expansion valve (V-4) is 45%. The liquid yield is therefore 55% in the process shown in Fig. 7.20 and operating with the refrigerant mixtures shown in Table 7.13.

The flow rate of the high-pressure refrigerant reduces progressively as the high boilers are separated in the phase separators. This results in a smaller heat load for the final heat exchangers, as shown in Fig. 7.20. It can also be seen from Fig. 7.20 that

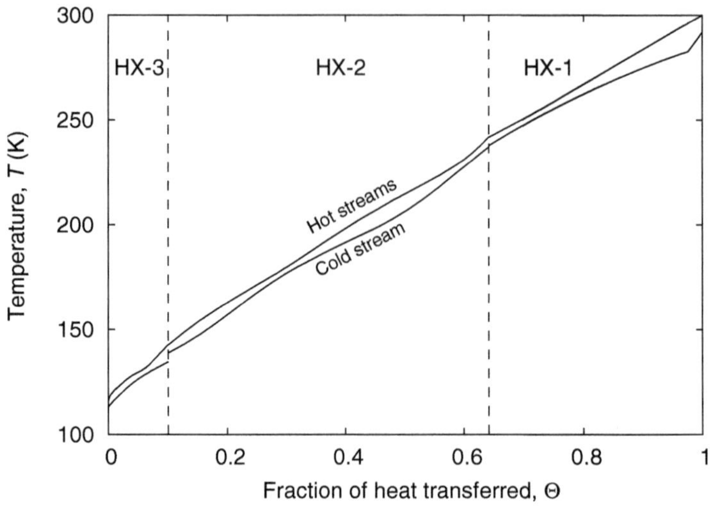

Fig. 7.20. Temperature profiles in the three heat exchangers of a Kapitza process nitrogen liquefier from Fig. 7.19 and operating with the mixture shown in Table 7.15.

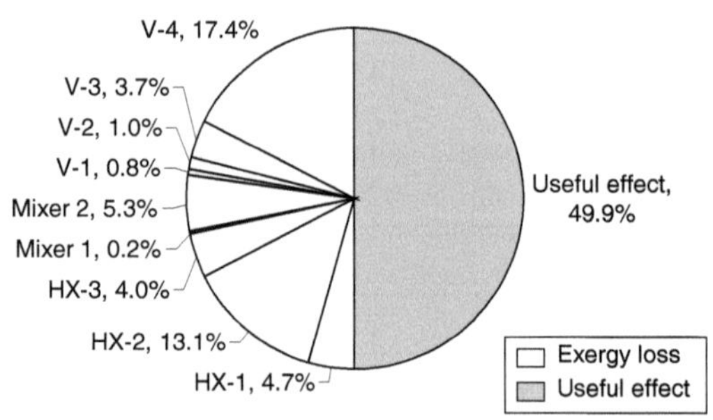

Fig. 7.21. Exergy utilization in the cold box of the Kleemenko process nitrogen liquefier shown in Fig. 7.19 and operating with the mixture shown in Table 7.13.

the temperature approach between the streams is small in the coldest heat exchanger (HX-3), while that in the second heat exchanger (HX-2) is somewhat high, particularly close to the warm end, resulting in considerable exergy loss in the second heat exchanger.

Figure 7.21 shows the exergy utilization in the cold box of Fig. 7.19. The exergy efficiency of the cold box is 49.9%, with the maximum of the exergy loss occurring in the second heat exchanger (HX-2) and the refrigerant expansion valve (V-3).

As seen in the case of a precooled Linde–Hampson process, the refrigeration available with the flash vapor can be used effectively to improve the exergy efficiency of the process. Figure 7.22 shows the Kleemenko process for the liquefaction of nitrogen with a flash nitrogen recycle. As in the case of the Kleemenko process without flash nitrogen recycle, the exergy efficiency of the cold box was used as the

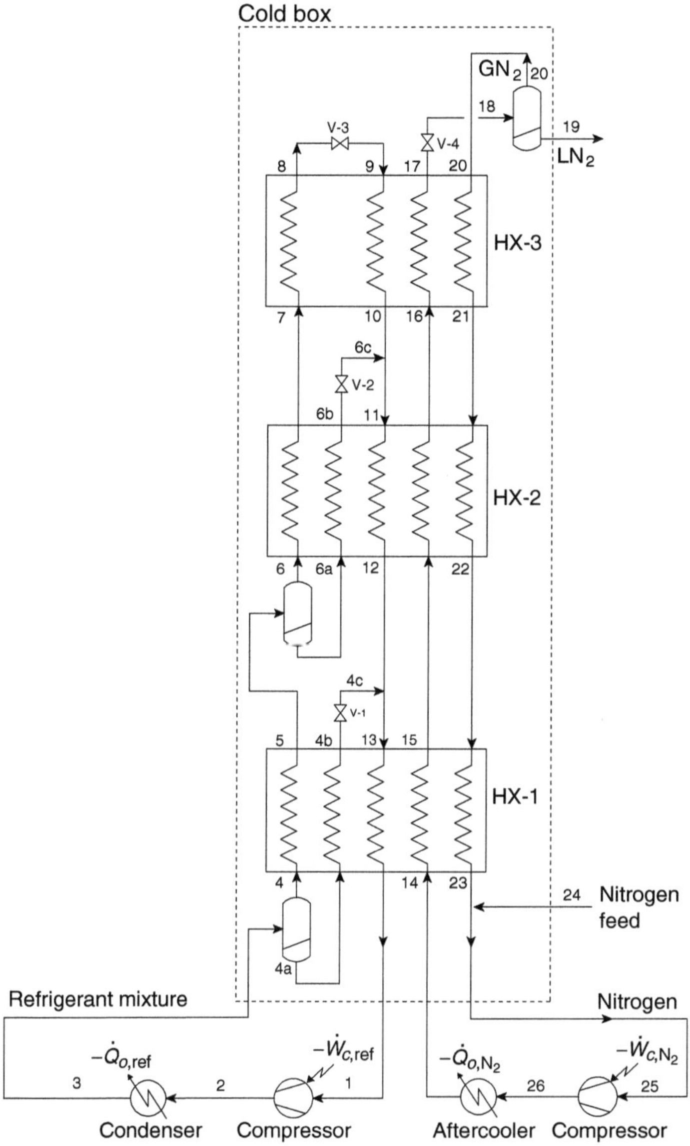

Fig. 7.22. Kapitza process for the liquefaction of nitrogen with flash nitrogen recycle.

objective function in the optimization study. The cold box in this case includes the nitrogen expansion valve (V-4) and the nitrogen phase separator (PS-3).

The exergy efficiency of the cold box of Fig. 7.22 is given by the expression

$$\eta_{\text{ex, cb}} = \frac{\text{minimum power for liquefaction}}{\text{exergy expenditure}} = \frac{\dot{n}_{24}(\text{ex}_{19} - \text{ex}_{24})}{\dot{n}_3(\text{ex}_3 - \text{ex}_1) + \dot{n}_{14}(\text{ex}_{14} - \text{ex}_{25})}. \tag{7.17}$$

The overall exergy efficiency is given by the expression

$$\eta_{\text{ex},o} = \frac{\text{minimum power for liquefaction}}{\text{compressor power input}} = \frac{\dot{n}_{24}(\text{ex}_{19} - \text{ex}_{24})}{(-\dot{W}_{c,\text{ref}}) + (-\dot{W}_{c,\text{N}_2})}. \tag{7.18}$$

Table 7.15 shows the optimum composition of the mixture. Though large amounts of butane and pentane are used in the refrigerant mixture, only negligible quantities of high boilers are present in the refrigerant passing through the final heat exchanger (stream 6), thus preventing the possibility of freezing of the high boilers at low temperatures.

It can be seen from Table 7.16 that the vapor fraction at the exit of the nitrogen expansion valve (stream 18) reduces from 45% in the case of the Kleemenko process without flash nitrogen recycle to a value of 28%. The temperature of the streams at the exit of the three heat exchangers is also lower in the present case when compared with a conventional Kleemenko process without flash gas recycle.

A comparison of Tables 7.16 and 7.14 also shows that the ratio of mole flow rate of liquid nitrogen product (stream 19) to the mole flow rate of refrigerant through the compressor (stream 1) increases from 17.0% in the case of the Kleemenko process without flash gas recycle to 27.7% with the use of flash gas recycle. This results in a substantial improvement in the exergy efficiency of the cold box of the process to 73.2%, as shown in Fig. 7.23. This includes an exergy loss of 6.2% in the nitrogen expansion valve. The exergy efficiency of the cold box of the Kleemenko process with flash nitrogen recycle therefore is of the same order as that obtained in a very efficient natural gas liquefaction process discussed in the previous chapter.

The temperature of the streams at the exit of the three heat exchangers (Fig. 7.24) is also lower in the present case when compared with a conventional Kleemenko process without flash nitrogen recycle. The temperature approach between the hot and cold fluid streams is small in all the heat exchangers at the cold end, as shown in Fig. 7.24, resulting in a small exergy loss in the three heat exchangers.

Table 7.15. Composition of high-pressure refrigerant at different locations of the Kapitza nitrogen liquefaction process with flash gas recycle shown in Fig. 7.22

	Mol %				
	Stream 3	Stream 4	Stream 4a	Stream 6	Stream 6a
Neon	1.08	1.25	0.03	1.83	0.04
Nitrogen	6.94	7.99	0.33	11.49	0.60
Methane	41.50	47.22	5.62	62.76	14.40
Ethane	30.87	32.88	18.25	23.44	52.82
Propane	0.22	0.20	0.33	0.05	0.53
*n*Butane	12.74	8.16	41.51	0.41	24.52
*n*Pentane	6.65	2.30	33.93	0.03	7.09

Table 7.16. Temperature, pressure, vapor fraction, and mole flow rate of different streams of the Kleemenko process nitrogen liquefaction process with flash nitrogen recycle shown in Fig. 7.22 and operating with refrigerant mixtures shown in Table 7.15

	Stream					
	1	3	4	5	6	7
Temperature, K	295.8	300.0	300.0	230.0	230.0	148.2
Pressure, bar	3.0	20.0	20.0	20.0	20.0	20.0
Vapor fraction	1.000	0.860	1.000	0.680	1.000	0.080
Flow rate, mol/s	1.000	1.000	0.860	0.860	0.590	0.590
	8	9	10	11	12	13
Temperature, K	107.3	103.7	135.2	141.5	224.8	225.5
Pressure, bar	20.0	3.0	3.0	3.0	3.0	3.0
Vapor fraction	0.000	0.067	0.584	0.395	0.898	0.782
Flow rate, mol/s	0.585	0.585	0.585	0.862	0.862	1.000
	14	15	16	17	18	19
Temperature, K	300.0	230.0	148.2	103.6	77.4	77.4
Pressure, bar	40.0	40.0	40.0	40.0	1.0	1.0
Vapor fraction	1.000	1.000	1.000	0.000	0.283	0.000
Flow rate, mol/s	0.363	0.363	0.363	0.363	0.363	0.260
	20	21	22	23	6a	6b
Temperature, K	77.4	134.2	209.4	297.0	230.0	148.2
Pressure, bar	1.0	1.0	1.0	1.0	20.0	20.0
Vapor fraction	1.000	1.000	1.000	1.000	0.000	0.000
Flow rate, mol/s	0.103	0.103	0.103	0.103	0.277	0.277
	6c	4a	4b	4c		
Temperature, K	148.9	300.0	230.0	225.9		
Pressure, bar	3.0	20.0	20.0	3.0		
Vapor fraction	0.003	0.000	0.000	0.065		
Flow rate, mol/s	0.277	0.138	0.138	0.138		

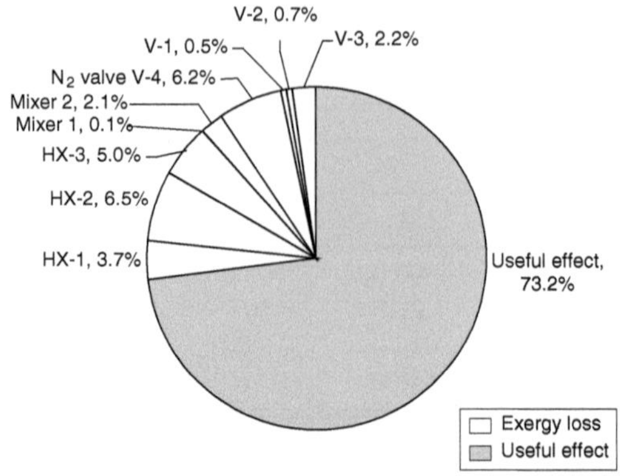

Fig. 7.23. Exergy utilization in the cold box for the Kleemenko process nitrogen liquefier operating with mixtures shown in Table 7.15.

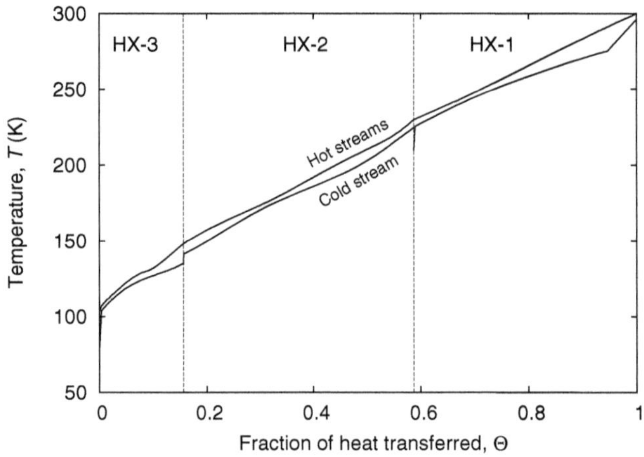

Fig. 7.24. Temperature profiles in the three heat exchangers of a Kapitza process nitrogen liquefier with flash nitrogen recycle operating with mixtures shown in Table 7.15.

Kleemenko process nitrogen liquefiers operating with optimum nitrogen-hydrocarbon mixtures can effectively compete with turbine-based liquefiers because of their high exergy efficiency.

7.6 Other liquefaction processes and refrigerants

Any of the natural gas liquefaction processes such as the propane precooled process (C3-MR) or the dual mixed refrigerant process (DMR) process can also be used for the liquefaction of nitrogen. The flash nitrogen also needs to be recycled to the compressor, as in the case of mixed refrigerant precooled Linde–Hampson, Kapitza, and Kleemenko liquefiers.

A number of non flammable refrigerants such as R14 (CF_4), R23 (CHF_3), R116 (C_2F_6), R218 (C_3F_8), etc. can be used instead of methane, ethane, propane, and so forth. The properties of mixtures of these refrigerants and the binary interaction parameters for different equations of state are, however, not known in the open literature. No attempt has therefore been made to present examples of the performance of liquefaction or refrigeration processes operating with nonflammable refrigerants. Hopefully, this monograph will be revised and such examples presented when experimental data for these refrigerant mixtures become available in the open literature.

7.7 Summary

Most natural gas liquefaction processes discussed in the previous chapter can be used for the liquefaction of nitrogen and other pure fluids with minor modification. Similarly, conventional processes such as the Kaptiza process can also be used with additional mixed refrigerant precooling. As in conventional liquefaction processes such as the Linde–Hampson process, only a part of the nitrogen gas liquefies on expansion. The unliquefied nitrogen gas therefore needs to be recycled in all mixed refrigerant nitrogen liquefaction processes.

The exergy efficiency of mixed refrigerant precooled Linde–Hampson liquefaction process is slightly higher than that of the mixed refrigerant precooled Kapitza liquefaction process. The exergy efficiency of the Kleemenko nitrogen liquefaction process (with flash nitrogen recycle) is higher than that of the precooled Linde–Hampson and Kapitza liquefaction processes.

The mixed refrigerant precooled Linde–Hampson process is preferable for small plants because of its simplicity, and the Kleemenko process is preferable for large plants because of its higher exergy efficiency.

References

[1] Acharya, A., Arman, B., Olszewski, W. J., and Bonaquist, D. P. (2000a). Variable load refrigeration system particularly for cryogenic temperatures. U.S. Patent 6,076,372.

[2] Acharya, A., Arman, B., Weber, J. A., Srinivasan, V., Nowobliski, J. J., Smolarek, J., and Nenov, N. T. (2000b). Multiple circuit cryogenic liquefaction of industrial gas. U.S. Patent 6,105,388.

[3] Akharov, A., Marfenina, I., and Mikulin, Y. (1981). *Theory and Design of Cryogenic Systems*. MIR Publishers, Moscow.

[4] Alexeev, A. and Klipping, G. (2005). Cryogenics in Russia. In *Proc. of the Twentieth International Cryogenic Engineering Conference (ICEC 20): Beijing, China, 11–14 May 2004*, pages 1–8. Elsevier.

[5] Alexeev, A. and Quack, H. (2000). Vorrichtung und verfahren zur kryolagerung biologischer stoffe. German Patent 19922364.

[6] Alexeev, A. and Quack, H. (2003). Refrigerant mixture for a mixture-throttling process. U.S. Patent 6,513,338.

[7] Alfeev, V. N., Brodyanskii, V., Yagodin, V., Nikolsky, V., and Ivantsov, A. (1973). Refrigerant for a cryogenic throttling unit. U.K. Patent 1,336,892.

[8] Alfeev, V. N., Yagodin, V. M., and Nikol'skii, V. A. (1972). Cooling articles to cryogenic temperatures. S.U. Patent 362,978.

[9] ASHRAE (1998). *Chapter 39, ASHRAE Handbook-Refrigeration*. ASHRAE, USA.

[10] Aspentech (2004). *Aspen Plus and Aspen Plus Optimizer simulation programs, Ver. 12.1*. Aspen Tech Inc., USA.

[11] Aspentech (2007). *Aspen Plus and Aspen Plus Optimizer simulation programs, Ver. 20.0*. Aspen Tech Inc., USA.

[12] Barclay, M. and Denton, N. (2005). Selecting offshore LNG processes. *LNG Journal*, pages 34–36.

[13] Barron, R. F. (1983). *Cryogenic Systems*. Oxford Press, second edition.

[14] Berger, E. (2005). Small scale LNG production in Europe from the early seventies until today — experience of a global LNG plant contractor. In *Proc. of the*

First International Conference on Small Scale LNG, 29-30 September 2005, Oslo, Norway.

[15] Berger, E., Förg, W., Heiersted, R. S., and Paurola, P. (2003). The MFC® (mixed fluid cascade) process for the first European baseload LNG production plant. *Linde Technology*, 1:12–23.

[16] Boiarskii, M., Yudin, B., Longsworth, R. C., and Khatri, A. (1998a). High efficiency throttle cryogenic refrigerator based on one stage compressor. U.S. Patent 5,706,663.

[17] Boiarskii, M., Brodyanskii, V. M., and Longsworth, R. C. (1998b). Retrospective of mixed-refrigerant technology and modern status of cryocoolers based on one-stage, oil-lubricated compressors. *Advances in Cryogenic Engineering*, 43:1701–1708.

[18] Boiarskii, M., Khatri, A., and Kovalenko, V. (1999). Design optimization of the throttle cycle cooler with mixed refrigerant. In *Cryocoolers 10*, pages 457–465.

[19] Boiarskii, M., Khatri, A., and Podchernieav, O. (2000). Enhanced refrigeration performance of the throttle-cycle coolers operating with mixed refrigerants. *Advances in Cryogenic Engineering*, 45:291–297.

[20] Boiarskii, M., Mogorychmy, V. L., and Klusmier, L. (1995). Cryogenic mixed gas refrigerant for operation within temperature ranges of 80°K– 100°K. U.S. Patent 5,441,658.

[21] Boiarskii, M., Podcherniaev, O., and Lunin, A. (2003). Optimal design and generalized performance of throttle cycle coolers operating with mixed refrigerants. In *Proc. of the International Congress of Refrigeration, Aug. 17–22, 2003, Washington D.C., paper ICR 0352.*

[22] Boiarskii, M. Y. and Lapshin, V. A. (1979). Determination of refrigerating capacity of regnerating plants operating on mixtures at variable temperatures. *Kholod. Tekh.*, (10):23–26.

[23] Boiarskii, M. Y., Lukin, A. I., Potapov, G. G., and Ladokhin, S. D. (1993a). Optimization of parameters in installations having a cooling mixture cycle for production of liquid nitrogen. *Chemical and Petroleum Engineering*, 29(12):20–23.

[24] Boiarskii, M. Y., Lunin, A. I., Potapov, G. G., and Ladokhin, S. D. (1993b). Optimization of parameters of installations with recycle of mixtures for manufacture of liquid nitrogen. *Chemical and Petroleum Engineering*, 29(12):623–627.

[25] Bonaquist, D. P., Prosser, N. M., and Arman, B. (2003). Advances in refrigeration for air separation — using mixed refrigerant cycles to reduce cost and improve efficiency. In *Paper ICR0524, Proc. of the International Congress of Refrigeration 2003, Washington, D.C.*, pages 1–7.

[26] Brent, R. P. (1973). *Algorithms for Minimization Without Derivatives*. Prentice-Hall, Englewood Cliffs, NJ.

[27] Brodyanskii, V. M., Gramov, E. A., Grezin, A. K., Yagodin, V. M., Nikol'skii, V. A., and Tashchina, A. G. (1971). Efficient throttling cryogenic refrigerators which operate on mixtures. *Chemical and Petroleum Engineering*, 8(12):1057–1061.

[28] Brodyanskii, V. M., Sorin, M. V., and Goff, P. L. (1994). *The Efficiency of Industrial Processes: Exergy Analysis and Optimization*. Elsevier Science B.V., Amsterdam.

[29] Brostow, A. A., Agrawal, R., Herron, D. M., and Roberts, M. J. (2001). Process for nitrogen liquefaction. U.S. Patent 6,298,688.

[30] Chakravarthy, V., Weber, J., Rashad, A., Acharya, A., and Bonaquist, D. (2003). Oxygen liquefier using a mixed gas refrigeration cycle. In *Proc. of the ASME Heat Transfer Division — 2003 Volume 2: Gas Turbine Heat Transfer, Heat Transfer in Electronic Equipment, Heat Transfer in Energy Systems, Low-Temperature Heat Transfer*, pages 491–495.

[31] Coers, D. H., and Sudduth, J. W. (1976). Refrigerant apparatus and process using multicomponent refrigerant. U.S. Patent 3,932,154.

[32] Dobak, J. D., Radebaugh, R., Huber, M. L., and Marquardt, E. D. (1998). Mixed gas refrigeration method. U.S.Patent 5,787,715.

[33] Dubar, C. A. T., and Tu, O. L. M. (2001). Liquefaction apparatus. U.S. Patent 6,250,244.

[34] Fedele, L., Bobboa, S., Camporese, R., and Stryjek, R. (2004). VLLE measurements and correlation for the pentafluoroethane (R125) + n-butane (R600) system. *Fluid Phase Equilibria*, 222–223:283–289.

[35] Fischer, B., Martin, P.-Y., and Rojey, A. (2004). Liquefaction of natural gas with natural gas recycling. U.S. Patent 6,763,680.

[36] Flynn, T. M. (1996). *Cryogenic Engineering*. Marcel Dekker, New York.

[37] Foglietta, J. H. (2002). LNG production using dual independent expander refrigeration cycles. U.S. Patent 6,412,302.

[38] Förg, W. (2003). Natural gas trade routes and liquefaction processes. *Linde Technology*, 1:4–11.

[39] Förg, W., Bach, W., and Stockmann, R. (1999). A new LNG baseload process and manufacturing of the main heat exchanger. *Linde Reports on Science and Technology*, 61:3–11.

[40] Fuderer, A. (1965). Compression process for refrigeration. U.S. Patent 3,203,194.

[41] Garcia, D. C., and Luks, K. D. (1999). Patterns of solid-fluid phase equilibria: New possibilities? *Fluid Phase Equilibria*, 161:91–106.

[42] Garier, C., and Paradowski, H. (1981). Method and plant for liquefying a gas with low boiling temperature. U.S. Patent 4,274,849.

[43] Gaumer Jr., S., Lee, S., and Charles, N. L. (1973). Combined cascade and multicomponent refrigeration system and method. U.S. Patent 3,763,658.

[44] Gerdsmeyer, K.-D., and Isalski, W. H. (2005). On-board reliquefaction for LNG ships. In *Proc. of the Gas Processors Association Europe Conference, London, May 2005*.

[45] Golubev, D. (2003). Kühlung eines resistiven HTSL-Kurzschlussstrom-begrenzers mit einer Gemisch-Joule-Thomson-Kältemaschine. Ph.D. thesis, TU Dresden, Germany.

[46] Golubev, D., Alexeev, A., Haberstroh, C., and Quack, H. (2000). Efficiency study of a nitrogen-mixed gas cascade for cooling the current leads. In *Proc. of ICEC 18, February 21–25, 2000 Mumbai.*

[47] Gong, M., Wu, J., Luo, E., Qi, Y., and Zhou, Y. (2004). Study of the single-stage mixed-gases refrigeration cycle for cooling temperature-distributed heat loads. *Cryogenics*, 43(1):31–41.

[48] Gong, M. Q., Luo, E. C., Zhou, Y., Liang, J. T., and Zhang, L. (2000). Optimum composition calculation for multicomponent cryogenic mixture used in Joule-Thomson refrigerators. *Advances in Cryogenic Engineering*, 45:283–290.

[49] Grezin, A. K., and Zacharov, N. D. (1979). General principles of formation and optimization of multicomponent working fluids for cryogenic systems. In *Proc. of the 15th International Congress of Refrigeration*, volume 1, pages 169–172.

[50] Kleemenko, A. P. (1959). One flow cascade cycle. In *Proc. of the 10th International Congress of Refrigeration*, volume 1, pages 34–39.

[51] Knapp, H., Döring, R., Oellrich, L., Plöcker, U., and Prausnitz, J. M. (1981). *Vapor-Liquid equilibria for mixtures of low boiling substances*, volume VI. Dechema Chemistry Data Series.

[52] Krieger, H. (1976). Process and arrangement for cooling fluids. U.S. Patent 3,964,891.

[53] Kumar, K. R., Somayajulu, N., Venkatarathnam, G., Mukherjee, D., and Sarangi, S. (1994). CRESP: A steady state simulator for cryogenic process plants. *Advances in Cryogenic Engineering*, 39:745–750.

[54] Kumar, P. S. (2004). Studies on Joule-Thomson cryogenic refrigerators operating with refrigerant mixtures. Ph.D. thesis, Indian Institute of Technology, Madras.

[55] Lavrenchenko, G. K., Trotsenko, A. B., Ruvinsky, G. Y., Tabachnik, E. U., Balkin, B. N., and Balavachenko, B. U. (1983). Investigation of phase equilibria in binary mixtures of refrigerants with limits of mixing. *Kholod. Tekh. Tekhnolog.*, 37:38–44.

[56] Lawrence, C. T., Zhou, J. L., and Tits, A. L. (1997). User's Guide for CFSQP Version 2.5: A C Code for Solving (Large Scale) Constrained Nonlinear (Minimax) Optimization Problems, Generating Iterates Satisfying All Inequality Constraints. Institute for Systems Research, University of Maryland, Technical Report TR-94-16r1, College Park, MD.

[57] Little, W. A. (1997). Method for efficient counter-current heat exchange using optimized mixtures. U.S. Patent 5,644,502.

[58] Maher, J. B., and Sudduth, J. W. (1975). Method and apparatus for liquefying gases. U.S. Patent 3,914,949.

[59] Missimer, D. J. (1972). Self-balancing low temperature refrigeration system. U.S. Patent 3,768,273.

[60] Missimer, D. J. (1997). Refrigerant conversion of auto-refrigerating cascade (arc) systems. *Int. J. Refrigeration*, 20:201–207.

[61] Newton, C. L. (1983). Dual mixed refrigerant natural gas liquefaction with staged compression. U.S. Patent 4,525,185.

[62] Nisenoff, M., Patten, F., and Wolf, S. (1996). ...And What About Cryogenic Refrigeration. In *Proc. of International Cryocooler Conference, June 25–27, 1996, Waterville, USA*, pages 25-28.

[63] NIST (2000). *NIST-12 Thermodynamic and transport properties of pure fluids Database: Version 5.0.*

[64] Olszewski, W. J., Arman, B., Acharya, A., Weber, J. A., and Rashad, M. A.-A. (2000). Cryogenic industrial gas liquefaction with hybrid refrigeration generation. U.S. Patent 6,041,620.

[65] Paradowski, H., and Hagyard, P. (2003). Comparing five LNG processes. *Hydrocarbon Engineering*, 8(10):32–37.

[66] Paradowski, H., and Rojey, A. (2000). Method and device for liquefying a natural gas without phase separation of the coolant mixtures. U.S. Patent 6,105,389.

[67] Pasad, G. V., and Venkatarathnam, G. (1999). A method for avoiding trivial roots in isothermal flash calculations using cubic equations of state. *Ind. Engg. Chem Res*, 38:3530–3534.

[68] Petersson, B., and Thorsell, H. (1990). Comparison of the refrigerants HFC 134a and CFC 12. *Int. J. Refrigeration*, 13:176–180.

[69] Podbielniak, W. (1936). Art of refrigeration. U.S. Patent 2,041,725.

[70] Prausnitz, J., Lichtenthaler, R., and de Azevedo, E. (1985). *Molecular Thermodynamics of Fluid-Phase Equilibria*, chapter 9. Prentice-Hall, Englewood Cliffs, New Jersey.

[71] Price, B. C. (2003). Small-scale LNG facility development. *Hydrocarbon Processing*, 82(1):37–40.

[72] Radebaugh, R. (2000). Pulse tube cryocoolers for cooling infrared sensors. In *Proc. of SPIE, The International Society for Optical Engineering, Infrared Technology and Applications XXVI*, volume 4130, pages 363–379.

[73] Radebaugh, R. (2004). Refrigeration for superconductors. *Proc. of the IEEE*, 92(10):1719–1734.

[74] Ramasubramanian, N. V. (2005). A program for simulation and optimization of cryogenic systems. Master's thesis, Indian Institute of Technology Madras.

[75] Ransbarger, W. (2007). A fresh look at LNG process efficiency. *LNG Industry*, Spring 2007 issue.

[76] Reddy, V. R. (2004). Experimental studies on mixed refrigerant cryocoolers with and without phase separators. Master's thesis, Indian Institute of Technology Madras.

[77] Roberts, J. (2006). Integrated multiloop refrigeration process for gas liquefaction. U.S. Patent 7,086,251.

[78] Roberts, M. J., and Agrawal, R. (2000). Dual mixed refrigerant cycle for gas liquefaction. U.S. Patent 6,119,479.

[79] Roberts, M. J., and Agrawal, R. (2001). Hybrid cycle for the production of liquefied natural gas. U.S. Patent 6,308,531.

[80] Roberts, M. J., Agrawal, R., and Daugherty, T. L. (2002). Single mixed refrigerant gas liquefaction process. U.S. Patent 6,347,531.

[81] Sarsten, J. A. (1977). Natural gas liquefaction process. G.B. Patent 1,472,196, US 4,057,972.

[82] Sato, N. (2004). *Chemical Energy and Exergy*. Elsevier, Burlington, MA.

[83] Shell (1962). Improvements in or relating to process and apparatus for liquefying natural gas. G.B. Patent 895,094.

[84] Siva Sankar, M., and Venkatarathnam, G. Studies on a small mixed refrigerant cycle nitrogen gas cooler. *In Proc. of International Cryogenic Engineering Conference, July 17–21, 2006 Prague*, pages 633–636.

[85] Somayajulu, N. V. (1991). CRYOSIM: A process simulator for cryogenic process. Master's thesis, Indian Institute of Technology Kharagpur.

[86] Stockmann, R., Forg, W., Bolt, M., Steinbauer, M., Pfeiffer, C., Paurola, P., Fredheim, A. O., and Sorensen, O. (2001). Method for liquefying a stream rich in hydrocarbons. U.S. Patent 6,253,574.

[87] Streng, A. G. (1971). Miscibility and compatibility of some liquefied and solidified gases at low temperatures. *J. Chem. Eng. Data*, 16(3):357–359.

[88] Swenson, L. K. (1977). Single mixed refrigerant, closed loop process for liquefying natural gas. U.S. Patent 4,033,735.

[89] Thorp, N., and Scott, R. L. (1956). Fluorocarbon solutions at low temperatures. I. The liquid mixtures of CF4-CHF3, CF4-CH4, CF4-Kr, CH4-Kr. *J. Physi. Chemi.*, 60:670–680.

[90] Timmerhaus, K. D., and Flynn, T. M. (1989). *Cryogenic Process Engineering*. Plenum Press, New York.

[91] Varma, G. V., Lau, K. H., and Ulrichson, D. L. (1993). A new algorithm for process flowsheeting. *Computers Chem. Eng.*, 17(4):355–360.

[92] Venkatarathnam, G. (2007). A refrigerant composition for refrigeration systems. India patent 211,627.

[93] Venkatarathnam, G., Mokashi, G., and Murthy, S. S. (1996). Occurrence of pinch points in the condensers and evaporators for zeotropic refrigerant mixtures. *Int. J. Refrigeration*, 19:361–368.

[94] Venkatarathnam, G. and Murthy, S. S. (1999). Effect of mixture composition on the formation of pinch points in condensers and evaporators for zeotropic refrigerant mixtures. *Int. J. Refrigeration*, 22:205–215.

[95] Walas, S. M. (1985). *Phase Equilibria in Chemical Engineering*. Butterworth-Heinemann, New York.

[96] Yudin, B. V. (1993). Study of phase transformations at low temperatures in multicomponent working bodies containing oil mixtures, synopsis (Autoreferat) of D.Sc. thesis, Moscow Power Engineering Institute, Moscow.

Index

MIX
Papier aus verantwortungsvollen Quellen
Paper from responsible sources
FSC® C105338

If you have any concerns about our products,
you can contact us on
ProductSafety@springernature.com

In case Publisher is established outside the EU,
the EU authorized representative is:
Springer Nature Customer Service Center GmbH
Europaplatz 3, 69115 Heidelberg, Germany

Printed by Libri Plureos GmbH
in Hamburg, Germany